COMPUTATIONAL FLUID DYNAMICS FOR MECHANICAL ENGINEERING

COMPUTATIONAL FLUID DYNAMICS FOR MECHANICAL ENGINEERING

George Qin

CRC Press
Taylor & Francis Group
Boca Raton London New York

CRC Press is an imprint of the
Taylor & Francis Group, an **informa** business

First edition published 2022
by CRC Press
6000 Broken Sound Parkway NW, Suite 300, Boca Raton, FL 33487-2742

and by CRC Press
2 Park Square, Milton Park, Abingdon, Oxon, OX14 4RN

Library of Congress Cataloging-in-Publication Data
A catalog record has been requested for this book

ISBN: 978-0-367-68729-8 (hbk)
ISBN: 978-0-367-68730-4 (pbk)
ISBN: 978-1-003-13882-2 (ebk)

DOI: 10.1201/9781003138822

Typeset in Times
by MPS Limited, Dehradun

Access the Support Materials: https://www.routledge.com/9780367687298

Contents

Preface

This manuscript is an outcome of preparation for the computational fluid dynamics (CFD) elective course on the mechanical engineering curriculum of Cedarville University. At first it was only intended to be notes facilitating students' learning of the subject but as its scope grew it became more and more like a book.

Most students who take this course do not have much previous exposure to CFD. For this reason many great CFD classics are not very suitable as their first CFD textbook: some are a little too deep; some are a little too mathematical; some are a little too "Aero," and some provide a little too few worked examples. I wish students find the current book relatively easy-to-follow and pave their way to those more advanced texts. To this end, most important points are explained by using relevant examples with (hopefully) enough details. All numerical procedures are presented with MATLAB® as the default programming language. Topics are selected with a mechanical engineering bias.

The students should complete numerical methods and a fluid mechanics course before they can understand this book. Having taken heat transfer and thermodyanmics is also helpful to appreciate related contents.

Like many CFD guys, I have spent a lot of hours writing CFD codes, debugging them, and scratching my head when they do not converge. I would like to encourage students reading this book to be willing to do the same from day one. I know it is intimidating at times, but it will pay off.

For my fellow teachers, I recommend you to focus on the first five chapters and cover some topics selected from the last three.

I would like to thank my colleagues at the School of Engineering and Computer Science of Cedarville University for graciously taking extra workload so that I can focus on writing this book. I appreciate the contribution of my student Matthew P. Earl who kindly reviewed a draft of the first three chapters and made many helpful suggestions. The generous permission of Dr. Persson (UC Berkeley) and Dr. Tryggvason (Johns Hopkins) to include several of their programs in this book and in the solutions manual is gratefully acknowledged. I would like to dedicate this text to Dr. Richard H. Pletcher, the late professor of Iowa State University, who taught me much more than just CFD.

MATLAB® is a trademark of The MathWorks, Inc. and is used with permission. The MathWorks does not warrant the accuracy of the text or exercises in this book. This book's use or discussion of MATLAB® software or related products does not constitute endorsement or sponsorship by The MathWorks of a particular pedagogical approach or particular use of the MATLAB® software.

George Qin
Cedarville University, Cedarville, Ohio

Author biography

Dr. George Qin is currently an associate professor at the School of Engineering and Computer Science of Cedarville University at Cedarville, Ohio. George obtained his B.S. and M.S. degrees in Mechanical Engineering from Shanghai Jiaotong University of China. He carried out research on large eddy simulation (LES) of turbulent flows in rotating ducts under supervision of Dr. Richard Pletcher of Iowa State University for his Ph.D. degree. Upon receiving his degree in 2007, George began to work as a lecturer and post-doc research fellow in Iowa State University. He moved to Cedarville University in 2012. George teaches and researches in the general thermal-fluids area including Thermodynamics, Fluid Mechanics, Heat Transfer, Computational Fluid Dynamics, and Turbulent Flows.

1 Essence of Fluid Dynamics

1.1 INTRODUCTION

Fluid flows are ubiquitous in nature and are an integral part of our daily-life experiences with fluids. Fluid flows or fluid dynamics can be explained with the examples of gentle movement of air as well as strong winds. We can see fluid flow mechanics in work in the flowing streams of water as well as in rough sea waves. The list of such examples is infinite as we are literally surrounded by natural phenomena that involve fluid flow mechanics; we are surrounded by air and water, while at the same time blood circulation within our body is based on fluid flow mechanics.

Fluid flows are categorized on the basis of various criteria. If the flow is bounded by solid surfaces or has at least one free surface, fluid flows can be classified as internal flows or external flows. According to whether the flow behavior is regular or irregular, fluid flows may be differentiated into laminar flows and turbulent flows. Also, fluid flows can be classified as single-phase and multiphase flows depending on the number of phases involved e.g. one phase or many phases co-existing in the flow.

Although fluid flows are of various types, almost each type can be determined accurately using the same equations, namely the Navier–Stokes equations. However, except for some simple flows, there are no known exact solutions to the Navier–Stokes equations and we, hence, have to rely on computers to solve them numerically to predict how fluid flows. This type of fluid dynamics study is called computational fluid dynamics (CFD) and is the focus of the present book.

The past few decades have witnessed the significant growth and penetration of computer technology. As a consequence, CFD has flourished and is increasingly gaining popularity in academia as well as industries. Today, CFD is a billion-dollar business that gives rise to a large number of employment opportunities. A simple Google search can provide evidence to support this fact. CFD is being employed in diverse fields as a result of its broad applications. In addition to practical values, CFD is interesting and a beautiful concept. It is the author's personal experience that working on CFD and obtaining results that match the analytical solution or experimental data were some of the most satisfactory moments of life.

Learning CFD without the programming is like learning to swim without entering the pool. Therefore, to gain maximum out of this book, readers are urged to write their own codes while solving examples and exercises.

DOI: 10.1201/9781003138822-1

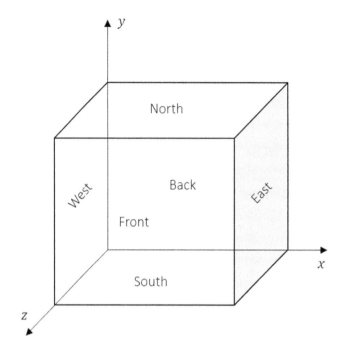

FIGURE 1.1 Control volume.

1.2 GENERAL FORM OF TRANSPORT EQUATIONS

1.2.1 DERIVATION OF GENERAL FORM OF TRANSPORT EQUATIONS

Fluid flows form the basis of many transport processes. Consider a certain extensive property Φ (e.g. mass, momentum, or energy) of a fluid, whose specific value (the amount of Φ per unit mass of fluid) is ϕ. Let us focus on a rectangular control volume in the fluid field as shown in Figure 1.1. The change in Φ in this volume is usually due to three mechanisms: convection, diffusion, and generation. That is, Φ can be carried into/out of the volume passively by the fluid flow (convection); it can migrate into/out of the volume from/into an adjacent region because of a higher/lower level of ϕ in that region (diffusion); and it may be generated/destroyed in the volume because of certain mechanisms (generation). Notice that convection has a favorite direction, which is the fluid flow direction, but diffusion usually does not have such a preference. Instead, Φ simply diffuses from a higher to a lower ϕ region, no matter which direction it may be.

Suppose the fluid density is ρ and its velocity in the x direction is u. The net amount of Φ carried into the volume through the west and east faces due to convection per unit time is then

$$(\rho u A \phi)_w - (\rho u A \phi)_e \qquad (1.1)$$

where the subscript w denotes the west face and e the east, and A is the face area. The net amount of Φ brought into the volume via the west and east faces due to diffusion per unit time is

$$\left(\Gamma A \frac{\partial \phi}{\partial x}\right)_e - \left(\Gamma A \frac{\partial \phi}{\partial x}\right)_w \tag{1.2}$$

where Γ is the fluid diffusion coefficient of property ϕ. It has the same unit as that of the product of density, velocity, and length, which is $kg/(m{\cdot}s)$ or $Pa{\cdot}s$. In fact, diffusion is essentially the mixing due to random motion of fluid molecules, and one can show that the diffusion coefficient is proportional to the product of fluid density, mean molecular velocity, and mean free path of fluid molecules (Boltzmann, 2011).

Suppose the amount of Φ generated per unit volume per unit time is S, the change in Φ in this control volume ΔV in Δt time interval is then

$$\Delta\Phi = \Delta t \left\{ [(F_x)_w - (F_x)_e]A_{we} + \left[\left(F_y\right)_s - \left(F_y\right)_n\right]A_{sn} \right.$$
$$\left. + \left[(F_z)_b - (F_z)_f\right]A_{bf} + S\Delta V \right\} \tag{1.3}$$

where the flux of Φ across a face perpendicular to the jth direction is

$$F_j = \rho u_j \phi - \Gamma \frac{\partial \phi}{\partial x_j} \tag{1.4}$$

Equation (1.3) can be written in differential form by dividing both sides by $\Delta V{\cdot}\Delta t$ and then let ΔV and $\Delta t \to 0$. The equation becomes

$$\frac{\partial(\rho\phi)}{\partial t} + \sum_{j=1}^{3} \frac{\partial F_j}{\partial x_j} = S \tag{1.5}$$

or

$$\frac{\partial(\rho\phi)}{\partial t} + \sum_{j=1}^{3} \frac{\partial\left(\rho u_j \phi\right)}{\partial x_j} = \sum_{j=1}^{3} \frac{\partial}{\partial x_j}\left(\Gamma \frac{\partial \phi}{\partial x_j}\right) + S \tag{1.6}$$

Equation (1.6) is the general form of equations that govern a great many transport phenomena in fluid flows under the Cartesian coordinate system. It is also called the general convection-diffusion equation. For obvious reasons, the first, second, third, and fourth terms of this equation are called transient (or unsteady) term, convection term, diffusion term, and generation (or source) term.

We, probably, have already learned about material derivative in fluid mechanics

$$\frac{Dq}{Dt} = \frac{\partial q}{\partial t} + \sum_{j=1}^{3} u_j \frac{\partial q}{\partial x_j} = \frac{\partial q}{\partial t} + \sum_{j=1}^{3} \frac{\partial (u_j q)}{\partial x_j} \tag{1.7}$$

which denotes the time derivative of a property q of a moving fluid particle. So indeed the general convection–diffusion equation simply provides us the reasons (terms on the right side of the equation) for the change in property $\rho\phi$ of a moving fluid particle (terms on the left side of the equation).

1.2.2 Maximum Principle, Conservativeness, and Boundedness

Equation (1.6) is essentially the conservation law of property Φ, which may be thought of as an extension of the first law of thermodynamics (the energy conservation principle). There is also a similar generalization of the second law of thermodynamics. According to the second law, heat (which is a form of energy) can only diffuse spontaneously from a higher to a lower temperature region. As a result of such heat diffusion, the temperature difference between these two regions decreases. The same law holds for any extensive property Φ, which only diffuses spontaneously from a higher to a lower ϕ position and such diffusion results in a decrease of the ϕ difference between the two involved positions.

It can be proved (Wesseling, 2001) that for a flow with constant fluid properties (ρ and Γ), if the source term $S \leq 0$ and at portions of the flow boundary where the derivative of ϕ is specified this derivative is nonpositive as we go out of the boundary, the maximum of ϕ inside the flow will never increase with time. This is the so-called maximum principle. Physical intuition should convince us about its correctness: convection mixes fluid particles of different levels of ϕ, which only leads to a more uniform ϕ distribution; diffusion tends to smooth out the differences of ϕ in the flow field, which again lowers the maximum ϕ value. Hence, the ϕ maximum can be increased only by a positive source, or a positive outward ϕ derivative at the boundary. These two possibilities are excluded by the given conditions of the principle, therefore the conclusion follows.

If the source term $S \geq 0$ and the ϕ outward derivative along the flow boundary is specified as nonnegative, the minimum of ϕ cannot decrease with time. Finally, if $S = 0$ and along the boundary the ϕ derivative, if prescribed, is zero, ϕ inside the flow will always be restricted between the initial extrema of ϕ.

The maximum principle at steady state, that is as $\partial(\rho\phi)/\partial t = 0$, is as follows. If $S \leq 0$, local maxima of ϕ can occur only at the boundary; if $S \geq 0$, local minima of ϕ can appear only at the boundary; and if $S = 0$, local extrema of ϕ cannot present in interior, which means ϕ must be monotonic and bounded by the boundary values.

Notice that we have two versions of the maximum principle: the unsteady version and steady version. The former is "temporal" and "global": a flow property in an unsteady process must always stay in the range defined by the initial global extrema (conditions apply). It does not rule out the possibility of local extrema in space and/or in time, as long as such local extrema do not exceed the initial global extrema[1]. For example, suppose we have a long water channel in which water is still. We then inject ink into the water at several spots along the channel. Water begins to flow after the ink injection is over. If we focus on a fixed point downstream the ink injection sites, we will observe the ink concentration at this point fluctuates with time; the ink concentration also has some local maxima and minima in the channel at a given time moment. But, all these local extrema can never exceed the initial extrema when the water begins to flow. On the other hand, the steady version of the maximum principle is "spatial" and "local": a flow property at the steady state cannot have any local extrema except at the boundary (if $S = 0$). One way to unify these two versions of the maximum principle is tracking the local extrema as they advect with the fluid: the magnitude of these extrema never increases with time. Instead, they gradually diminish and eventually vanish at the steady state. Moreover, no new local extrema can arise during this process (if $S = 0$). This property is termed total variation diminishing (TVD, see Section 3.3.6, Chapter 3), which will be used interchangeably with boundedness in this book.

The conservativeness and boundedness of ϕ entailed by the conservation law and the maximum principle are among the most important physical properties. Generally, we will want to preserve these two properties in our numerical procedures.

1.3 NAVIER–STOKES EQUATIONS

1.3.1 Continuity Equation and Momentum Equations

By feeding Equation (1.6) with desired properties and corresponding source terms, we can determine the governing equations of these properties. For example, by setting $\phi = 1$, the corresponding extensive property $\Phi = m\phi = m$ is mass and the source term in its transport equation is zero because mass can neither be created nor be destroyed (the mass conservation principle). We thence have the continuity equation

$$\frac{\partial \rho}{\partial t} + \frac{\partial (\rho u)}{\partial x} + \frac{\partial (\rho v)}{\partial y} + \frac{\partial (\rho w)}{\partial z} = 0 \tag{1.8}$$

where u, v, and w are fluid velocities in the x, y, and z directions, respectively. For an incompressible flow in which every fluid particle has a constant density, Equation (1.8) can be simplified to

$$\frac{\partial u}{\partial x} + \frac{\partial v}{\partial y} + \frac{\partial w}{\partial z} = 0 \tag{1.9}$$

By substituting $\phi = u_i$ into Equation (1.6), we obtain the momentum equation in the ith direction:

$$\frac{\partial(\rho u_i)}{\partial t} + \frac{\partial(\rho u u_i)}{\partial x} + \frac{\partial(\rho v u_i)}{\partial y} + \frac{\partial(\rho w u_i)}{\partial z} = \left[\frac{\partial}{\partial x}\left(\mu\frac{\partial u_i}{\partial x}\right) + \frac{\partial}{\partial y}\left(\mu\frac{\partial u_i}{\partial y}\right)\right.$$

$$\left. + \frac{\partial}{\partial z}\left(\mu\frac{\partial u_i}{\partial z}\right)\right] + S_i \qquad (1.10)$$

where μ is the diffusion coefficient of momentum, usually known as the dynamic viscosity. Its unit, like the other diffusion coefficients, is $Pa \cdot s$. S_i is the source term which is the time rate of momentum $\Phi = mu_i$ being generated per unit volume of fluid. Can momentum be generated? Of course. As the second law of Newton tells us, force equals the time rate of change in momentum. S_i is therefore nothing but the forces acting on per unit volume of fluid along the ith direction. Viscous force, pressure force, and gravity force are the three most commonly encountered forces in fluid flows. Taking into account these forces, the momentum equation reads

$$\frac{\partial(\rho u_i)}{\partial t} + \frac{\partial(\rho u u_i)}{\partial x} + \frac{\partial(\rho v u_i)}{\partial y} + \frac{\partial(\rho w u_i)}{\partial z} = \left[\frac{\partial}{\partial x}\left(\mu\frac{\partial u_i}{\partial x}\right) + \frac{\partial}{\partial y}\left(\mu\frac{\partial u_i}{\partial y}\right)\right.$$

$$\left. + \frac{\partial}{\partial z}\left(\mu\frac{\partial u_i}{\partial z}\right)\right] - \frac{\partial p}{\partial x_i} + \rho g_i + f_i \qquad (1.11)$$

in which p is the fluid pressure and g_i is the gravitational acceleration component along the ith direction. The diffusion terms are indeed the viscous force per unit volume of fluid; they are followed by the two terms representing the pressure and gravity forces. All the other forces are included in the last term f_i.

It turns out this form of momentum equation is only valid for incompressible fluids with constant viscosity. For an incompressible fluid with variable viscosity due to e.g. multiple phases coexisting in the fluid, some changes have to be made to the diffusion terms (see Section 5.6.3 for the reason). The momentum equation then becomes

$$\frac{\partial(\rho u_i)}{\partial t} + \frac{\partial(\rho u u_i)}{\partial x} + \frac{\partial(\rho v u_i)}{\partial y} + \frac{\partial(\rho w u_i)}{\partial z} = \left\{\frac{\partial}{\partial x}\left[\mu\left(\frac{\partial u_i}{\partial x} + \frac{\partial u}{\partial x_i}\right)\right]\right.$$

$$\left. + \frac{\partial}{\partial y}\left[\mu\left(\frac{\partial u_i}{\partial y} + \frac{\partial v}{\partial x_i}\right)\right] + \frac{\partial}{\partial z}\left[\mu\left(\frac{\partial u_i}{\partial z} + \frac{\partial w}{\partial x_i}\right)\right]\right\}$$

$$- \frac{\partial p}{\partial x_i} + \rho g_i + f_i \qquad (1.12)$$

You may verify that Equation (1.12) reduces to Equation (1.11) if viscosity is constant.

The continuity equation and the momentum equation together form the Navier–Stokes equations, which govern the behavior of most fluid flows. Notice that the momentum equation in fact stands for three separate equations for u, v, and w. Because the Navier–Stokes equations are first-order in time and second-order in space, we typically need one initial condition and two boundary conditions in each spatial direction to complement these equations.

For a 2-D incompressible flow with constant density and viscosity with negligible body forces, the Navier–Stokes equations can be greatly simplified. They become

$$\frac{\partial u}{\partial x} + \frac{\partial v}{\partial y} = 0 \tag{1.13}$$

$$\frac{\partial u}{\partial t} + \frac{\partial (u^2)}{\partial x} + \frac{\partial (vu)}{\partial y} = \nu\left(\frac{\partial^2 u}{\partial x^2} + \frac{\partial^2 u}{\partial y^2}\right) - \frac{1}{\rho}\frac{\partial p}{\partial x} \tag{1.14}$$

$$\frac{\partial v}{\partial t} + \frac{\partial (uv)}{\partial x} + \frac{\partial (v^2)}{\partial y} = \nu\left(\frac{\partial^2 v}{\partial x^2} + \frac{\partial^2 v}{\partial y^2}\right) - \frac{1}{\rho}\frac{\partial p}{\partial y} \tag{1.15}$$

where $\nu = \mu/\rho$ is the kinematic viscosity.

1.3.2 DIMENSIONLESS FORM OF EQUATIONS AND REYNOLDS NUMBER

It is sometimes desirable to render the governing equations of a transport phenomenon, e.g. the Navier–Stokes equations, into a nondimensional form by normalizing each variable in the equation with a reference value.

If in a fluid flow, e.g. a channel flow, the reference length is L e.g. the channel height; and reference velocity is U e.g. the mean flow velocity, then the compatible reference time is L/U and reference pressure is ρU^2. Ignoring body forces and assuming constant fluid density and viscosity, the 2-D incompressible Navier–Stokes equations can be normalized to

$$\frac{\partial \tilde{u}}{\partial \tilde{x}} + \frac{\partial \tilde{v}}{\partial \tilde{y}} = 0 \tag{1.16}$$

$$\frac{\partial \tilde{u}}{\partial \tilde{t}} + \frac{\partial (\tilde{u}^2)}{\partial \tilde{x}} + \frac{\partial (\tilde{v}\tilde{u})}{\partial \tilde{y}} = \frac{1}{Re}\left(\frac{\partial^2 \tilde{u}}{\partial \tilde{x}^2} + \frac{\partial^2 \tilde{u}}{\partial \tilde{y}^2}\right) - \frac{\partial \tilde{p}}{\partial \tilde{x}} \tag{1.17}$$

$$\frac{\partial \tilde{v}}{\partial \tilde{t}} + \frac{\partial (\tilde{u}\tilde{v})}{\partial \tilde{x}} + \frac{\partial (\tilde{v}^2)}{\partial \tilde{y}} = \frac{1}{Re}\left(\frac{\partial^2 \tilde{v}}{\partial \tilde{x}^2} + \frac{\partial^2 \tilde{v}}{\partial \tilde{y}^2}\right) - \frac{\partial \tilde{p}}{\partial \tilde{y}} \tag{1.18}$$

where variables with a tilde are the normalized variables e.g. $\tilde{u} = u/U$; Re is the Reynolds number

$$Re = \frac{\rho U L}{\mu} \tag{1.19}$$

The initial and boundary conditions can be normalized in the same way. The reason why we sometimes prefer such nondimensional forms of equations is because the normalized flow field solved from such equations is determined by only a few dimensionless parameters like the Reynolds number.

At this point, differential equations like the Navier–Stokes equations seem to be just something we have to solve with numerical methods. But as you will see in Chapters 6 and 7, differential equations are thermselves very powerful numerical methods.

Exercises
1. Show that Equation (1.14) can also be written as

$$\frac{\partial u}{\partial t} + u\frac{\partial u}{\partial x} + v\frac{\partial u}{\partial y} = \nu\left(\frac{\partial^2 u}{\partial x^2} + \frac{\partial^2 u}{\partial y^2}\right) - \frac{1}{\rho}\frac{\partial p}{\partial x} \tag{1.20}$$

Although mathematically this equation is equivalent to Equation (1.14), they may have different numerical consequences. Equations in the form of Equation (1.6) like Equation (1.14) are the so-called conservative form equations; equations in the form of (1.20) are the nonconservative form equations.
2. Derive Equation (1.17).
3. Derive a pressure Poisson equation from Equations (1.13) through (1.15):

$$\frac{\partial^2 p}{\partial x^2} + \frac{\partial^2 p}{\partial y^2} = 2\rho\left(\frac{\partial u}{\partial x}\frac{\partial v}{\partial y} - \frac{\partial v}{\partial x}\frac{\partial u}{\partial y}\right) \tag{1.21}$$

4. For a 2-D incompressible flow we can define the stream function ϕ by requiring

$$u = \frac{\partial \phi}{\partial y}; \quad v = -\frac{\partial \phi}{\partial x} \tag{1.22}$$

We also can define a flow variable called vorticity

$$\omega = \frac{\partial v}{\partial x} - \frac{\partial u}{\partial y} \tag{1.23}$$

Show that

$$\omega = -\left(\frac{\partial^2 \phi}{\partial x^2} + \frac{\partial^2 \phi}{\partial y^2}\right) \tag{1.24}$$

Stream function and vorticity can replace the familiar u, v, and p to be used in 2-D incompressible flow simulations; they also can be used to visualize the flow field, see Section 5.3.4.

NOTE

1 A global extremum is the extremum of the whole flow domain; a local extremum is the one in a small neighborhood around a specific point in the domain.

2 Finite Difference and Finite Volume Methods

2.1 FINITE DIFFERENCE METHOD

2.1.1 FINITE DIFFERENCES

The finite difference method involves replacing derivatives in a differential equation with finite differences so that the differential equation is transformed into a system of algebraic equations, which can be solved with the help of computers.

The methods of converting differential equations into algebraic equations are called "numerical schemes" and the techniques used for solving such algebraic equations are termed "numerical algorithms."

As shown in Figure 2.1, the derivative of a smooth function $\phi(x)$ at a specific $x = x_i$, that is $(d\phi/dx)_i$ is numerically equal to the slope of the line tangent to the ϕ curve at point (x_i, ϕ_i).

This slope can be approximated by the slope of a secant line passing through the point (x_i, ϕ_i) and a neighboring point (x_{i+1}, ϕ_{i+1}) on the curve i.e.

$$\left(\frac{d\phi}{dx}\right)_i \approx \frac{\phi_{i+1} - \phi_i}{\delta x} \tag{2.1}$$

where $\delta x = x_{i+1} - x_i$. Notice that we use notations like ϕ_i to denote $\phi(x_i)$, etc. The quotient on the right side of Equation (2.1) is a finite difference approximation of the derivative on the left side. This finite difference is called a forward difference.

If we use the slope of the secant line passing through points (x_i, ϕ_i) and (x_{i-1}, ϕ_{i-1}) to approximate the same derivative, we end up with the so-called backward difference:

$$\left(\frac{d\phi}{dx}\right)_i \approx \frac{\phi_i - \phi_{i-1}}{\delta x} \tag{2.2}$$

And if we use the slope of the line passing points (x_{i-1}, ϕ_{i-1}) and (x_{i+1}, ϕ_{i+1}) to approximate the derivative, we have the so-called central difference:

$$\left(\frac{d\phi}{dx}\right)_i \approx \frac{\phi_{i+1} - \phi_{i-1}}{2\delta x} \tag{2.3}$$

DOI: 10.1201/9781003138822-2

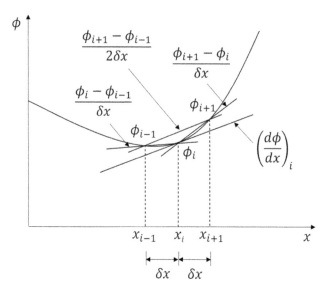

FIGURE 2.1 Forward, backward, and central differences.

These finite difference formulas, of course, do not approximate the derivative to the same degree of accuracy. If we expand ϕ_{i+1} and ϕ_{i-1} into Taylor's series about ϕ_i, we have

$$\phi_{i+1} = \phi(x_i + \delta x) = \phi_i + \left(\frac{d\phi}{dx}\right)_i \delta x + \frac{1}{2!}\left(\frac{d^2\phi}{dx^2}\right)_i \delta x^2 + \ldots$$

$$\phi_{i-1} = \phi(x_i - \delta x) = \phi_i - \left(\frac{d\phi}{dx}\right)_i \delta x + \frac{1}{2!}\left(\frac{d^2\phi}{dx^2}\right)_i \delta x^2 - \ldots$$

(2.4)

so that

$$\frac{\phi_{i+1} - \phi_i}{\delta x} = \left(\frac{d\phi}{dx}\right)_i + \frac{1}{2!}\left(\frac{d^2\phi}{dx^2}\right)_i \delta x + \ldots$$

(2.5)

As can be seen, the left side of this equation is the forward difference and the first term on the right side is the derivative we want to approximate. All the remaining terms on the right side are neglected in the forward difference approximation of the derivative, and therefore determines how accurate the forward difference is. We will call such terms the truncation error. The forward difference is said to be first-order accurate, or of first-order accuracy in the sense that the truncation error is roughly proportional to δx as $\delta x \to 0$. Notice that $\delta x \gg \delta x^2 \gg \delta x^3 \ldots$ as $\delta x \to 0$. Similarly, the backward difference is also first-order accurate because

$$\frac{\phi_i - \phi_{i-1}}{\delta x} = \left(\frac{d\phi}{dx}\right)_i - \frac{1}{2!}\left(\frac{d^2\phi}{dx^2}\right)_i \delta x + \dots \tag{2.6}$$

The central difference, on the other hand, is second-order accurate as its truncation error is approximately proportional to δx^2 as $\delta x \to 0$:

$$\frac{\phi_{i+1} - \phi_{i-1}}{2\delta x} = \left(\frac{d\phi}{dx}\right)_i + \frac{1}{3!}\left(\frac{d^3\phi}{dx^3}\right)_i \delta x^2 + \dots \tag{2.7}$$

Now let us formally define the truncation error. If we use L_Δ to denote a finite difference and L the corresponding derivative that L_Δ approximates, the truncation error involved in the finite difference of a $\phi(x)$ function is

$$E = L_\Delta \phi - L\phi \tag{2.8}$$

For example the truncation error of the central difference at point x_i is

$$E_i = \frac{\phi_{i+1} - \phi_{i-1}}{2\delta x} - \left(\frac{d\phi}{dx}\right)_i = \frac{1}{3!}\left(\frac{d^3\phi}{dx^3}\right)_i \delta x^2 + \dots = \mathcal{O}(\delta x^2) \tag{2.9}$$

The symbol $\mathcal{O}$ reads "of the order of," which means "roughly proportional to": the truncation error of the central difference is of the order of δx^2 i.e. it is roughly proportional to δx^2 (as $\delta x \to 0$).

A higher-order accurate finite difference is typically a more accurate approximation to the derivative than a lower-order accurate difference. For example, to evaluate the derivative of $\phi(x) = \sin x$ at $x_i = 1$, we may use the forward difference and $\delta x = 0.1$:

$$\left(\frac{d \sin x}{dx}\right)_i \approx \frac{\sin x_{i+1} - \sin x_i}{\delta x} = \frac{\sin(1.1) - \sin(1)}{0.1} = 0.497364 \tag{2.10}$$

or the backward difference:

$$\left(\frac{d \sin x}{dx}\right)_i \approx \frac{\sin x_i - \sin x_{i-1}}{\delta x} = \frac{\sin(1) - \sin(0.9)}{0.1} = 0.58144 \tag{2.11}$$

or the central difference:

$$\left(\frac{d \sin x}{dx}\right)_i \approx \frac{\sin x_{i+1} - \sin x_{i-1}}{2\delta x} = \frac{\sin(1.1) - \sin(0.9)}{2 \times 0.1} = 0.53940 \tag{2.12}$$

TABLE 2.1

Common Finite Difference Formulas

Derivative	First-Order Accurate Finite Differences	Second-Order Accurate Finite Differences
$\left(\dfrac{d\phi}{dx}\right)_i$	$\dfrac{\phi_{i+1}-\phi_i}{\delta x}$	$\dfrac{\phi_{i+1}-\phi_{i-1}}{2\delta x}$
	$\dfrac{\phi_i-\phi_{i-1}}{\delta x}$	$\dfrac{-3\phi_i+4\phi_{i+1}-\phi_{i+2}}{2\delta x}$
		$\dfrac{3\phi_i-4\phi_{i-1}+\phi_{i-2}}{2\delta x}$
$\left(\dfrac{d^2\phi}{dx^2}\right)_i$	$\dfrac{\phi_i-2\phi_{i+1}+\phi_{i+2}}{\delta x^2}$	$\dfrac{\phi_{i-1}-2\phi_i+\phi_{i+1}}{\delta x^2}$
	$\dfrac{\phi_i-2\phi_{i-1}+\phi_{i-2}}{\delta x^2}$	

The exact value of this derivative is 0.54030. The central difference result is clearly more accurate than both forward and backward difference results.

Some commonly used finite differences are listed in Table 2.1. Central differences are shown in bold font since they will be used extensively in this book. These finite difference formulas can be derived in many different ways. We have already shown that this can be done using Taylor's series. Another means to construct finite differences of desired order of accuracy is the polynomial fitting method.

For example, if we want to develop a second-order accurate finite difference for $d\phi/dx$ at point i using ϕ values at points i, $i + 1$, and $i + 2$ along a uniform mesh (which means the points are equally spaced), we can assume $\phi(x)$ can be represented by a polynomial of x:

$$\phi(x) = a + bx + cx^2 + dx^3 + \dots \qquad (2.13)$$

Without losing generality, we may set $x = 0$ at point i, then easily we see the derivative we try to approximate

$$\left(\frac{d\phi}{dx}\right)_i = b \qquad (2.14)$$

at $x = 0$. To find this coefficient b, we only need to write down the ϕ values at points i, $i + 1$, and $i + 2$ with the help of the polynomial (keep in mind that $x_i = 0$, $x_{i+1} = \delta x$, and $x_{i+2} = 2\delta x$):

$$\phi_i = a$$

$$\phi_{i+1} = a + b\delta x + c\delta x^2 + d\delta x^3 + \dots$$

$$\phi_{i+2} = a + 2b\delta x + 4c\delta x^2 + 8d\delta x^3 + \dots \tag{2.15}$$

By canceling the $c\delta x^2$ terms from these equations, we end up with

$$\left(\frac{d\phi}{dx}\right)_i = b = \frac{-3\phi_i + 4\phi_{i+1} - \phi_{i+2}}{2\delta x} + 2d\delta x^2 + \mathcal{O}(\delta x^3) \tag{2.16}$$

That is

$$\left(\frac{d\phi}{dx}\right)_i = \frac{-3\phi_i + 4\phi_{i+1} - \phi_{i+2}}{2\delta x} + \mathcal{O}(\delta x^2) \tag{2.17}$$

which is the second-order forward difference that can be found in Table 2.1.

There are other methods we can use to derive finite difference equations e.g. the finite volume method, which will be discussed later.

By replacing derivatives with finite differences, a differential equation is approximated by a system of algebraic equations, which can be solved numerically. This is the so-called finite difference method. Let us explore one example that shows the basic steps of this method.

2.1.2 EXAMPLE: LAMINAR CHANNEL FLOW

As seen in Figure 2.2, a fluid with constant density and viscosity flows steadily in a channel formed by two infinitely large parallel plates separated by a distance $H = 2h = 2\,m$. The flow direction is set to be the x-direction and the origin of y-coordinates is at the lower plate. The flow is fully developed, which means that the velocity does not change in the x-direction i.e.

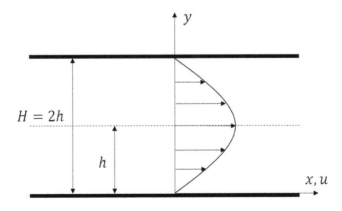

FIGURE 2.2 Laminar channel flow example setup.

$$\frac{du}{dx} = 0 \tag{2.18}$$

By substituting Equation (2.18) into the continuity Equation (1.13), we can see

$$\frac{\partial v}{\partial y} = 0 \tag{2.19}$$

which means the v-velocity does not change along the y-direction. Since $v = 0$ at the channel surface and it is not going to change when we move upward from the surface, we conclude that $v = 0$ everywhere in the channel. So the u-momentum equation, Equation (1.14) simplifies to

$$\frac{d^2u}{dy^2} = \frac{1}{\mu}\frac{dp}{dx} \tag{2.20}$$

To maintain this flow, the pressure gradient dp/dx cannot be zero but a constant (because the flow is the same at every x-location, so there is no reason for dp/dx to vary with x). Here we will simulate such a flow with $dp/\mu dx = -1$ $(m{\cdot}s)^{-1}$.

Since the flow is symmetric to the channel centerline, we only need to calculate the lower half of the channel i.e. $0 \le y \le 1$. The two boundary conditions are

$$\begin{cases} u = 0 & y = 0 \\ \frac{du}{dy} = 0 & y = 1 \end{cases} \tag{2.21}$$

At the lower boundary the u value is given, and we call such a boundary condition a Dirichlet boundary condition; at the upper boundary the derivative of u is specified, and such a condition is called Neumann boundary condition.

We want to solve the velocity distribution $u(y)$ with the finite difference method. To this end, we first set up a mesh (grid) system, which is nothing but a certain number of points (nodes) we select along the y-axis.

Let us use 5 nodes evenly distributed in the y range with two nodes at the two ends, as shown in Figure 2.3. The distance between two neighboring nodes is therefore $\delta y = 0.25$.

Typically, we use i, j, k as indices in the x, y, and z-directions, respectively. Therefore, the y-coordinate of the jth node is y_j and u at the same node is u_j. By replacing the derivatives in (2.20) with a central difference, we have the following finite difference equation for the jth node:

$$\frac{u_{j-1} - 2u_j + u_{j+1}}{\delta y^2} = -1 \tag{2.22}$$

Written in a more symbolic form, the finite difference equation is

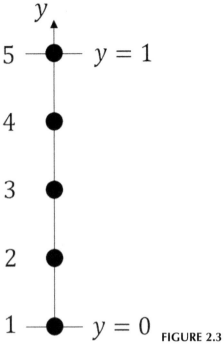

FIGURE 2.3   5-node mesh.

$$as_j u_{j-1} + ap_j u_j + an_j u_{j+1} = b_j \tag{2.23}$$

in which

$$as_j = \frac{1}{\delta y^2}, \; ap_j = -\frac{2}{\delta y^2}, \; an_j = \frac{1}{\delta y^2}, \; b_j = -1 \tag{2.24}$$

Notice that for the jth node, the $(j-1)^{th}$ node is its south neighboring node and the $(j+1)^{th}$ node is its north neighboring node.

Equation (2.23) is only valid for interior nodes, which are nodes with both south and north neighboring points i.e. $2 \le j \le 4$. For the two boundary nodes, node 1 and node 5, we will have to apply the two boundary conditions in Equation (2.21) to establish the finite difference equations needed at these two nodes. The Dirichlet condition at node 1 is simply

$$u_1 = 0 \tag{2.25}$$

If we view this condition in light of Equation (2.23) the coefficients and constant are

$$as_1 = 0, \; ap_1 = 1, \; an_1 = 0, \; b_1 = 0 \tag{2.26}$$

The first-order derivative in the Neumann condition at $y = 1$, that is node 5, should be approximated with a finite difference formula. Since this node does not have a north neighboring point, we cannot use a central difference. Instead, a backward difference e.g. the first-order backward difference is used:

$$\frac{u_5 - u_4}{\delta y} = 0 \qquad (2.27)$$

Again if we view this condition in light of Equation (2.23) the coefficients and constant are

$$as_5 = -\frac{1}{\delta y}, \quad ap_5 = \frac{1}{\delta y}, \quad an_5 = 0, \quad b_5 = 0 \qquad (2.28)$$

Now we can write down the finite difference equations at all five nodes

$$
\begin{aligned}
u_1 &= 0 \\
\frac{1}{\delta y^2} u_1 - \frac{2}{\delta y^2} u_2 + \frac{1}{\delta y^2} u_3 &= -1 \\
\frac{1}{\delta y^2} u_2 - \frac{2}{\delta y^2} u_3 + \frac{1}{\delta y^2} u_4 &= -1 \\
\frac{1}{\delta y^2} u_3 - \frac{2}{\delta y^2} u_4 + \frac{1}{\delta y^2} u_5 &= -1 \\
-\frac{1}{\delta y} u_4 + \frac{1}{\delta y} u_5 &= 0
\end{aligned}
\qquad (2.29)
$$

This is a tridiagonal system of linear equations because when written in a matrix form,

$$
\begin{bmatrix}
ap_1 & an_1 & 0 & 0 & 0 & 0 \\
as_2 & ap_2 & an_2 & 0 & 0 & 0 \\
0 & as_3 & ap_3 & an_3 & 0 & 0 \\
0 & 0 & \ddots & \ddots & \ddots & 0 \\
0 & 0 & \ddots & \ddots & \ddots & 0 \\
0 & 0 & 0 & as_{N-1} & ap_{N-1} & an_{N-1} \\
0 & 0 & 0 & 0 & as_N & ap_N
\end{bmatrix}
\times
\begin{pmatrix}
u_1 \\ u_2 \\ u_3 \\ \vdots \\ u_{N-1} \\ u_N
\end{pmatrix}
=
\begin{pmatrix}
b_1 \\ b_2 \\ b_3 \\ \vdots \\ b_{N-1} \\ b_N
\end{pmatrix}
\qquad (2.30)
$$

or more concisely

$$Au = b \qquad (2.31)$$

the coefficient matrix A only has nonzero elements along three diagonals. Here the total number of nodes, which is also the total number of unknowns and the total number of finite difference equations is $N = 5$.

Once we create the coefficient matrix A and the constant vector b in a MATLAB program, the u values can be found easily by using MATLAB left division, $u = A\backslash b$. Since most of the elements of A are zeros, we can create it as a sparse matrix which only stores the nonzero elements to save the computer memory and speed up the data access. A MATLAB function for this purpose can be found in Appendix A.1 Assemble Diagonal Vectors to Form Coefficient Matrix.

Although the left division is very quick to solve small- to medium-size systems of equations, its calculation speed becomes less satisfactory as the number of node points becomes large. A very efficient method (numerical algorithm) of solving large tridiagonal systems of equations is the tridiagonal matrix algorithm (TDMA) also known as Thomas algorithm.

2.1.3 TDMA ALGORITHM

The idea behind the TDMA algorithm is to pursue a solution in the form of

$$\phi_i = Q_i - P_i\phi_{i+1} \tag{2.32}$$

For example, the first of a tridiagonal system of finite difference equations is

$$ap_1 \cdot \phi_1 + an_1 \cdot \phi_2 = b_1 \tag{2.33}$$

hence $\phi_1 = Q_1 - P_1\phi_2$ in which

$$Q_1 = \frac{b_1}{ap_1}, \quad P_1 = \frac{an_1}{ap_1} \tag{2.34}$$

Substitute this result into the second finite difference equation

$$as_2 \cdot \phi_1 + ap_2 \cdot \phi_2 + an_2 \cdot \phi_3 = b_2 \tag{2.35}$$

we will be able to find the constants Q_2 and P_2 in $\phi_2 = Q_2 - P_2\phi_3$. Then we go ahead to substitute this formula into the next finite difference equation to find Q_3 and P_3, and so on.

One may find that P_i and Q_i satisfy the following recursive formulae

$$P_i = \frac{an_i}{ap_i - as_i P_{i-1}} \tag{2.36}$$

and

$$Q_i = \frac{b_i - as_i Q_{i-1}}{ap_i - as_i P_{i-1}} \tag{2.37}$$

The last Q, say Q_N, is equal to the last unknown ϕ_N

$$\phi_N = Q_N \tag{2.38}$$

Then the other ϕs can be found with a back-substitution process: first $\phi_{N-1} = Q_{N-1} - P_{N-1}\phi_N$, then ϕ_{N-2} and so on until we find ϕ_1. A MATLAB function that realizes the TDMA algorithm is given in Appendix A.2 TDMA Algorithm (Thomas Algorithm).

The numerical and exact solutions are compared in Table 2.2 and Figure 2.4. The exact solution is

TABLE 2.2

Numerical Results with a 5-node Mesh

Node	y	Numerical Solution	Exact Solution	Error
2	0.25	0.1875	0.2188	0.0313
3	0.50	0.3125	0.3750	0.0625
4	0.75	0.3750	0.4688	0.0938
5	1.00	0.3750	0.5000	0.1250

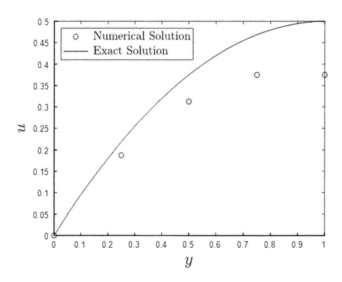

FIGURE 2.4 Numerical results with a 5-node mesh.

TABLE 2.3

Numerical Results with a 9-node Mesh

Node	y	Numerical Solution	Exact Solution	Error
3	0.25	0.2031	0.2188	0.0156
5	0.50	0.3438	0.3750	0.0313
7	0.75	0.4219	0.4688	0.0469
9	1.00	0.4375	0.5000	0.0625

$$u(y) = y - \frac{y^2}{2} \qquad (2.39)$$

To explore the mesh dependency of the discrepancy between the numerical solution and the exact solution, which is also called the (global) error in the numerical solution, we recalculate the flow using 9 nodes so that the mesh size δy is reduced to 1/2 of the original value by using 5 nodes. The results are shown in Table 2.3 and Figure 2.5.

Compared with the results shown in Table 2.2, the error at every interior point on the refined mesh is very close to 1/2 of the error at the same location on the coarser mesh. This observation implies that the error is proportional to the mesh size δy, that is, our numerical scheme is only first-order accurate everywhere (excluding the Dirichlet boundary of course).

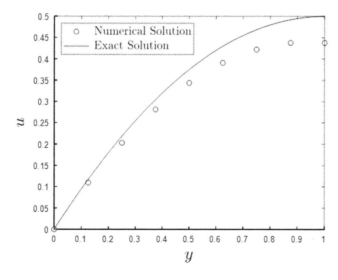

FIGURE 2.5 Numerical results with a 9-node mesh.

This observation is a little surprising since we used the second-order accurate central difference at all nodes except at the boundary $y = 1$ where a first-order accurate backward difference was adopted. So a lower-order accurate finite difference at even one single point is going to pull down the accuracy of the whole solution. Is this always the case?

Let us answer this question by using a different method, the finite volume method, to solve the same channel flow problem.

2.2   FINITE VOLUME METHOD

The finite difference method replaces the derivatives in a differential equation with finite differences. The finite volume method, on the other hand begins with integrating the differential equation over a control volume.

For the channel flow example, we first split the y-range into 5 equal control volumes (or "cells"), with one node at the center of each control volume, as shown in Figure 2.6.

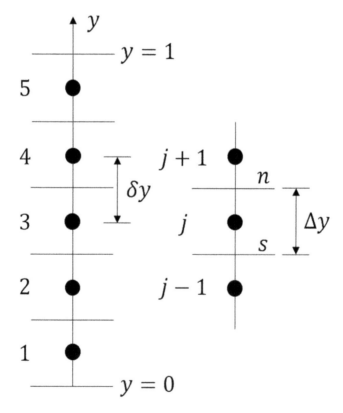

FIGURE 2.6   5-node finite volume mesh.

By integrating (2.20) over the jth control volume from its south (s) face to its north (n) face (see Figure 2.6) we have

$$\int_s^n \frac{d^2u}{dy^2}dy = \int_s^n \frac{1}{\mu}\frac{dp}{dx}dy \tag{2.40}$$

That is

$$\left(\frac{du}{dy}\right)_n - \left(\frac{du}{dy}\right)_s = \frac{1}{\mu}\frac{dp}{dx}\Delta y \tag{2.41}$$

where Δy is the size of the control volume, that is the distance between the north and the south surfaces of the control volume.

If we add up equations like Equation (2.41) over all control volumes, we will have

$$\mu\left(\frac{du}{dy}\right)_{y=1} - \mu\left(\frac{du}{dy}\right)_{y=0} = \frac{dp}{dx}(1) \tag{2.42}$$

This is the exact force balance in the channel if we multiply both sides of the equation by the channel surface area: the pressure force balances the viscous forces. So a good feature of the finite volume method is that it preserves the conservativeness of the governing differential equation if the conservative form of governing equation is used (see problem 1 of exercises at the end of Chapter 1).

Now we will have to make some approximations to the two face derivatives in Equation (2.41). A simple choice is using the central difference

$$\left(\frac{du}{dy}\right)_n \approx \frac{u_{j+1} - u_j}{\delta y}; \quad \left(\frac{du}{dy}\right)_s \approx \frac{u_j - u_{j-1}}{\delta y} \tag{2.43}$$

We have

$$\frac{u_{j+1} - 2u_j + u_{j-1}}{\delta y} = \frac{1}{\mu}\frac{dp}{dx}\Delta y \rightarrow \frac{u_{j-1} - 2u_j + u_{j+1}}{\delta y^2} = -1 \tag{2.44}$$

Notice $\delta y = \Delta y$ for a uniform mesh (see Figure 2.6) and the pressure gradient value has been given in the problem statement. This is the finite difference equation we need. You may find it is exactly the same as the equation derived by using the finite difference method, Equation (2.22).

For the two volumes next to the boundaries ($j = 1$ and $j = 5$), special treatment has to be practiced. At $y = 0$, that is the south face of control volume 1 (denoted by CV 1 in the equation below), the derivative cannot be found with Equation (2.43) because there is no node 0. We then simply use a first-order accurate forward difference

$$\left(\frac{du}{dy}\right)_{s \text{ of } CV\ 1} \approx \frac{u_1 - u(y = 0)}{0.5\Delta y} = \frac{2u_1}{\Delta y} = \frac{2u_1}{\delta y} \tag{2.45}$$

since we have the boundary condition $u(y = 0) = 0$; and the finite difference equation for control volume 1 then becomes

$$\frac{u_2 - u_1}{\delta y^2} - \frac{2u_1}{\delta y^2} = \frac{u_2 - 3u_1}{\delta y^2} = -1 \tag{2.46}$$

At $y = 1$, the upper boundary, instead of using Equation (2.43) we directly apply the boundary condition

$$\left(\frac{du}{dy}\right)_{n \text{ of } CV\ 5} = 0 \tag{2.47}$$

So the finite difference equation for control volume 5 is

$$0 - \frac{u_5 - u_4}{\delta y^2} = \frac{u_4 - u_5}{\delta y^2} = -1 \tag{2.48}$$

Now we can write down finite difference equations for all five control volumes

$$
\begin{array}{lllll}
-\frac{3}{\delta y^2}u_1 & +\frac{1}{\delta y^2}u_2 & & & = -1 \\
\frac{1}{\delta y^2}u_1 & -\frac{2}{\delta y^2}u_2 & +\frac{1}{\delta y^2}u_3 & & = -1 \\
& \frac{1}{\delta y^2}u_2 & -\frac{2}{\delta y^2}u_3 & +\frac{1}{\delta y^2}u_4 & = -1 \\
& & \frac{1}{\delta y^2}u_3 & -\frac{2}{\delta y^2}u_4 & +\frac{1}{\delta y^2}u_5 & = -1 \\
& & & \frac{1}{\delta y^2}u_4 & -\frac{1}{\delta y^2}u_5 & = -1
\end{array}
\tag{2.49}
$$

which again can be written in the same matrix form as Equations (2.30) and (2.31). They can be solved by using the MATLAB left division or the TDMA algorithm.

The numerical results are summarized in Table 2.4 of which the upper half is for a 5-node mesh and the lower half for a 15-node mesh.

These results show that the error on the coarser mesh is almost exactly $9 = 3^2$ times the error at the same physical position on a mesh three times finer. Therefore, we reached the second-order accuracy since the error is $\mathcal{O}(\delta y^2)$. Previously, we used a first-order accurate backward difference, Equation (2.45), in deriving the finite

TABLE 2.4
Numerical Results with the Finite Volume Method

Node	y	Numerical Solution	Exact Solution	Error
1	0.1	0.10	0.095	−0.005
2	0.3	0.26	0.255	−0.005
3	0.5	0.38	0.375	−0.005
4	0.7	0.46	0.455	−0.005
5	0.9	0.50	0.495	−0.005
Node	y	Numerical Solution	Exact Solution	Error
2	0.1	0.0956	0.0950	−0.000555
5	0.3	0.2556	0.2550	−0.000555
8	0.5	0.3756	0.3750	−0.000555
11	0.7	0.4556	0.4550	−0.000555
14	0.9	0.4956	0.4950	−0.000555

difference equation for control volume 1, Equation (2.46). Let us examine the truncation error thus introduced.

Equation (2.46), repeated below, is used to approximate the governing equation, (2.20), also repeated blow. At node 1.

$$\frac{u_2 - 3u_1}{\delta y^2} = -1$$
$$\left(\frac{d^2u}{dy^2}\right)_1 = -1 \tag{2.50}$$

Using Taylor's series

$$u_2 = u_1 + u_1'\delta y + \frac{1}{2}u_1''\delta y^2 + \frac{1}{6}u_1'''\delta y^3 + \dots$$
$$u(y=0) = 0 = u_1 - \frac{1}{2}u_1'\delta y + \frac{1}{8}u_1''\delta y^2 - \frac{1}{48}u_1'''\delta y^3 + \dots \tag{2.51}$$

so

$$u_2 - 3u_1 = -2u_1 + u_1'\delta y + \frac{1}{2}u_1''\delta y^2 + \frac{1}{6}u_1'''\delta y^3 + \dots$$
$$0 = 2u_1 - u_1'\delta y + \frac{1}{4}u_1''\delta y^2 - \frac{1}{24}u_1'''\delta y^3 + \dots \tag{2.52}$$

Adding up the two formulas in Equation (2.52), we have

$$u_2 - 3u_1 = \frac{3}{4}u_1''\delta y^2 + \frac{1}{8}u_1'''\delta y^3 + \dots \tag{2.53}$$

Therefore, the truncation error at the first node point is

$$\frac{u_2 - 3u_1}{\delta y^2} - u_1'' = E_1 = -\frac{1}{4}u_1'' + \frac{1}{8}u_1'''\delta y + \ldots \qquad (2.54)$$

Surprisingly the truncation error at node 1 is $\mathscr{O}(1)$. That is, no matter how fine a mesh we use, the truncation error at node 1 is always finite (the value is 0.25 in fact). However, the solution is still of the second-order accuracy. Therefore, the conjecture that using a lower-order accurate finite difference at the boundary always pulls down the accuracy of the whole solution is not correct.

To understand why this is the case, let us first introduce some concepts[1].

2.3   ERROR ANALYSIS

2.3.1   BOUNDED SCHEME AND POSITIVE SCHEME

A bounded finite difference scheme L_Δ is one which generates bounded numerical solutions i.e. solutions that satisfy the maximum principle (see Section 1.2.2).

For example, if we approximate a steady-state transport equation of ϕ with a nonpositive source term

$$L\phi = S \leq 0 \qquad (2.55)$$

by using a bounded scheme at interior points:

$$L_\Delta\phi_i = S_i, \ \ 1 < i < N \qquad (2.56)$$

the numerical solution thus obtained at the interior points will be bounded by the larger one of the numerical solutions at the two boundaries:

$$\phi_i \leq \max(\phi_1, \phi_N), \ \ 1 < i < N \qquad (2.57)$$

Notice that this statement is independent of the boundary condition treatment practices. If, moreover, one of the boundary conditions, for instance the condition at $i = N$, can be approximated by

$$L_\Delta'\phi_N = S_N \qquad (2.58)$$

in which L_Δ' is also a bounded finite difference and $S_N \leq 0$, then we will have a stronger conclusion:

$$\phi_i \leq \phi_1, \ \ 1 < i \leq N \qquad (2.59)$$

A finite difference scheme L_Δ

$$L_\Delta \phi_i = \sum_k a_k \phi_{i+k} \qquad (2.60)$$

is a positive scheme if

$$\begin{cases} \sum_k a_k = 0 \\ a_k < 0 \, if \, k \neq 0 \end{cases} \qquad (2.61)$$

where a_ks are constant coefficients and ks are certain integers. For example, the central difference used to approximate a second-order derivative (more precisely, a negative second-order derivative)

$$L_\Delta u_i = \sum_{k=-1}^{1} a_k u_{i+k} = \frac{-u_{i-1} + 2u_i - u_{i+1}}{\delta x^2} \qquad (2.62)$$

is positive since

$$a_{-1} = a_1 = -\frac{1}{\delta x^2} < 0, \, a_0 = \frac{2}{\delta x^2}, \, a_{-1} + a_0 + a_1 = 0 \qquad (2.63)$$

A simple way to ensure if a scheme is positive is to apply this scheme to equation $L_\Delta \phi_i = 0$, then move all ϕ terms, except ϕ_i to the right side of the equation. The scheme is positive if all coefficients before these ϕs are positive and the summation of coefficients on the right side is equal to the coefficient before ϕ_i on the left. Applying this method to the central difference we get

$$\frac{2}{\delta x^2} u_i = \frac{1}{\delta x^2} u_{i-1} + \frac{1}{\delta x^2} u_{i+1} \qquad (2.64)$$

and it is obviously a positive scheme.

2.3.2 Properties of Positive Scheme

A good feature of positive schemes is that they are bounded schemes. The reason is as follows. If L_Δ is a positive scheme as defined in Equation (2.60), and we use such a scheme to approximate a transport equation of ϕ with a zero source term at interior points i.e. $L_\Delta \phi_i = S_i = 0$ for $1 < i < N$, then the interior point solution is

$$\phi_i = (-a_{-1}\phi_{i-1} - a_1\phi_{i+1} - \ldots)/a_0 \qquad (2.65)$$

Since all coefficients are positive in this equation and

$$-a_{-1} - a_1 - \ldots = a_0 \tag{2.66}$$

ϕ_i is a weighted average of ϕ_{i-1}, ϕ_{i+1}, etc., so it cannot exceed the maximum value of these ϕ values. That is, ϕ_i refuses to be a maximum. Since none of the interior point solutions are maximum, the maximum can only be obtained at the boundary nodes i.e. $\phi_i \le \max\left(\phi_1, \phi_N\right)$. If the positive scheme is also used at one of the boundary nodes, say at point N, then ϕ_N also refuses to be the maximum. Therefore, $\phi_i \le \phi_1$. We have the same conclusion when $S_i < 0$. Comparing this conclusion with the definition of the bounded scheme, it is clear that a positive linear scheme i.e. a scheme that can be written in the form of Equation (2.60), is a bounded scheme. If a linear scheme is not positive, it is always possible for such a scheme to produce unbounded results. You may verify this conclusion by changing the coefficients in Equation (2.64) to make it a nonpositive scheme, and feeding it different numbers. You will find that sometimes u_i is not bounded by u_{i-1} and u_{i+1}. So an (unconditionally) bounded linear scheme has to be positive. Therefore, boundedness and positiveness are interchangeable terms for linear schemes.

Because it is desirable to preserve boundedness in our numerical calculations, positive schemes were recommended by many authorities, such as (Patankar, 1980) and (Versteeg & Malalasekera, 2007). However, an unbounded scheme does not necessarily always produce unbounded solutions. We will lose many valuable schemes if we exclude all nonpositive schemes from consideration. This point will be expounded in Chapter 3.

For a positive scheme L_Δ, we have (check it!)

$$\left|L_\Delta\right|\phi_i\left| = \sum_k a_k \left|\phi_{i+k}\right| \le \left|\sum_k a_k \phi_{i+k}\right| = \left|L_\Delta \phi_i\right| \tag{2.67}$$

If we use a positive scheme L_Δ to approximate the transport equation of ϕ with source term S at the interior nodes and one of the boundary nodes, say $L_\Delta\phi_i = S_i$ for $1 < i \le N$, the absolute value of the solution at these points should satisfy

$$L_\Delta|\phi_i| \le |L_\Delta\phi_i| = |S_i| \tag{2.68}$$

We may even be able to estimate the magnitude of $|\phi_i|$ without solving such finite difference equations. For example, if we can find a nonnegative function f so that

$$|S_i| \le L_\Delta f_i \tag{2.69}$$

for $1 < i \leq N$, we then have

$$L_\Delta |\phi_i| \leq |S_i| \leq L_\Delta f_i \tag{2.70}$$

or

$$L_\Delta (|\phi_i| - f_i) = L_\Delta \phi_i' = S_i' \leq 0 \tag{2.71}$$

where we define $\phi' = |\phi| - f$.

Since L_Δ is a positive scheme, it is a bounded scheme. For a bounded scheme with a nonpositive source term S' applied to interior points and one of the boundaries $(1 < i \leq N)$, we must have

$$\phi_i' \leq \phi_1' \tag{2.72}$$

that is

$$|\phi_i| - f_i \leq |\phi_1| - f_1 \tag{2.73}$$

Notice that f is a function we choose, so we know its values at all points; ϕ_1 is a boundary ϕ value. Let us assume it is known e.g. a Dirichlet condition is imposed at point 1. Then we can estimate the magnitude of ϕ_i without solving the finite difference equations:

$$|\phi_i| \leq f_i + |\phi_1| - f_1 \tag{2.74}$$

2.3.3 GLOBAL ERROR AND TRUNCATION ERROR

Suppose a differential equation

$$L\phi = b \tag{2.75}$$

is approximated by a finite difference equation

$$L_\Delta \phi = b \tag{2.76}$$

where b is a known function.

The exact solution of the differential equation is denoted by ϕ^{exact}, which of course satisfies the differential equation:

$$L\phi^{exact} = b \tag{2.77}$$

Similarly, the solution we obtain from the finite difference equation, $\phi^{numerical}$, satisfies the finite difference equation

$$L_\Delta \phi^{numerical} = b \tag{2.78}$$

We define the global error as the difference between the exact solution and the numerical solution:

$$e = \phi^{exact} - \phi^{numerical} \tag{2.79}$$

Obviously, we are really interested in the global error.

Truncation error, on the other hand, is the difference between $L_\Delta \phi$ and $L\phi$. Notice ϕ is the function whose derivatives are approximated by finite differences (see Figure 2.1). Since the exact solution ϕ^{exact} is a function whose differential equation is approximated by a finite difference equation, the truncation error is indeed

$$E = L_\Delta \phi^{exact} - L\phi^{exact} \tag{2.80}$$

Since

$$L\phi^{exact} = b = L_\Delta \phi^{numerical} \tag{2.81}$$

we have

$$\begin{aligned} E = L_\Delta \phi^{exact} - L\phi^{exact} &= L_\Delta \phi^{exact} - L_\Delta \phi^{numerical} \\ &= L_\Delta (\phi^{exact} - \phi^{numerical}) = L_\Delta e \end{aligned} \tag{2.82}$$

as long as the finite difference operator L_Δ is linear. The truncation error is, therefore, related to the global error but they are neither same nor are they necessarily of the same order of magnitude.

This relationship can be readily verified with the data we obtained for the channel flow example.

The finite difference equations used in Section 2.2 and the leading terms of their truncation errors are summarized below (I violate the convention of using j as the index along the y-direction for the sake of presentation consistency):

$$L_\Delta u_i = \begin{cases} \dfrac{u_2 - 3u_1}{\delta y^2} = -1 & i = 1 \\[2mm] \dfrac{u_{i-1} - 2u_i + u_{i+1}}{\delta y^2} = -1 & 1 < i < N, \\[2mm] \dfrac{u_{N-1} - u_N}{\delta y^2} = -1 & i = N \end{cases}$$

(2.83)

$$E_i \approx \begin{cases} -\dfrac{1}{4}u_1'' & i = 1 \\[2mm] \dfrac{1}{12}u_i''''\delta y^2 & 1 < i < N \\[2mm] -\dfrac{1}{24}u_N'''\delta y & i = N \end{cases}$$

The truncation error then can be evaluated with the help of the exact solution, Equation (2.39). You may find $u_i'' = -1$ and $u_i''' = u_i'''' = ...=0$ for this flow. The truncation error at node 1 is then

$$E_1 \approx -\frac{1}{4}u_i'' = 0.25$$

(2.84)

At the same node,

$$L_\Delta e_1 = \frac{e_2 - 3e_1}{\delta y^2}$$

(2.85)

which can be evaluated with the global error values listed in Table 2.4: for the 5-node mesh, $e_2 = e_1 = -0.005$ and $\delta y = 0.2$. Therefore, $L_\Delta e_1 = 0.25$, which equals E_1. For the 15-node mesh, $e_2 = e_1 = -0.000555$ and $\delta y = 0.06667$, you may find $L_\Delta e_1 = 0.25 = E_1$ as well.

Although truncation error and global error can differ by many orders of magnitude, we may still estimate the global error based on the truncation error, thanks to the properties of positive schemes discussed in the previous section.

2.3.4 Global Error Estimation

Assume that a positive scheme L_Δ is used at the interior as well as at one of the boundary points ($1 < i \le N$) in a certain numerical calculation. Since

$$L_\Delta e_i = E_i$$

(2.86)

if we can find a nonnegative function f so that for $1 < i \le N$

$$|E_i| \le L_\Delta f_i$$

(2.87)

we must have

$$|e_i| \leq f_i + |e_1| - f_1 \tag{2.88}$$

in virtue of Equation (2.74). So to evaluate the magnitude of global error $|e_i|$, the key is to find a proper function f that satisfies Equation (2.87). Such a function is called barrier function (Wesseling, 2001).

I find that the following form of barrier function seems to always work:

$$f(y) = C\delta y^p (a + by + cy^2) \tag{2.89}$$

where C, a, b, c and p are constants to be determined. I will call this function the universal barrier function.

Let us estimate the magnitude of global error due to the finite volume method formulas used in Section 2.2. The finite difference equations and the leading terms of their truncation errors are summarized blow. This time, I changed the sign on both sides of these finite difference equations so that they are closer to the form of Equations (2.60) and (2.61).

$$L_\Delta u_i = \begin{cases} \frac{3u_1 - u_2}{\delta y^2} = 1 & i = 1 \\ \frac{-u_{i-1} + 2u_i - u_{i+1}}{\delta y^2} = 1 & 1 < i < N, \\ \frac{u_N - u_{N-1}}{\delta y^2} = 1 & i = N \end{cases}$$

$$E_i \approx \begin{cases} \frac{1}{4}u_1'' & i = 1 \\ -\frac{1}{12}u_i''''\delta y^2 & 1 < i < N \\ \frac{1}{24}u_N'''\delta y & i = N \end{cases} \tag{2.90}$$

We use the universal barrier function $f = C\delta y^p (a + by + cy^2)$ and require that

$$|E_i| \leq L_\Delta f_i \tag{2.91}$$

for $1 < i \leq N$ (notice that the finite difference schemes are positive at these points), that is

$$\begin{cases} \frac{1}{12}|u_i''''|\delta y^2 \leq \frac{-f_{i-1} + 2f_i - f_{i+1}}{\delta y^2} & 1 < i < N \\ \frac{1}{24}|u_N'''|\delta y \leq \frac{f_N - f_{N-1}}{\delta y^2} & i = N \end{cases} \tag{2.92}$$

You may find that

$$
\begin{cases}
\dfrac{-f_{i-1} + f_i - f_{i+1}}{\delta y^2} = -2Cc\delta y^p & 1 < i < N \\[4mm]
\dfrac{f_N - f_{N-1}}{\delta y^2} = C[(b + 2c)\delta y^{p-1} - 2c\delta y^p] & i = N
\end{cases}
\tag{2.93}
$$

Obviously the inequities in (2.92) can be satisfied for no matter how small a δy if we select a $c < 0$, a $b > -2c$, a p with $p \leq 2$ and a large enough C. To reach the most accurate estimate of the global error, we should choose $p = 2$.

Based on Equation (2.88), we have

$$
|e_i| \leq f_i + |e_1| - f_1
\tag{2.94}
$$

for $1 < i \leq N$. We are now very close to an estimate of the global error magnitude, except that we do not know $|e_1|$.

This difficulty can be addressed if we notice that we can make $L_\Delta |e|_1 \leq |L_\Delta e_1| = |E_1| \leq L_\Delta f_1$ i.e.

$$
\frac{3|e_1| - |e_2|}{\delta y^2} \leq \frac{|3e_1 - e_2|}{\delta y^2} = |E_1| = \frac{1}{4}|u_1''| \leq \frac{3f_1 - f_2}{\delta y^2} = C\left(2a - \frac{3}{2}c\delta y^2\right)
\tag{2.95}
$$

if we choose a positive a and a large enough C (notice that the barrier function f is positive in the y interval considered with the selected C, a, b, and c values). Therefore, we have $L_\Delta (|e| - f)_1 \leq 0$, that is

$$
\frac{3(|e_1| - f_1) - (|e_2| - f_2)}{\delta y^2} \leq 0
\tag{2.96}
$$

or

$$
3\left(|e_1| - f_1\right) \leq |e_2| - f_2
\tag{2.97}
$$

Now since for any interior point, we have (see Equation (2.94))

$$
|e_i| - f_i \leq |e_1| - f_1
\tag{2.98}
$$

The second point, point 2 is an interior point, so

$$
|e_2| - f_2 \leq |e_1| - f_1
\tag{2.99}
$$

Combine this result with Equation (2.97), we conclude that

$$|e_1| - f_1 \le 0 \tag{2.100}$$

Therefore, Equation (2.98) infers that

$$|e_i| - f_i \le 0 \tag{2.101}$$

whence

$$|e_i| \le f_i = \mathcal{O}(\delta y^2) \tag{2.102}$$

The numerical scheme derived in Section 2.2 is therefore of the second-order accuracy.

Based on these results, we find that the finite volume method has some unique advantages over the finite difference method. For example, the accuracy of the finite volume method does not deteriorate because of the one-order-less accurate treatment at the boundaries[2]. The finite volume method can be easily applied to the unstructured mesh, which is usually challenging in finite difference method (see Chapter 6). The finite volume method also automatically preserves conservativeness, since it recovers the conservation law by integrating the conservative form of governing equations.

2.4   FINITE VOLUME METHOD (CONTINUED)

2.4.1   VIRTUAL CONTROL VOLUME AND VIRTUAL NODE

Although finite volume method has a number of beneficial features, we noticed that the two boundary control volumes need special treatments. This can be inconvenient, especially, for multidimensional problems. Hereby we introduce another way to handle boundary conditions, namely using virtual control volumes and virtual nodes, which are also known as ghost control volumes and ghost nodes. The channel flow example will be employed again for this purpose.

As shown in Figure 2.7, two virtual control volumes with a virtual node at the center of each volume are added beyond the two boundaries. Therefore, if there are N real control volumes in between $y = 0$ and 1, the two virtual control volumes make this number $N + 2$. So the numbering of nodes is such that the "real" nodes begin with node 2 and end with node $N + 1$.

With the virtual nodes, the real nodes next to boundaries, nodes 2 and $N + 1$, now have both south and north neighboring nodes so that the finite difference equation for the interior nodes can be applied to these two nodes.

The boundary conditions can be very easily handled by setting proper equations at the virtual nodes.

At the lower boundary $y = 0$, $u = 0$. To implement this condition we can assume that this velocity equals the average of velocities at the two nodes, nodes 1 and 2, that sandwich the boundary:

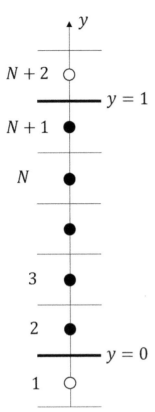

FIGURE 2.7   Virtual control volumes and virtual nodes.

$$u(y = 0) \approx \frac{u_1 + u_2}{2} = 0 \tag{2.103}$$

This provides an extra finite difference equation for virtual node 1.

If we combine the finite difference equation at node 2,

$$\frac{u_1 - 2u_2 + u_3}{\delta y^2} = -1 \tag{2.104}$$

with Equation (2.103), we obtain the original finite difference equation for the first real node derived in Section 2.2:

$$\frac{-3u_2 + u_3}{\delta y^2} = -1 \tag{2.105}$$

At the upper boundary $y = 1$, the boundary condition is $du/dy = 0$, which can be approximated by a central difference:

$$\left(\frac{du}{dy}\right)_{y=1} \approx \frac{u_{N+2} - u_{N+1}}{\delta y} = 0 \qquad (2.106)$$

Again if one combines this equation with the one being used at node $N + 1$

$$\frac{u_N - 2u_{N+1} + u_{N+2}}{\delta y^2} = -1 \qquad (2.107)$$

one has

$$\frac{u_N - u_{N+1}}{\delta y^2} = -1 \qquad (2.108)$$

which again is exactly the same as the finite difference equation for the last real node derived in Section 2.2. So this method of virtual nodes, while much simpler to implement, is essentially the same as the boundary treatments without using virtual nodes. Let us solve another example with the finite volume method and virtual nodes.

2.4.2 EXAMPLE: CHANNEL FLOW CONSISTING OF TWO IMMISCIBLE FLUIDS

As seen in Figure 2.8, two immiscible fluids occupy the upper and lower halves of a horizontal channel. The density and viscosity of fluid 1 (ρ_1 and μ_1) are different from those of fluid 2 (ρ_2 and μ_2). So it is a very simple two-phase flow. We want to determine the steady-state, fully developed flow velocity distribution in these two fluids driven by a constant pressure gradient dp/dx.

The x-momentum equation (Equation (1.12); Chapter 1) that takes into account the variable viscosity reduces to

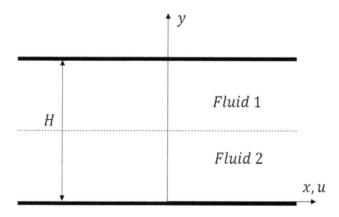

FIGURE 2.8   Channel flow consisting of two immiscible fluids.

$$-\frac{\partial}{\partial y}\left[\mu\left(\frac{\partial u}{\partial y}+\frac{\partial v}{\partial x}\right)\right]=-\frac{d}{dy}\left(\mu\frac{du}{dy}\right)=-\frac{dp}{dx} \tag{2.109}$$

The boundary conditions are the no-slip conditions at the two channel surfaces

$$u = 0 \; at \; y = 0 \; and \; y = H \tag{2.110}$$

The viscosity distribution in the channel is

$$\mu = \begin{cases} \mu_1 & y > \frac{H}{2} \\ \mu_2 & y < \frac{H}{2} \end{cases} \tag{2.111}$$

It is advantageous to normalize these equations since it is much easier to write the exact solution in the dimensionless form. We define

$$\begin{cases} \tilde{u} = \frac{u}{U}; \; U = -\frac{H^2}{\mu_1}\frac{dp}{dx} \\ \tilde{\mu} = \frac{\mu}{\mu_1} \\ \tilde{y} = \frac{y}{H} \end{cases} \tag{2.112}$$

Then the governing equation, the boundary conditions, and the viscosity distribution can be rendered dimensionless:

$$-\frac{d}{d\tilde{y}}\left(\tilde{\mu}\frac{d\tilde{u}}{d\tilde{y}}\right) = 1 \tag{2.113}$$

$$\tilde{u} = 0 \; at \; \tilde{y} = 0 \; and \; \tilde{y} = 1 \tag{2.114}$$

$$\tilde{\mu} = \begin{cases} 1 & \tilde{y} > \frac{1}{2} \\ K = \frac{\mu_2}{\mu_1} & \tilde{y} < \frac{1}{2} \end{cases} \tag{2.115}$$

The exact solution to this problem is

$$\begin{cases} \tilde{u}_1 = -\frac{\tilde{y}^2}{2} + \frac{3K+1}{4(K+1)}\tilde{y} - \frac{K-1}{4(K+1)} \\ \tilde{u}_2 = \frac{1}{K}\left[-\frac{\tilde{y}^2}{2} + \frac{3K+1}{4(K+1)}\tilde{y}\right] \end{cases} \tag{2.116}$$

We will simulate a case with viscosity ratio $K = 20$. From now on we will drop those tildes over variables as long as we understand they are dimensionless variables.

To set up the mesh, we split the y range into equal control volumes, and it is obviously a good choice to use the interface of the two fluids as a control volume surface. This requires us to use even number of control volumes. To facilitate the implementation of boundary conditions, two virtual control volumes are added as shown in Figure 2.9. Therefore, if there are $2N$ control volumes in between $y = 0$ and 1, the two virtual control volumes make this number $2N + 2$. So the node numbering is such that the "real" nodes begin with node 2 and end with node $2N + 1$. The fluid interface is in between control volume $N + 1$ and volume $N + 2$.

By integrating Equation (2.113) over the jth control volume from its south face to its north face, we have

$$\int_s^n \frac{d}{dy}\left(\mu\frac{du}{dy}\right)dy = \int_s^n 1\,dy \tag{2.117}$$

Remember all variables are dimensionless. Therefore, we have

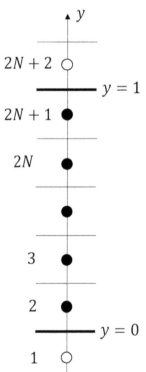

FIGURE 2.9 Finite volume mesh for the channel flow consisting of two immiscible fluids.

TABLE 2.5

Numerical Results with a 10-Interior-Node Mesh

Node	y	Numerical Solution	Exact Solution	Percentage Difference
2	0.05	0.00187	0.00175	−6.44
3	0.15	0.00510	0.00488	−4.38
4	0.25	0.00783	0.00751	−4.19
5	0.35	0.01006	0.00965	−4.31
6	0.45	0.01179	0.01128	−4.58
7	0.55	0.01414	0.02196	35.63
8	0.65	0.02878	0.03458	16.80
9	0.75	0.03341	0.03720	10.19
10	0.85	0.02805	0.02982	5.95
11	0.95	0.01268	0.01244	−1.94

$$\mu_n \frac{u_{j+1} - u_j}{\Delta y} - \mu_s \frac{u_j - u_{j-1}}{\Delta y} = \Delta y \qquad (2.118)$$

Now we face a problem: how to evaluate μ at control volume surfaces? I guess your response probably would be the same as mine, which is using the average of μ values at the two nodes straddling the surface. With this intuition, Equation (2.118) becomes

$$- \left[\left(\frac{\mu_{j+1} + \mu_j}{2} \right) \frac{u_{j+1} - u_j}{\Delta y} - \left(\frac{\mu_j + \mu_{j-1}}{2} \right) \frac{u_j - u_{j-1}}{\Delta y} \right] = \Delta y \qquad (2.119)$$

which is

$$- \left(\frac{\mu_j + \mu_{j-1}}{2} \right) u_{j-1} + \left(\frac{\mu_{j-1} + 2\mu_j + \mu_{j+1}}{2} \right) u_j - \left(\frac{\mu_{j+1} + \mu_j}{2} \right) u_{j+1} = \Delta y^2 \quad (2.120)$$

We can thus see that this scheme is positive. The μ values are determined by

$$\mu_j = \begin{cases} 1 & j > N + 1 \\ K & j \leq N + 1 \end{cases} \qquad (2.121)$$

At the boundaries we simply use

$$u_1 + u_2 = 0; \ u_{2N+1} + u_{2N+2} = 0 \qquad (2.122)$$

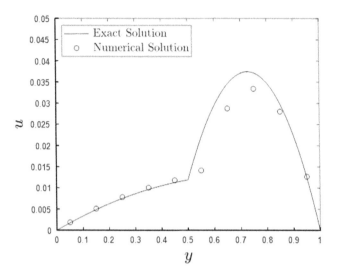

FIGURE 2.10   Numerical results with a 10-interior-node mesh.

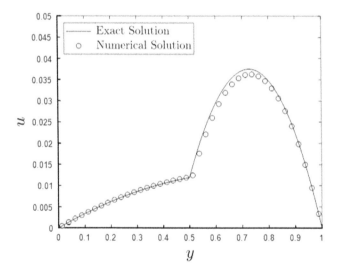

FIGURE 2.11   Numerical results with a 40-interior-node mesh.

which serves as finite difference equations for the two virtual nodes, node 1 and node $2N + 2$. In this way, node 2 and node $2N + 1$ do not need any special treatment, but use the same equations as the other interior nodes i.e. Equation (2.120).

The finite difference equations again form a tridiagonal system and can be solved with the MATLAB left division or TDMA. We use 10 interior nodes, and the results are shown in Table 2.5 and Figure 2.10.

Obviously the numerical solution is not satisfactory. You may suspect it is because the mesh is not fine enough and therefore we have significant errors. Well, let

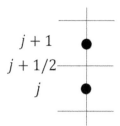

$j + 1$

$j + 1/2$

j

FIGURE 2.12   Interface between two control volumes.

us try using 40 interior nodes. The numerical solution does improve but still clearly deviates from the exact solution as shown in Figure 2.11, which is quite unusual compared with the channel flow with one single fluid (see Figure 2.5). It implies that probably there is something wrong with our numerical scheme.

The reason for the unsatisfactory results is our intuitive treatment of viscosity at the two fluid interface. We should make use of harmonic instead of algebraic average of viscosities of the two fluids at this interface.

As seen in Figure 2.12, the shear stress at the interface $j + 1/2$ between control volume j and control volume $j + 1$ can be evaluated in two ways

$$\tau_{j+\frac{1}{2}} \approx \mu_{j+1} \frac{u_{j+1} - u_{j+\frac{1}{2}}}{\frac{\Delta y}{2}}, \; also \; \tau_{j+\frac{1}{2}} \approx \mu_j \frac{u_{j+\frac{1}{2}} - u_j}{\frac{\Delta y}{2}} \tag{2.123}$$

According to Newton's third law, these two shear stresses should be equal, so we have

$$\tau_{j+\frac{1}{2}} \approx \frac{u_{j+1} - u_{j+\frac{1}{2}}}{\frac{\Delta y}{2\mu_{j+1}}} = \frac{u_{j+\frac{1}{2}} - u_j}{\frac{\Delta y}{2\mu_j}} = \frac{u_{j+1} - u_j}{\frac{\Delta y}{2\mu_{j+1}} + \frac{\Delta y}{2\mu_j}} \tag{2.124}$$

Here we have made use of an identity: if $a/b = c/d$, then

$$\frac{a}{b} = \frac{c}{d} = \frac{a + c}{b + d} \tag{2.125}$$

You might verify this identity yourselves.

Now since

$$\tau_{j+\frac{1}{2}} = \mu_{j+\frac{1}{2}} \left(\frac{du}{dy} \right)_{j+\frac{1}{2}} \approx \mu_{j+\frac{1}{2}} \frac{u_{j+1} - u_j}{\Delta y} = \frac{u_{j+1} - u_j}{\frac{\Delta y}{\mu_{j+\frac{1}{2}}}} \tag{2.126}$$

We therefore have

TABLE 2.6

Numerical Results with Improved Interface Viscosity Evaluation Method

Node	y	Numerical Solution	Exact Solution	Percentage Difference
2	0.05	0.00182	0.00175	−3.57
3	0.15	0.00495	0.00488	−1.28
4	0.25	0.00758	0.00751	−0.83
5	0.35	0.00971	0.00965	−0.65
6	0.45	0.01134	0.01128	−0.55
7	0.55	0.02321	0.02196	−5.69
8	0.65	0.03583	0.03458	−3.61
9	0.75	0.03845	0.03720	−3.36
10	0.85	0.03107	0.02982	−4.19
11	0.95	0.01369	0.01244	−10.05

$$\frac{1}{\mu_{j+\frac{1}{2}}} = \frac{1}{2}\left(\frac{1}{\mu_{j+1}} + \frac{1}{\mu_j}\right) \tag{2.127}$$

That is

$$\mu_{j+\frac{1}{2}} = \frac{2}{\dfrac{1}{\mu_{j+1}} + \dfrac{1}{\mu_j}} = \frac{2\mu_j\mu_{j+1}}{\mu_j + \mu_{j+1}} \tag{2.128}$$

This is the correct way to calculate the interface viscosity, which is not our intuitive algebraic average, but a harmonic average of two neighboring nodes' viscosities. This is also the correct way to calculate the other varying diffusion coefficients at control volumes' surfaces.

By using this formula, the finite difference equation for the interior nodes reads

$$-\frac{2\mu_j\mu_{j-1}}{\mu_j + \mu_{j-1}}u_{j-1} + \left(\frac{2\mu_j\mu_{j-1}}{\mu_j + \mu_{j-1}} + \frac{2\mu_j\mu_{j+1}}{\mu_j + \mu_{j+1}}\right)u_j - \frac{2\mu_j\mu_{j+1}}{\mu_j + \mu_{j+1}}u_{j+1} = \Delta y^2 \tag{2.129}$$

Calculation results using 10 interior nodes are given in Table 2.6 and the left graph in Figure 2.13. Now the results are much more accurate. When we use 40 interior nodes, the numerical results are almost indistinguishable from the exact solution, as shown in the right graph of Figure 2.13.

This is a very simple multiphase flow example. This subject will be expounded in more detail in Chapter 7.

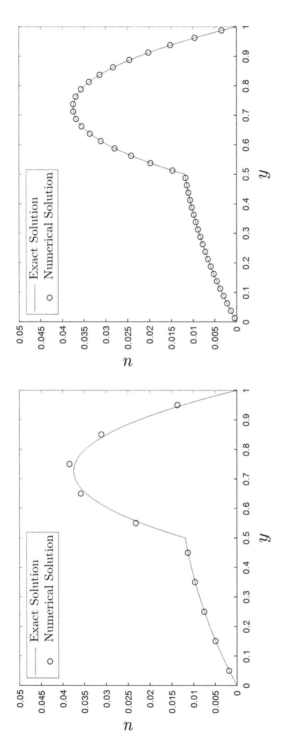

FIGURE 2.13 Numerical results with the improved interface viscosity evaluation method.

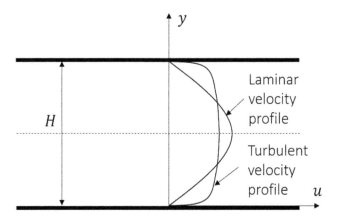

FIGURE 2.14   Velocity profiles of laminar and turbulent flows.

2.5   A GLIMPSE OF TURBULENCE

2.5.1   INTRODUCTION

In the channel flow discussed in Section 2.1.2, if the Reynolds number exceeds 1300 (Patel & Head, 1969) the flow becomes turbulent: the flow is no longer steady and regular, instead velocity at each point in the flow field continuously fluctuates in an apparently random manner. If we "zoom out" a little to observe a region instead of a point in the flow, we may find the velocity fluctuations are caused by swirls or "eddies" of various sizes, life-spans, orientations, and kinetic energy levels. It is, therefore, usually more practical to pursue the time-averaged or ensemble-averaged flow field (also called the mean flow field) rather than the instantaneous flow field of a turbulent flow. The time-averaged flow field is obtained by averaging the flow field over a long enough time period while the ensemble-averaged flow field is the mean of the flow fields realized in a large number of repetitions of the same experiment. For a steady turbulent flow, the time-average and ensemble-average are equivalent (Pope, 2000).

If we compare the time-averaged velocity distribution of a turbulent channel flow with that of a laminar channel flow, we may readily find that the former is no longer a parabolic profile as the latter. Instead, the turbulent mean velocity profile is "flatter" in most of the channel, while being "steeper" close to the channel walls compared with its laminar counterpart, as shown in Figure 2.14.

Due to the steep velocity gradient at the wall, a turbulent flow experiences more significant wall friction and pressure drop than a laminar flow does at the same Reynolds number. For example, at Reynolds number $Re = 10^5$, the wall friction of a turbulent channel flow is about 33 times that of a laminar channel flow at the same Reynolds number, if the laminar flow status could be maintained at such a high Reynolds number.

The characteristics of the turbulent mean velocity profile are due to the turbulent eddy motion, which transports fluid particles from one region to another. Since fluid particles with different velocities mix together, a rather uniform mean velocity distribution arises in most of the channel except in close proximity to the walls where the mean velocity drops to zero in a short distance and results in a large velocity gradient at the wall.

Notice that the turbulent eddy motion and regular diffusion have very similar effects: they both promote mixing and smooth out fluid property distributions in a flow; they both contribute to the wall friction.

The resemblance between the turbulent eddy motion and the diffusion motivated the concept of so-called eddy viscosity, or turbulent viscosity μ_T, which is the counterpart of the molecular viscosity μ but usually is much larger than μ since the turbulent eddy motion mixes the fluid particles at a rate much greater than that of the molecular motion.

With the eddy viscosity included in the picture, the Navier–Stokes equation for the steady-state fully developed turbulent channel flow shown in Figure 2.14 is then

$$-\frac{d}{dy}\left[(\mu + \mu_T)\left(\frac{d\bar{u}}{dy} + \frac{d\bar{v}}{dx}\right)\right] = -\frac{d}{dy}\left[(\mu + \mu_T)\frac{d\bar{u}}{dy}\right] = -\frac{d\bar{p}}{dx} \quad (2.130)$$

where $\bar{u}$ and $\bar{v}$ are the time-averaged velocities ($\bar{v} = 0$) and $\bar{p}$ the time-averaged pressure. Notice that the flow is steady in the statistical sense i.e. the time-averaged velocity is time-independent although the instantaneous velocity fluctuates all the time. To solve such equations we need the eddy viscosity values, which unlike the regular viscosity, cannot be found in a fluid property table. This is because the eddy viscosity is a direct consequence of the turbulent motion, which of course is not a fluid physical property. Determination of the turbulent eddy viscosity is the central task of turbulent models.

2.5.2 MIXING LENGTH MODEL

Just like the regular molecular viscosity (μ) is proportional to the product of fluid density, mean molecular velocity, and mean free path of molecules (Boltzmann, 2011), the eddy viscosity μ_T is usually modeled as the product of fluid density, the mean eddy velocity scale, and mean eddy length scale at the point where the eddy viscosity is evaluated:

$$\mu_T = \rho u_{eddy} l_{eddy} \quad (2.131)$$

Prandtl (Goldstein, 1938) proposed the mixing length model in 1925 to estimate u_{eddy} and l_{eddy}. According to this model

$$u_{eddy} \approx l_{eddy}\left|\frac{d\bar{u}}{dy}\right| \quad (2.132)$$

Therefore,

$$\mu_T = \rho l_{eddy}^2 \left| \frac{d\bar{u}}{dy} \right| \tag{2.133}$$

Prandtl further argued that the mean eddy size should be restricted by its distance from the nearest solid wall y_{wall}:

$$l_{eddy} \approx \kappa y_{wall} \tag{2.134}$$

where κ is a constant. Based on experimental measurements, $\kappa \approx 0.41$. This argument makes sense since an eddy close to a wall cannot be too big, otherwise it will penetrate the wall. It has been found that this model only works in a certain range of y_{wall}. For very small y_{wall} values, that is in the region very close to a wall, the eddies are impaired by viscosity. One way to take into account this effect is adding the van Driest damping function (van Driest, 1956) to Equation (2.134):

$$l_{eddy} = \kappa y_{wall} \left(1 - e^{-\frac{y^+}{26}} \right) \tag{2.135}$$

where

$$y^+ = \frac{\rho u_\tau y_{wall}}{\mu} ; \; u_\tau = \sqrt{\frac{\tau_{wall}}{\rho}} ; \; \tau_{wall} = \mu \left(\frac{d\bar{u}}{dy} \right)_{y=0} \tag{2.136}$$

Here τ_{wall} is simply the wall shear stress (friction force acting on per unit wall area); u_τ is the so-called friction velocity since it is based on the wall friction and has the same unit of a velocity. y^+ is a dimensionless distance from the wall. You might have noticed that the definition of y^+ is similar to the definition of a Reynolds number. Hence, y^+ is dimensionless.

I know your feeling when you see these formulas for the first time. They are out of blue, frustrating, and seem to be intractable. You may feel a little better once you try to put them into a program.

Now the situation is like this: the time-averaged velocity has to be solved from the Navier–Stokes equation, Equation (2.130), which depends on the eddy viscosity; but eddy viscosity depends on the time-averaged velocity, see Equation (2.133). So the solution of turbulent flows is intrinsically iterative.

2.5.3 EXAMPLE: TURBULENT CHANNEL FLOW

Let us simulate a turbulent channel flow with the following parameters. The fluid is water, whose density is $\rho = 1000 \; kg/m^3$, viscosity $\mu = 0.00112 \; Pa{\cdot}s$. The height of the channel is $H = 0.2\ m$. Because of symmetry, we only need to simulate half the channel. The pressure gradient that drives the flow is $d\bar{p}/dx = -0.21\ Pa/m$.

We want to determine

1. The time-averaged velocity distribution $\bar{u}(y)$.
2. The turbulent shear stress τ_{xy}^{tur} distribution. It is

$$\tau_{xy}^{tur} = \mu_t \frac{d\bar{u}}{dy} \tag{2.137}$$

3. The u^+ versus y^+ distribution. u^+ is the dimensionless velocity normalized by u_τ:

$$u^+ = \frac{\bar{u}}{u_\tau} \tag{2.138}$$

4. The Reynolds number based on the bulk velocity U_b and channel height H. The bulk velocity U_b is

$$U_b = \frac{1}{H} \int_0^H \bar{u}dy \tag{2.139}$$

5. The friction coefficient c_f, which is defined as

$$c_f = \frac{\tau_{wall}}{\frac{1}{2}\rho U_b^2} \tag{2.140}$$

We will use the same mesh as the one shown in Figure 2.7, this time with 300 interior nodes. Integrating Equation (2.130) over the jth control volume from its south surface to its north surface, we have

$$-\mu_s^{tot}\bar{u}_{j-1} + \left(\mu_s^{tot} + \mu_n^{tot}\right)\bar{u}_j - \mu_n^{tot}\bar{u}_{j+1} = \left(-\frac{d\bar{p}}{dx}\right)\Delta y^2 \tag{2.141}$$

where the total viscosity

$$\mu^{tot} = \mu + \mu_T = \mu + \rho\kappa^2 y^2\left(1 - e^{-\frac{y^+}{26}}\right)^2\left|\frac{d\bar{u}}{dy}\right| \tag{2.142}$$

The procedure of solving this problem is

1. Assign an initial guess of the time-averaged velocity distribution to $\bar{u}$. The laminar flow velocity distribution can be used.
2. Calculate the wall shear stress and u_τ, then calculate the eddy viscosity at each face of every control volume.
3. Use the TDMA algorithm to solve the system of equations given by Equation (2.141) so that we have a new velocity distribution.
4. Repeat steps (2) and (3) until the difference between two consecutive solutions is less than a prescribed small tolerance.

The calculated mean velocity normalized by its maximum, the normalized turbulent shear stress, and u^+ distributions are compared with the direct numerical simulation (DNS) data (Mansour, Kim, & Moin, 1988) in Figures 2.15, 2.16, and 2.17.

As you can see, the agreement between the current computational fluid dynamics (CFD) results and DNS data is very good in most of the channel.

You may observe from Figure 2.17 that there is a region close to the wall in which u^+ is roughly a linear function of $\ln y^+$:

$$u^+ = \frac{1}{\kappa} \ln y^+ + C \tag{2.143}$$

where $\kappa \approx 0.41$; $C \approx 5$. This is the so-called log-law of the wall, which is found to be approximately valid in a wide range of wall-bounded turbulent flows. The log-law region usually begins at $y^+ \approx 30$. In the zone really close to the wall ($y^+ \lesssim 5$), we also have a very nice formula:

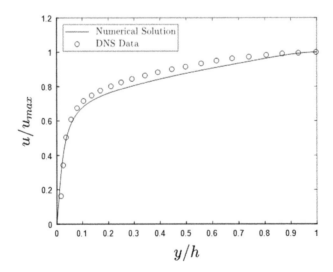

FIGURE 2.15   Mean velocity distribution.

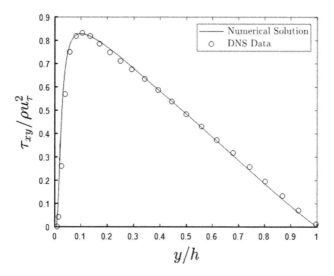

FIGURE 2.16 Turbulent shear stress distribution.

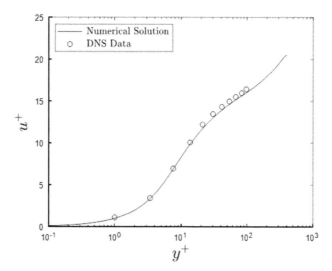

FIGURE 2.17 u^+ versus y^+ distribution.

$$u^+ = y^+ \tag{2.144}$$

More information about these regions will be given in Chapter 8.

Based on the mean velocity profile obtained, we can evaluate the bulk velocity according to (2.139). Then we may determine the Reynolds number based on the bulk velocity U_b and channel height H, which is about 1.38×10^4. This value agrees well with the exact value 1.398×10^4 (Abe, Kawamura, & Matsuo, 2001).

The calculated friction coefficient $c_f = 2\tau_w/(\rho U_b^2) \approx 0.0067$, which compares favorably with the exact value 0.0065 (Abe, Kawamura, & Matsuo, 2001).

The mixing length turbulence model works well in the current channel flow example. However, it might not be able to handle more complicated turbulent flows with flow separation, complex boundaries, rotation, and so on. More advanced turbulence models like the k-ϵ model, usually, have to be used for better prediction of such flows. We will discuss this important topic in Chapter 8.

2.6 THE CFD PROCEDURE

At this point, you should have already had some rough ideas on how to use CFD to solve a fluid flow problem. What is the starting point of the CFD procedure? Does it begin with the mesh generation? Or maybe the code writing?

Well, it is hard to tell what the exact starting point of a CFD project is. My personal experience is that a CFD work usually does not begin with a computer, but a stack of paper. I have to write down the governing equations, the mesh arrangements, the numerical schemes used to discretize each term of the governing differential equation, and the numerical algorithms for solving the finite difference equations on paper sheets. This paperwork has to be done carefully and in detail to an extent that I can compare my computer code with it to figure out what is wrong with the code (or what is wrong with my numerical method). I recommend you to do the same.

Although this step may be a little time-consuming at first, it will save you enormous time eventually. In fact you will want to make almost all the important decisions at this step because many such decisions are mutually dependent. If you skip this step, you may end up with having to start all over again when your code is almost done because you find out, for instance, that there are no efficient ways to solve the finite difference equations based on the very good numerical scheme you choose.

Once you complete this documentation, the rest of the CFD procedure is only an implementation of the plan laid out in this document with a computer code: creating the mesh, setting up the finite difference equations according to the numerical schemes, and solving these equations using the numerical algorithms. Of course, some extra work is needed to verify the correctness of the numerical results (see Section 5.7 in Chapter 5).

Exercises

1. Develop a second-order accurate finite difference approximation for $(d\phi/dx)_i$ on a nonuniform mesh, as shown in Figure 2.18, using the information (ϕ and x values) of mesh points $i - 1$, i and $i + 1$. Suppose $\delta x_i = x_{i+1} - x_i = \alpha \delta x_{i-1} = \alpha (x_i - x_{i-1})$.

2. Use the scheme you developed for problem 1 to evaluate the derivative of $\phi(x) = \sin(x - x_i + 1)$ at point i. Suppose $\Delta x_{i-1} = 0.02$ and $\Delta x_i = 0.01$.

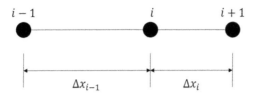

FIGURE 2.18 Nonuniform mesh.

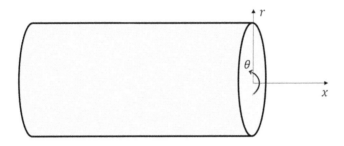

FIGURE 2.19 Turbulent pipe flow.

Compare your numerical result with the exact solution, which is $\cos(1)$. Then halve both Δx_{i-1} and Δx_i, and redo the calculation. Is the scheme truly second-order accurate? You will find later that we do not need such a complex interpolation to reach second-order accuracy on a nonuniform mesh.

3. Reproduce the calculation presented in Section 2.1.2.
4. Show that the finite difference method used in Section 2.1.2 is first-order accurate by using the global error estimate technique.
5. Show that although the method used in Section 2.1.2 is only first-order accurate, the global error at a fixed index, say $j = 3$, is $\mathcal{O}(\delta y^2)$ when we refine the mesh.
6. Reproduce the numerical results presented in Section 2.4.2 by using correct interfacial viscosity treatment.
7. Reproduce the CFD results presented in Section 2.5.3.
8. Use the mixing length model to simulate the turbulent flow in a horizontal pipe. For such a flow, Nikuradse (Nikuradse, 1950) suggested the following expression for the mixing length

$$\frac{l_{mix}}{R} = 0.14 - 0.08\left(1 - \frac{y}{R}\right)^2 - 0.06\left(1 - \frac{y}{R}\right)^4 \qquad (2.145)$$

where R is the pipe radius and y the distance from the nearest pipe surface. You may need to add the van Driest damping function to this formula.

Notice that in the cylindrical coordinates (see Figure 2.19) the x-momentum equation is

$$0 = -\frac{d\bar{p}}{dx} + \frac{1}{r}\frac{d}{dr}\left[r\left(\mu_{mix} + \mu\right)\frac{d\bar{u}}{dr}\right] \qquad (2.146)$$

for this steady-state fully developed turbulent flow.

The fluid density is $\rho = 1.23 \, kg/m^3$; viscosity is $\mu = 1.79 \times 10^{-5} \, Pa\cdot s$; the pressure gradient along the flow direction is $d\bar{p}/dx = -1.112 \times 10^3 \, Pa/m$; the pipe radius is $2 \times 10^{-3} \, m$.

You should produce the following results:

1. The mean velocity distribution along a radius from the pipe wall to the pipe center.
2. The friction factor, which is

$$f = -\frac{D}{\frac{1}{2}\rho U_b^2}\frac{d\bar{p}}{dx} \qquad (2.147)$$

and the bulk velocity is defined as

$$U_b = \frac{2\int_0^R \bar{u}rdr}{R^2}. \qquad (2.148)$$

3. The Reynolds number based on the bulk velocity and pipe diameter.

4. The u^+ vs. r^+ plot on a semi-logarithmic graph

9. An interesting fluid flows through a horizontal channel. The flow is a fully developed laminar flow and the pressure gradient that drives the flow is $dp/dx = -0.01 \, Pa/m$. The fluid is interesting because its viscosity is not constant but changing with its distance from the lower surface of the channel: $\mu = (0.4 - 0.1y) \, Pa\cdot s$. The height of the channel is $H = 2 \, m$. What is the maximum flow speed? Give four accurate significant digits after the decimal point.

10. A long uranium rod of radius $R = 0.1 \, m$ is stored in a water pool. The rod temperature distribution satisfies the following equation:

$$\frac{1}{r}\frac{d}{dr}\left(kr\frac{dT}{dr}\right) + S = 0 \qquad 2.149$$

in which $k = 20/Wm\cdot K$ is the thermal conductivity of uranium and $S = 1000W/m^3$ is the rate of heat generation due to nuclear decay. The boundary condition at the rod center is $dT/dr = 0$ because of symmetry; the condition at the rod surface is $-k(dT/dr) = h(T - T_w)$. Here $h = 50\,W/m^2{\cdot}K$ is the convective heat transfer coefficient due to the natural convection of water. The water temperature is $T_w = 20°C$. What is the temperature at the rod center?

NOTES

1 You can skip most of the Section 2.3 if you are not interested in the math. You should, however, read the paragraphs about boundedness, positiveness, and their relationship; the concept of global error and its relation to the truncation error are also important.
2 This conclusion is invalid for the compact finite volume method schemes, which are sensitive to the boundary treatments (Kobayashi, 1999).

3 Numerical Schemes

3.1 SCHEMES FOR TIME ADVANCING

3.1.1 Example: Start-Up of Couette Flow

As seen in Figure 3.1, a channel formed by two infinitely large parallel plates is filled with an incompressible viscous fluid. The fluid is initially at rest. At a certain moment, say $t = 0$, the upper plate suddenly begins to move at a constant speed u_s to the right. Fluid in the channel then begins to flow. We want to calculate the fluid velocity distribution in the channel at a later time moment.

For this flow, the u-momentum equation reduces to

$$\frac{\partial u}{\partial t} - \nu \frac{\partial^2 u}{\partial y^2} = 0 \tag{3.1}$$

where $\nu = \mu/\rho$ is the kinematic viscosity whose unit is m^2/s. This is a 1-D unsteady diffusion equation. Since the flow is driven by the boundary movement instead of a pressure gradient, the pressure gradient is zero in this flow.

Obviously, the velocity distribution $u(t, y)$ is affected by factors including channel height, upper boundary velocity, and fluid viscosity. To reduce the number of parameters involved and make the final results more general, we can define some dimensionless variables:

$$\tilde{t} = \frac{t}{H/u_s}; \; \tilde{y} = \frac{y}{H}; \; \tilde{u} = \frac{u}{u_s} \tag{3.2}$$

Using these definitions, the governing equation is normalized as

$$\frac{\partial \tilde{u}}{\partial \tilde{t}} - \frac{1}{Re} \frac{\partial^2 \tilde{u}}{\partial \tilde{y}^2} = 0 \tag{3.3}$$

where

$$Re = \frac{u_s H}{\nu} \tag{3.4}$$

is the Reynolds number.

The initial and boundary conditions are

$$\tilde{u}(0, \tilde{y}) = 0 \tag{3.5}$$

DOI: 10.1201/9781003138822-3

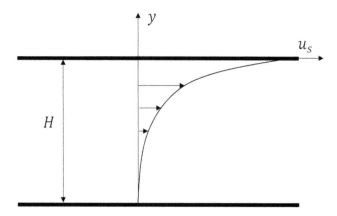

FIGURE 3.1   Start-up of Couette flow example setup.

and

$$\begin{cases} \tilde{u}(\tilde{t},\ 0) = 0 \\ \tilde{u}(\tilde{t},\ 1) = 1 \end{cases} \tag{3.6}$$

You can see the evolution of the dimensionless velocity is only controlled by one single parameter, namely the Reynolds number.

We will drop these tildes above variables from now on, but we should keep in mind that these variables are dimensionless.

We will use the finite difference method instead of finite volume method for this example.

We first discretize the y-range from $y = 0$ to $y = 1$ with $N + 1$ equally spaced node points, as seen in Figure 3.2. The grid spacing is therefore $\Delta y = 1/N$.

The temporal derivative in Equation (3.3) is approximated with a one-sided difference and the spatial derivative with the familiar central difference

$$\frac{u_j^n - u_j^{n-1}}{\Delta t} - \frac{1}{Re} \frac{u_{j-1} - 2u_j + u_{j+1}}{\Delta y} = 0 \tag{3.7}$$

Here the superscripts n and $n - 1$ are not exponents, instead they denote the u values at the present, n^{th} time moment, and the previous, $(n - 1)^{th}$ time moment, respectively. The interval between two consecutive time moments is Δt. The initial and boundary conditions are discretized as

$$u_j^1 = 0; \begin{cases} u_1^n = 0 \\ u_{N+1}^n = 1 \end{cases}; j = 1,\ 2,\ \dots,\ N;\ n = 1,\ 2,\ \dots \tag{3.8}$$

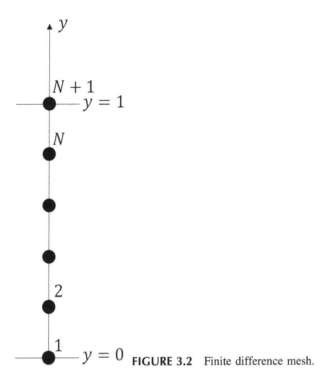

FIGURE 3.2 Finite difference mesh.

Now we face a problem: at which time moment should we evaluate u values in the central difference? Well, we have several options here and studying the different performance of these options is the main focus of this section.

3.1.2 Euler Explicit Scheme

If we use the u values at the last time moment in the diffusion term, we have the so-called Euler explicit scheme (or simply explicit scheme if no confusion is caused):

$$\frac{u_j^n - u_j^{n-1}}{\Delta t} - \frac{1}{Re} \frac{u_{j-1}^{n-1} - 2u_j^{n-1} + u_{j+1}^{n-1}}{\Delta y^2} = 0 \tag{3.9}$$

The truncation error of this scheme is $\mathcal{O}(\Delta t) + \mathcal{O}(\Delta y^2)$ since a first-order forward difference and a second-order central difference are utilized.

To solve the u distribution at a certain time moment with the explicit scheme, we begin with the known initial ($n = 1$ i.e. $t = 0$) velocity distribution u_j^1, then we march to the second time moment ($n = 2$, $t = \Delta t$) and calculate the u_j^2 values (notice it is not u_j squared) directly from Equation (3.9) since they are the only unknowns in such equations. That is, we can explicitly express the u values at the second time moment with the known u values at the first time moment:

$$u_j^2 = \frac{\Delta t}{Re\Delta y^2}u_{j-1}^1 + \left(1 - \frac{2\Delta t}{Re\Delta y^2}\right)u_j^1 + \frac{\Delta t}{Re\Delta y^2}u_{j+1}^1 \tag{3.10}$$

This is why we call this scheme explicit. After we obtain u_j^2, we can go ahead to compute u_j^3. So on and so forth we will eventually find the velocity distribution at the desired final time moment.

Let us solve the velocity distribution at the final moment $t = 0.1$ with $Re = 1$. We use $\Delta t = 0.01$ and $\Delta y = 0.1$ in our calculations. The result is shown in Figure 3.3.

The result is obviously very bad. In fact if we continue the calculation, the solution then keeps growing without limit. Such a solution is said to diverge. Why do we have such a bad result? Is it because we used too large Δt and Δy values that led to large truncation errors?

If you halve both Δt and Δy values and repeat the calculation, you may find the result becomes even worse: the velocity magnitude then reaches a few million.

The bad solution is actually due to the instability of the scheme with the chosen Δt and Δy values.

3.1.3 CONSISTENCY, STABILITY, AND CONVERGENCE

A numerical scheme is said to be unstable if the magnitude of the numerical solution grows without limit as the computation process continues[1]. An unstable scheme is therefore always unbounded. We hence immediately reach an important conclusion: a bounded scheme is always stable (otherwise it would be unbounded, but it is bounded). However, a stable scheme is not always bounded (in the sense of the maximum principle), as we will see in some examples later.

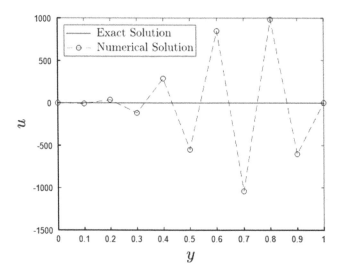

FIGURE 3.3 Numerical results with the Euler explicit method.

An unstable scheme, of course, always gives rise to divergent solutions. On the other hand, a consistent, stable scheme is generally convergent. That is, the numerical solution of the finite difference scheme approaches the exact solution of the differential equation which the numerical scheme approximates as the temporal and spatial mesh is refined. This observation is most clearly summarized in the famous Lax equivalence theorem (Richtmyer & Morton, 1967):

Given a properly posed initial-value problem and a finite difference approximation to it that satisfies the consistency condition, stability is the necessary and sufficient condition for convergence.

This theorem is applicable in the strict sense only to linear differential equations, yet it provides invaluable insight into nonlinear problems as well. This theorem tells us that we should only use consistent and stable schemes in numerical calculations.

A finite difference scheme is consistent if the truncation error approaches zero as the temporal and spatial mesh sizes approach zero[2]. Most finite difference schemes are consistent. Let us see how to determine the stability of a scheme. I will use the explicit scheme for the 1-D unsteady diffusion equation as an example.

3.1.4 VON NEUMANN AND MATRIX STABILITY ANALYSIS METHODS

It is possible to treat a numerical solution u_j^n of a finite difference equation as a superposition of a certain number of wave modes i.e. Fourier modes (Deen, 1998):

$$u_j^n = \sum_m A_m^n e^{i\xi_m y_j} = \sum_m A_m^n \left[\cos\left(\xi_m y_j\right) + i \sin\left(\xi_m y_j\right) \right] \tag{3.11}$$

where i is the imaginary unit i.e. $\sqrt{-1}$. ξ_m is the wave number of the mth wave mode and A_m^n is its amplitude at the nth moment.

Obviously if one or more of these wave modes keep growing in the computation process, the solution will eventually become infinite and the scheme is thus unstable.

Let us focus on the mth wave mode and see how it evolves:

$$u_{m,j}^{n-1} = A_m^{n-1} e^{i\xi_m y_j}; \quad u_{m,j}^n = A_m^n e^{i\xi_m y_j} \tag{3.12}$$

The equation that governs the evolution of this wave mode is the finite difference equation (since u_j^n is the solution to the finite difference equation):

$$u_{m,j}^n = u_{m,j}^{n-1} + r\left(u_{m,j-1}^{n-1} - 2u_{m,j}^{n-1} + u_{m,j+1}^{n-1}\right) \tag{3.13}$$

where $r = \Delta t/(Re\Delta y^2)$ is the so-called diffusion number.

Substituting Equation (3.12) into Equation (3.13) results in

$$A_m^n e^{i\xi_m y_j} = rA_m^{n-1} e^{i\xi_m y_{j-1}} + (1 - 2r)A_m^{n-1} e^{i\xi_m y_j} + rA_m^{n-1} e^{i\xi_m y_{j+1}} \qquad (3.14)$$

Notice that since

$$y_{j\mp1} = y_j \mp \Delta y \qquad (3.15)$$

Equation (3.14) is indeed

$$A_m^n e^{i\xi_m y_j} = rA_m^{n-1} e^{i\xi_m y_j}\left(e^{i\xi_m \Delta y} + e^{-i\xi_m \Delta y}\right) + (1 - 2r)A_m^{n-1} e^{i\xi_m y_j} \qquad (3.16)$$

Dividing both sides of this equation by $A_m^{n-1} e^{i\xi_m y_j}$ yields the ratio of wave amplitudes at two consecutive moments, or the so-called amplification factor of this specific wave mode

$$G_m = \frac{A_m^n}{A_m^{n-1}} = r(e^{i\xi_m \Delta y} + e^{-i\xi_m \Delta y}) + (1 - 2r) = 2r\cos\theta_m + (1 - 2r)$$
$$= 1 - 2r(1 - \cos\theta_m) = 1 - 4r\sin^2\frac{\theta_m}{2} \qquad (3.17)$$

where $\theta_m = \xi_m \Delta y$. Notice that

$$e^{i\theta_m} + e^{-i\theta_m} = 2\cos\theta_m \qquad (3.18)$$

and

$$1 - \cos\theta_m = 2\sin^2\frac{\theta_m}{2} \qquad (3.19)$$

The scheme is stable if the magnitude of the wave amplitude does not grow:

$$|G_m| = \left|\frac{A_m^n}{A_m^{n-1}}\right| = \left|1 - 4r\sin^2\frac{\theta_m}{2}\right| \le 1 \qquad (3.20)$$

There are two possible situations: if $1 - 4r\sin^2\theta_m/2 \ge 0$, then we require $1 - 4r\sin^2\theta_m/2 \le 1$, which is obviously always true; if $1 - 4r\sin^2\theta_m/2 < 0$, the scheme can only be stable when

$$-\left(1 - 4r\sin^2\frac{\theta_m}{2}\right) \le 1 \rightarrow 2r\sin^2\frac{\theta_m}{2} \le 1 \qquad (3.21)$$

This is guaranteed to be true for an arbitrary θ_m if

$$r \le \frac{1}{2} \qquad (3.22)$$

that is

$$\Delta t \leq \frac{Re}{2}\Delta y^2 \tag{3.23}$$

This is the stability condition of the explicit scheme when applied to the 1-D unsteady diffusion equation.

You might have a small concern: how can the velocity grow from 0 at the initial moment to a positive value at the final moment if none of its wave components grow? Well, it can happen. Say at a certain node, $u = (0.5) + (-0.5) = 0$ (suppose) initially, if the (0.5) mode does not change but the (-0.5) mode decays with time, then u increases with time until it reaches its steady-state value 0.5.

This stability analysis approach is the famous von Neumann stability analysis method (Tannehill, Anderson, & Pletcher, 1997). The main advantage of this method is that it gives us a physical picture (growth/decay of wave modes) to think of and work with. The main disadvantage of this method is that this physical picture may need careful interpretation in some cases e.g. when the transport phenomenon occurs in a finite physical or numerical domain. This disadvantage can be largely avoided by using the matrix stability analysis method (Tannehill, Anderson, & Pletcher, 1997), which is introduced now.

The finite difference equations of the current example can be written in a matrix form

$$\begin{pmatrix} u_1^n \\ u_2^n \\ u_3^n \\ \vdots \\ u_N^n \\ u_{N+1}^n \end{pmatrix} = \begin{bmatrix} 1 & 0 & 0 & \cdots & 0 & 0 \\ r & 1-2r & r & \cdots & 0 & 0 \\ 0 & r & 1-2r & r & 0 & 0 \\ 0 & 0 & \ddots & \ddots & \ddots & 0 \\ 0 & 0 & \ddots & \ddots & \ddots & 0 \\ 0 & 0 & \cdots & r & 1-2r & r \\ 0 & 0 & \cdots & 0 & 0 & 1 \end{bmatrix} \times \begin{pmatrix} u_1^{n-1} \\ u_2^{n-1} \\ u_3^{n-1} \\ \vdots \\ u_N^{n-1} \\ u_{N+1}^{n-1} \end{pmatrix} \tag{3.24}$$

or more concisely

$$u^n = Au^{n-1} \tag{3.25}$$

I will call A scheme matrix. Notice that the Dirichlet boundary conditions imply that the two boundary u values never change. This fact has been included in the scheme matrix, and hence the boundary condition effects are included in the analysis.

As our respectable linear-algebra professors told us, an $N \times N$ square matrix has N eigenvalues (some of them may be repeated):

$$A\phi_m = \lambda_m \phi_m; \quad m = 1, 2, \ldots, N \tag{3.26}$$

Here, λ_m is the mth eigenvalue (a number) and ϕ_m the corresponding eigenvector (a vector of N numbers). u^n may be expanded into a linear combination of these eigenvectors:

$$u^n = a_1^n \phi_1 + a_2^n \phi_2 + \ldots; \; u^{n-1} = a_1^{n-1} \phi_1 + a_2^{n-1} \phi_2 + \ldots \qquad (3.27)$$

Substituting these expansions into Equation (3.25), we have

$$a_1^n \phi_1 + a_2^n \phi_2 + \ldots = a_1^{n-1} A\phi_1 + a_2^{n-1} A\phi_2 + \ldots = \lambda_1 a_1^{n-1} \phi_1 + \lambda_2 a_2^{n-1} \phi_2 + \ldots$$
$$(3.28)$$

Equating the coefficients before the same eigenvectors, we have

$$a_1^n = \lambda_1 a_1^{n-1}; \; a_2^n = \lambda_2 a_2^{n-1} \ldots \qquad (3.29)$$

The amplitude of each constitutive eigenvector in the solution thus grows at a rate equal to the corresponding eigenvalue. The situation is therefore very similar to what we had in the von Neumann method: a solution is stable if none of its eigenmodes grow i.e.

$$\max |\lambda_m| \leq 1 \qquad (3.30)$$

Since MATLAB has a built-in function "eig" that calculates matrix eigenvalues for us, we can easily determine stability conditions using Equation (3.30). For example, for the current explicit scheme, you may compose a MATLAB function, say f which receives r as the input and calculates $\max |\lambda_m| - 1$ as the output. In this way the searching for stability conditions becomes seeking roots of $f(r)$, which can be done by using the MATLAB built-in function "fzero".

For the matrix we are dealing with, we can find its eigenvalues in closed-form without the help of those MATLAB built-in functions.

If each diagonal of a matrix contains the same elements, the matrix is then a so-called Toeplitz matrix. For an $N \times N$ tridiagonal Toeplitz matrix

$$\begin{bmatrix} b & c & 0 & \cdots & 0 & 0 \\ a & b & c & \cdots & 0 & 0 \\ 0 & a & b & c & 0 & 0 \\ 0 & 0 & \ddots & \ddots & \ddots & 0 \\ 0 & 0 & \ddots & \ddots & \ddots & 0 \\ 0 & 0 & \cdots & a & b & c \\ 0 & 0 & \cdots & 0 & a & b \end{bmatrix} \qquad (3.31)$$

its eigenvalues are well-known (to mathematicians of course e.g. Noschese, Pasquini, & Reichel, 2013):

$$\lambda_m = b + 2\sqrt{ac} \cos\left(\frac{m\pi}{N+1}\right); \; m = 1, \, 2, \, ..., \, N \tag{3.32}$$

Our scheme matrix is not a Toeplitz matrix (see Equation (3.24)) but its eigenvalues are the same as those of an $(N-1) \times (N-1)$ Toeplitz matrix with $a = c = r$, $b = 1 - 2r$ plus two extra eigenvalues of the same unity value. This can be proved rather easily by using the common eigenvalue definition formula, $\det(A - \lambda I) = 0$, but I am not going to bother you with that. The stability condition then should be

$$|\lambda_m| = \left|1 - 2r + 2r \cos\left(\frac{m\pi}{N}\right)\right| \le 1; \; m = 1, \, 2, \, ..., \, N - 1 \tag{3.33}$$

This stability condition is exactly the same as the von Neumann condition, Equation (3.17). The matrix stability analysis method, hence, gives exactly the same stability condition predicted by the von Neumann method for the current case i.e. $r \le 1/2$.

You may find that the stability condition $r \le 1/2$ is also the boundedness (positiveness) condition for the current scheme. But this is only a coincidence, we will see for many schemes that the stability and boundedness conditions are different.

For the current start-up of Couette flow example, the Euler explicit scheme is stable when we use a time step size $\Delta t \le 0.5 Re\Delta y^2 = 0.005$ if we use a 11-node mesh i.e. when $\Delta y = 0.1$.

If we use a time step size complying with the stability condition, say $\Delta t = 0.004$, the numerical result at $t = 0.1$ agrees well with the exact solution, as shown in Table 3.1 and Figure 3.4.

The exact solution of this problem is

$$u(t, y) = y - 2 \sum_{n=1}^{\infty} \frac{e^{-n^2\pi^2 t} \sin[n\pi(1 - y)]}{n\pi} \tag{3.34}$$

TABLE 3.1

Numerical Results with the Stable Euler Explicit Method

Node	y	Numerical Solution	Exact Solution	Percentage Difference
2	0.1	0.0310	0.0303	−2.36
3	0.2	0.0679	0.0663	−2.35
4	0.3	0.1165	0.1139	−2.26
5	0.4	0.1816	0.1780	−2.07
6	0.5	0.2674	0.2628	−1.77
7	0.6	0.3759	0.3707	−1.39
8	0.7	0.5072	0.5022	−0.99
9	0.8	0.6586	0.6547	−0.61
10	0.9	0.8252	0.8230	−0.27

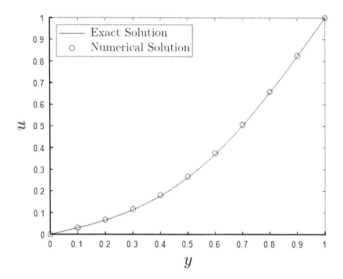

FIGURE 3.4 Numerical results with the stable Euler explicit method.

The stability analyses indicate that the explicit treatment of the diffusion term imposes a very stringent restriction on the time step size Δt: it has to be less than some constant times the grid spacing squared. For the current example, if one uses 101 nodes, the time step size Δt has to be less than 0.00005 and it takes at least 20,000 steps of calculation to obtain the solution at $t = 1$. The situation becomes even worse for 2-D and 3-D cases.

For example, if we are solving a 2-D unsteady diffusion equation

$$\frac{\partial u}{\partial t} = \frac{1}{Re}\left(\frac{\partial^2 u}{\partial x^2} + \frac{\partial^2 u}{\partial y^2}\right) \tag{3.35}$$

the finite difference formula will be

$$\frac{u_{i,j}^n - u_{i,j}^{n-1}}{\Delta t} - \frac{1}{Re}\left(\frac{u_{i-1,j}^{n-1} - 2u_{i,j}^{n-1} + u_{i+1,j}^{n-1}}{\Delta x^2} + \frac{u_{i,j-1}^{n-1} - 2u_{i,j}^{n-1} + u_{i,j+1}^{n-1}}{\Delta y^2}\right) = 0 \tag{3.36}$$

If we use a mesh with $\Delta x = \Delta y = h$, the formula becomes

$$u_{i,j}^n = u_{i,j}^{n-1} + r\left(u_{i-1,j}^{n-1} + u_{i+1,j}^{n-1} + u_{i,j-1}^{n-1} + u_{i,j+1}^{n-1} - 4u_{i,j}^{n-1}\right) \tag{3.37}$$

where the diffusion number $r = \Delta t/(Reh^2)$.

To analyze its stability, we may substitute a 2-D wave mode $u_{m,p}^n = A_{m,p}^n e^{i\left(\xi_m x_i + \xi_p y_j\right)}$ into this equation, and calculate the amplification factor of the wave mode. You may find the stability condition is

$$r \le \frac{1}{4} \tag{3.38}$$

i.e. $\Delta t \le Reh^2/4$. So, the time step size allowed for solving a 2-D flow with the Euler explicit scheme is only one-half (for 3-D, one-third) of the time step size permitted for a 1-D flow at the same Reynolds number.

For this reason the explicit scheme probably is not the best choice for multi-dimensional problems dominated by very long time scales. For such cases the implicit scheme, which will be discussed next, is likely to be more effective. One should be aware of, however, the many merits of the explicit scheme: it is easy to implement and typically demands less computer memory compared with implicit schemes; although the current simplest explicit scheme is only first-order time accurate, it can be easily transformed into second- or higher-order accurate schemes. For a problem governed by very short time scales, the explicit scheme is an attractive option since we have to use small time step sizes anyway. One example is the direct numerical simulation of turbulent flows, in which very small time step sizes have to be used to resolve the high-frequency components of turbulent fluctuations. Another example is the simulation of multiphase flows, which also requires small time step sizes to accurately capture the motion of the phase interface. These topics will be discussed in Chapters 7 and 8.

3.1.5 EULER IMPLICIT SCHEME

As mentioned previously, in many cases it is preferable to have an unconditionally stable scheme so that we can use larger time step sizes in the calculation. The Euler implicit scheme is such a scheme.

For our start-up of Couette flow example, if we use u values at the present instead of the last time moment in the diffusion term, the finite difference equation becomes

$$\frac{u_j^n - u_j^{n-1}}{\Delta t} - \frac{1}{Re}\frac{u_{j-1}^n - 2u_j^n + u_{j+1}^n}{\Delta y^2} = 0 \tag{3.39}$$

or

$$-ru_{j-1}^n + (1 + 2r)u_j^n - ru_{j+1}^n = u_j^{n-1} \tag{3.40}$$

This is an implicit scheme because multiple unknowns coexist on the left side of the finite difference equation, so they have to be solved simultaneously by using either a matrix or iterative method.

This scheme is positive and, therefore, is always bounded. A bounded scheme is always stable, therefore the scheme is unconditionally stable. This fact can be verified by using the von Neumann and matrix stability analysis methods as shown below.

Substituting a Fourier mode of the solution, $u_{m,j}^n = A_m^n e^{i\xi_m y_j}$, into Equation (3.40), one can find the amplification factor is

$$G_m = \frac{A_m^n}{A_m^{n-1}} = \frac{1}{1 + 2r - 2r \cos \theta_m} = \frac{1}{1 + 4r \sin^2 \frac{\theta_m}{2}} \tag{3.41}$$

where

$$\theta_m = \xi_m \Delta y \tag{3.42}$$

You can see since $4r \sin^2 \theta/2 \geq 0$, $|G_m| \leq 1$ always. The Euler implicit scheme is, hence, unconditionally stable.

If we use the matrix method to investigate the stability condition, we should write the finite difference equations in the matrix form

$$
\begin{bmatrix}
1 & 0 & 0 & \cdots & 0 & 0 \\
-r & 1+2r & -r & \cdots & 0 & 0 \\
0 & -r & 1+2r & -r & 0 & 0 \\
0 & 0 & \ddots & \ddots & \ddots & 0 \\
0 & 0 & \ddots & \ddots & \ddots & 0 \\
0 & 0 & \cdots & -r & 1+2r & -r \\
0 & 0 & \cdots & 0 & 0 & 1
\end{bmatrix}
\begin{pmatrix}
u_1^n \\
u_2^n \\
u_3^n \\
\vdots \\
\vdots \\
u_N^n \\
u_{N+1}^n
\end{pmatrix}
=
\begin{pmatrix}
u_1^{n-1} \\
u_2^{n-1} \\
u_3^{n-1} \\
\vdots \\
\vdots \\
u_N^{n-1} \\
u_{N+1}^{n-1}
\end{pmatrix}
\tag{3.43}
$$

or more briefly $Au^n = u^{n-1}$ i.e. $u^n = A^{-1} u^{n-1}$ where A^{-1} is the inverse matrix of A. The stability condition is then $\max |\lambda(A^{-1})| \leq 1$, or equivalently $\min |\lambda(A)| \geq 1$ because

$$\lambda(A^{-1}) = [\lambda(A)]^{-1} \tag{3.44}$$

Since $\lambda(A) = 1 + 2r - 2r \cos \theta_m$, you may find that this analysis again leads to exactly the same condition given by the von Neumann method.

The implicit scheme results at $t = 0.1$ using $\Delta t = 0.01$ and $\Delta y = 0.1$ are shown in Table 3.2 and Figure 3.5.

Because the Euler implicit scheme is unconditionally stable and bounded, the only concern we may have is its accuracy and efficiency. The truncation error of the

TABLE 3.2

Numerical Results with the Euler Implicit Method

Node	y	Numerical Solution	Exact Solution	Percentage Difference
2	0.1	0.0301	0.0303	0.65
3	0.2	0.0652	0.0663	1.70
4	0.3	0.1106	0.1139	2.87
5	0.4	0.1714	0.1780	3.70
6	0.5	0.2523	0.2628	3.97
7	0.6	0.3572	0.3707	3.66
8	0.7	0.4876	0.5022	2.91
9	0.8	0.6422	0.6547	1.91
10	0.9	0.8158	0.8230	0.88

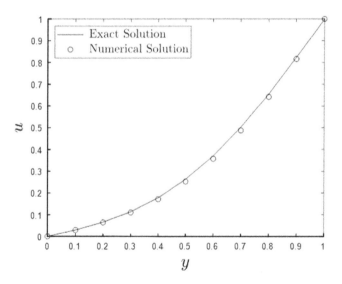

FIGURE 3.5 Numerical results with the Euler implicit method.

scheme is $\mathcal{O}(\Delta t) + \mathcal{O}(\Delta y^2)$, the same as that of the Euler explicit scheme. In terms of efficiency, carrying out one step of calculation with the implicit scheme is typically more time-consuming than using the explicit scheme, since all unknowns have to be solved simultaneously. But with the allowed larger time step sizes, the implicit scheme can reach the desired final time moment in fewer steps than the explicit scheme.

3.1.6 CRANK–NICOLSON SCHEME

If we average the explicit and implicit schemes we end up with the Crank–Nicolson scheme

$$\frac{u_j^n - u_j^{n-1}}{\Delta t} = \frac{1}{2}\left(\frac{u_{j-1}^n - 2u_j^n + u_{j+1}^n}{Re\Delta y^2} + \frac{u_{j-1}^{n-1} - 2u_j^{n-1} + u_{j+1}^{n-1}}{Re\Delta y^2}\right) \qquad (3.45)$$

which is

$$-\frac{r}{2}u_{j-1}^n + (1 + r)u_j^n - \frac{r}{2}u_{j+1}^n = \frac{r}{2}u_{j-1}^{n-1} + (1 - r)u_j^{n-1} + \frac{r}{2}u_{j+1}^{n-1} \qquad (3.46)$$

Using this scheme and $\Delta t = 0.01$ and $\Delta y = 0.1$, the result at $t = 0.1$ is shown in Table 3.3 and Figure 3.6.

You may find that the Crank–Nicolson scheme result is more accurate than both explicit and implicit scheme results. This is because, unlike the explicit and implicit schemes which are only first-order accurate in time, the Crank–Nicolson scheme is second-order accurate in both time and space: the truncation error involved is $\mathcal{O}(\Delta t^2) + \mathcal{O}(\Delta y^2)$ (see problem 4 in the exercises given at the end of this chapter).

The Crank–Nicolson scheme is unconditionally stable (see problem 5 in the exercises given at the end of this chapter). Is it also unconditionally bounded? Let us do another calculation using $\Delta t = 0.1$. The result at $t = 0.1$ (only one step of calculation is needed) is shown in Figure 3.7.

You can see that the solution is unbounded because the maximum u value exceeds the initial maximum, which is 1. That is, although the Crank–Nicolson scheme is unconditionally stable, using too large time step sizes still results in

---

TABLE 3.3
Numerical Results with the Crank–Nicolson Method

Node	y	Numerical Solution	Exact Solution	Percentage Difference
2	0.1	0.0305	0.0303	−0.80
3	0.2	0.0668	0.0663	−0.62
4	0.3	0.1143	0.1139	−0.41
5	0.4	0.1784	0.1780	−0.22
6	0.5	0.2630	0.2628	−0.09
7	0.6	0.3708	0.3707	−0.01
8	0.7	0.5021	0.5022	0.02
9	0.8	0.6545	0.6547	0.02
10	0.9	0.8229	0.8230	0.01

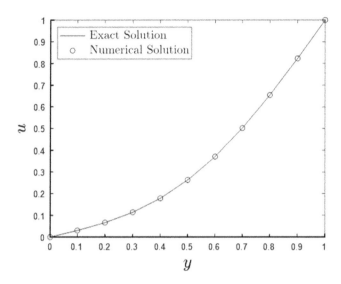

FIGURE 3.6 Numerical results with the Crank–Nicolson method.

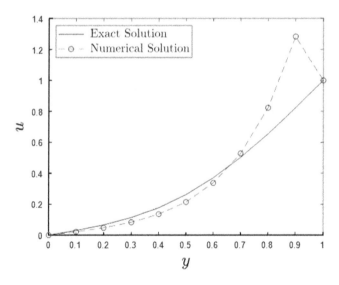

FIGURE 3.7 Unbounded Crank–Nicolson method results.

unphysical solutions. This is an example of a statement mentioned in Section 3.1.3: a bounded scheme is stable, but a stable scheme may be not bounded.

However, unlike an unstable solution, this unbounded solution remains finite all the time, as illustrated in Figure 3.8, which shows the velocity values at the point $y = 0.9$ in the channel at different time moments. This is because the Crank–Nicolson scheme is stable, so the solution cannot become infinite.

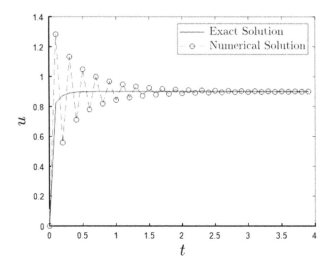

FIGURE 3.8   Velocity values at different time moments with the Crank–Nicolson method.

Under what conditions will the scheme become bounded? The Crank–Nicolson scheme is positive, therefore bounded when $r \leq 1$ (see Equation (3.46)). So if we want to use the Crank–Nicolson scheme in a time-accurate calculation, we have to make sure $r \leq 1$ to avoid wiggles in the solution[3] i.e. the time step size limit is $Re\Delta y^2$, which is only twice the maximum time step size allowed by the Euler explicit scheme (see Equation (3.23)). On the other hand, if we are only interested in the steady-state solution, we may use a much larger Δt to take advantage of the unconditional stability of the Crank–Nicolson scheme.

3.1.7 RUNGE–KUTTA SCHEMES

One disadvantage of the Crank–Nicolson scheme is that it is more difficult to implement than the explicit scheme. To ease our programming efforts but still enjoying a high-order accuracy, we can use one of the Runge–Kutta schemes.

The second-order Runge–Kutta scheme applied to our start-up of Couette flow example is as follows. We first use the Euler explicit method to predict the velocity at the new time moment:

$$u_j^{*n} = u_j^{n-1} + \Delta t \frac{u_{j-1}^{n-1} - 2u_j^{n-1} + u_{j+1}^{n-1}}{Re\Delta y^2} \tag{3.47}$$

The predicted velocities u_j^{*n}, etc. then replace u_j^n, etc. on the right side of Equation (3.45) to calculate u_j^n:

TABLE 3.4

Numerical Results with the Second-Order Runge–Kutta Method

Node	y	Numerical Solution	Exact Solution	Percentage Difference
2	0.1	0.0307	0.0303	-1.30
3	0.2	0.0670	0.0663	-0.97
4	0.3	0.1145	0.1139	-0.58
5	0.4	0.1784	0.1780	-0.24
6	0.5	0.2628	0.2628	-0.01
7	0.6	0.3703	0.3707	0.11
8	0.7	0.5015	0.5022	0.14
9	0.8	0.6539	0.6547	0.11
10	0.9	0.8226	0.8230	0.06

$$\frac{u_j^n - u_j^{n-1}}{\Delta t} = \frac{1}{2}\left(\frac{u_{j-1}^{*n} - 2u_j^{*n} + u_{j+1}^{*n}}{Re\Delta y^2} + \frac{u_{j-1}^{n-1} - 2u_j^{n-1} + u_{j+1}^{n-1}}{Re\Delta y^2}\right) \quad (3.48)$$

So in two steps we find the velocity at the new time moment and we have an explicit scheme in each step.

The solution at the time moment $t = 0.1$ obtained by using the second-order Runge–Kutta scheme with $\Delta t = 0.004$, $\Delta y = 0.1$ is summarized in Table 3.4. You may find that this result is more precise than the Euler explicit scheme result with the same Δt and Δy. This scheme, just like the Crank–Nicolson scheme, is second-order accurate in time.

The stability condition of the second-order Runge–Kutta scheme is the same as that of the Euler explicit scheme i.e. $r \le 0.5$. This conclusion can be made by using the von Neumann stability analysis method as follows.

If we substitute $u_{m,j}^n = A_m^n e^{i\xi_m y_j}$, etc. into Equations (3.47) and (3.48), we have

$$A_m^{*n} = A_m^{n-1} - \lambda_m^2 r A^{n-1} = (1 - \lambda_m^2 r)A^{n-1} \quad (3.49)$$

and

$$A_m^n - A_m^{n-1} = \frac{\Delta t}{2}\left(\frac{-\lambda_m^2 A^{*n}}{Re\Delta y^2} + \frac{-\lambda_m^2 A^{n-1}}{Re\Delta y^2}\right) = -\frac{\lambda_m^2 r}{2}(A^{*n} + A^{n-1}) \quad (3.50)$$

where $\lambda_m^2 = 4\sin^2\left(\xi_m \Delta y/2\right)$. Combining these two formulas we have

$$G_m = \frac{A_m^n}{A_m^{n-1}} = 1 - \lambda_m^2 r + \frac{(\lambda_m^2 r)^2}{2} \tag{3.51}$$

The stability condition $|G_m| \le 1$ entails $\lambda_m^2 r \le 2$, which is only possible for an arbitrary $\xi_m \Delta y$ as $r \le 0.5$.

$r \le 0.5$ is also the boundedness condition of this scheme, which can be established as follows.

Since an unbounded solution of the 1-D unsteady diffusion equation is usually characterized by oscillations in a sawtooth pattern along the time axis (see Figure 3.8), which implies that at least one of the constituent modes of the solution changes its sign in every time step: $(+, -, +, -...)$. The solution should be bounded if it is free of such modes, which requires the amplification factor to be nonnegative (Yankovskii, 2017):

$$0 \le G_m \le 1 \tag{3.52}$$

You can easily find that this is the case if $\lambda_m^2 r \le 2$ i.e. $r \le 0.5$ in Equation (3.51).

Notice that this requirement of being sawtooth-free in the time direction is usually more stringent than the condition for boundedness. A solution with sawtooth oscillations in time does not violate the maximum principle as long as the oscillations do not exceed the initial solution extrema. So if $r \in r_{awf}$ (reads "r belongs to range r_{awf}") denotes the all-time wiggle-free condition and $r \in r_b$ is the boundedness condition, while $r \in r_s$ the stability condition, we have

$$r_{awf} \in r_b \in r_s \tag{3.53}$$

That is, the all-time wiggle-free range should lie within the boundedness range, which is in turn enclosed by the stability range. For the second-order Runge–Kutta method, since $r_{awf} = r_s$, immediately we know this is also the boundedness range: $r_{awf} = r_b = r_s$.

You may feel the two-step procedure of the second-order Runge–Kutta scheme is still not very easy to implement in your code. In fact the two steps can be done in a very simple manner. In your program, you can repeat the explicit step

$$u_j^n = u_j^{n-1} + \Delta t \frac{u_{j-1}^{n-1} - 2u_j^{n-1} + u_{j+1}^{n-1}}{Re\Delta y^2} \tag{3.54}$$

twice with a `for` loop so that you have u_j^{n+1} values now. Then you simply average this u_j^{n+1} and u_j^{n-1}, you will have the same u_j^n predicted by the second-order Runge–Kutta scheme. You may want to verify that this procedure is indeed equivalent to the original one i.e. Equations (3.47) and (3.48).

There are higher-order Runge–Kutta schemes as well. For example, a well-known third-order Runge–Kutta scheme (Gottlieb & Shu, 1998) is as follows.

Suppose we want to solve

$$\frac{\partial u}{\partial t} = L(u) \tag{3.55}$$

where $L(u)$ stands for some differential operators like the second-order derivative in the 1-D unsteady diffusion equation. Then it might be solved in three steps:

Step 1:

$$U_j^2 = u_j^{n-1} + \Delta t L_\Delta\left(u_j^{n-1}\right) \tag{3.56}$$

Step 2:

$$U_j^3 = \frac{3}{4}u_j^{n-1} + \frac{1}{4}U_j^2 + \frac{\Delta t}{4}L_\Delta\left(U_j^2\right) \tag{3.57}$$

Step 3:

$$u_j^n = u_j^{n-1} + \frac{\Delta t}{6}L_\Delta\left(u_j^{n-1} + U_j^2 + 4U_j^3\right) \tag{3.58}$$

where L_Δ is the finite difference used to approximate the differential operator L. Again, we only have an explicit scheme in each step, which makes programming easier. This method is third-order accurate in time.

You may find the stability condition of this third-order Runge–Kutta scheme when applied to the 1-D unsteady diffusion equation is $r \leq 0.6282$. The all-time wiggle-free condition is $r \leq 0.3990$. The boundedness condition should be somewhere in between.

Among all Runge–Kutta schemes, probably the one we are most familiar with is the fourth-order Runge–Kutta scheme, which solves Equation (3.55) in four steps:

Step 1:

$$U_j^2 = u_j^{n-1} + \frac{\Delta t}{2}L_\Delta\left(u_j^{n-1}\right) \tag{3.59}$$

Step 2:

$$U_j^3 = u_j^{n-1} + \frac{\Delta t}{2}L_\Delta\left(U_j^2\right) \tag{3.60}$$

Step 3:

$$U_j^4 = u_j^{n-1} + \Delta t L_\Delta \left(U_j^3 \right) \tag{3.61}$$

Step 4:

$$u_j^n = u_j^{n-1} + \frac{\Delta t}{6} L_\Delta \left(u_j^{n-1} + 2U_j^2 + 2U_j^3 + U_j^4 \right) \tag{3.62}$$

This method is fourth-order accurate in time.

The stability condition of the fourth-order Runge–Kutta scheme when applied to the 1-D unsteady diffusion equation is $r \leq 0.696$. This is also its all-time wiggle-free as well as boundedness conditions.

3.1.8 SECOND-ORDER BACKWARD DIFFERENCE AND ADAMS–BASHFORTH SCHEMES

As you can see from the previous sections, most numerical schemes used for time advancing are only conditionally stable and bounded. The single unconditionally stable and bounded scheme is the Euler implicit scheme, which approximates the time derivative with a first-order backward difference. If we want the second-order accuracy in time yet still enjoying the unconditional stability, we may consider to use a second-order backward difference to approximate the unsteady term and treat the other terms implicitly. For the start-up of Couette flow example, this scheme reads

$$\frac{3u_j^n - 4u_j^{n-1} + u_j^{n-2}}{2\Delta t} - \frac{1}{Re} \frac{u_{j-1}^n - 2u_j^n + u_{j+1}^n}{\Delta y^2} = 0 \tag{3.63}$$

This scheme is unconditionally stable as you may verify using one of the stability analysis methods. This scheme needs known data at two previous time moments, the $(n-1)^{th}$ moment and the $(n-2)^{th}$ moment. This gives us a little trouble when we start the calculation, since we only have data at the initial time moment. We can circumvent this problem by using the Euler implicit scheme for the first step calculation, then transition to the second-order backward difference scheme afterwards.

How about its boundedness? This scheme is not positive because of the u_j^{n-2} term, boundedness is thus not guaranteed. However, this scheme usually works very well in realistic calculations. In fact, as long as we use proper initial values, e.g. data generated by the first-order implicit scheme, the second-order backward difference scheme never gives us unbounded results for the current example. An unbounded scheme does not always produce unbounded solutions.

discuss briefly is the Adams–Bashforth scheme, which in
...nes similar to the Runge–Kutta schemes. The most com-
...wo-step Adams–Bashforth scheme, which reads

$$u_j = u_j^{n-1} + \frac{3\Delta t}{2}L_\Delta\left(u_j^{n-1}\right) - \frac{\Delta t}{2}L_\Delta\left(u_j^{n-2}\right) \tag{3.64}$$

...use it to solve $\partial u/\partial t = L(u)$. This is another scheme that needs data at two
...ous time moments. This scheme is second-order accurate in time. It is stable
...hen applied to the current 1-D unsteady diffusion problem if $r \le 0.25$.
Boundedness is not guaranteed because it is not positive. The scheme, however,
works very well if the stability condition is met.

3.2 UNSTEADY CONVECTION–DIFFUSION EQUATION

3.2.1 EXAMPLE: MASS TRANSFER IN 1-D FLOW

The convection terms are absent in all governing equations we have solved hitherto.
However, in most fluid dynamics applications the convection terms are present. Let
us study an example, namely the mass transfer in a 1-D flow.

Suppose pure water flows at a constant speed u along the x-direction. At a certain
moment, dye is injected into a certain point in the flow and the dye concentration at
that point is maintained at a specific level, say C_0. We may conveniently choose this
point as the origin of the x-axis. An interesting question is how fast and how far the
dye is going to color the upstream water ($x < 0$)? The problem setup is sketched in
Figure 3.9.

This is a mass transfer problem and the dye concentration C (the amount of dye
per unit mass of water) is governed by an unsteady 1-D convection–diffusion
equation

$$\frac{\partial(\rho C)}{\partial t} + \frac{\partial(\rho u C)}{\partial x} - \frac{\partial}{\partial x}\left(\Gamma\frac{\partial C}{\partial x}\right) = 0 \tag{3.65}$$

in which ρ is the water density and Γ is the diffusion coefficient. Both ρ and Γ are
assumed to be constants. The initial and boundary conditions are

$$C = \begin{cases} 0 & t = 0; \, x \to -\infty \\ C_0 & x = 0 \end{cases} \tag{3.66}$$

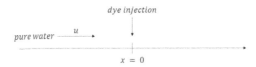

FIGURE 3.9 Mass transfer in a 1-D flow example setup.

Let us define a few dimensionless variables:

$$\tilde{x} = \frac{x}{L}; \quad \tilde{t} = \frac{tu}{L}; \quad \phi = \frac{C}{C_0}; \quad Pe = \frac{\rho u}{\Gamma}$$

where L is a certain reference distance and Pe is called Péclet number, which is the ratio of the strength of convection effects to the strength of diffusion effects. It is obviously very similar to the Reynolds number.

With these definitions, we can write down the dimensionless equation and boundary conditions:

$$\frac{\partial \phi}{\partial \tilde{t}} + \frac{\partial \phi}{\partial \tilde{x}} - \frac{1}{Pe} \frac{\partial^2 \phi}{\partial \tilde{x}^2} = 0 \tag{3.68}$$

and

$$\phi = \begin{cases} 0 & \tilde{t} = 0; \quad \tilde{x} = -\infty \\ 1 & \tilde{x} = 0 \end{cases} \tag{3.69}$$

From now on, the tildes above variables will be dropped although one should keep in mind that these variables are dimensionless.

The exact solution (Cabelli, 1977) to this problem is

$$\phi(t, x) = \frac{1}{2} \left[\mathrm{erfc} \left((t - x) \sqrt{\frac{Pe}{4t}} \right) + e^{Pex} \mathrm{erfc} \left(-(t + x) \sqrt{\frac{Pe}{4t}} \right) \right] \tag{3.70}$$

where

$$\mathrm{erfc}(x) = \frac{2}{\sqrt{\pi}} \int_x^\infty e^{-y^2} dy \tag{3.71}$$

is the complementary error function and MATLAB has a corresponding built-in function called "erfc".

3.2.2 FTCS Scheme

We will use the Euler explicit scheme for the unsteady term and the central difference for both convection and diffusion terms:

$$\frac{\phi_j^{n+1} - \phi_j^n}{\Delta t} + \frac{\phi_{j+1}^n - \phi_{j-1}^n}{2\Delta x} - \frac{1}{Pe} \frac{\phi_{j-1}^n - 2\phi_j^n + \phi_{j+1}^n}{\Delta x^2} = 0 \tag{3.72}$$

This is the so-called forward time, central space (FTCS) scheme.

I guess you would not directly set up a mesh, randomly select a time step size, and do the calculation right away after reading the previous section. Let us use the von Neumann method to determine the stability condition of this scheme first.

Assume the solution is a superposition of numerous wave modes, and each mode assumes the form $\phi_m^n = A_m^n e^{i\xi_m x}$, substituting this wave mode form into Equation (3.72), we end up with

$$G_m = \frac{A_m^{n+1}}{A_m^n} = 1 - 2r + 2r \cos \theta_m - ic \sin \theta_m \tag{3.73}$$

where

$$\theta_m = \xi_m \Delta x; \quad r = \frac{\Delta t}{Pe\Delta x^2}; \quad c = \frac{\Delta t}{\Delta x} \tag{3.74}$$

c is called Courant number or CFL number. Notice that here Δt and Δx are both dimensionless. If the original dimensional time step size and grid spacing are used, the Courant number is defined as

$$c = \frac{u\Delta t}{\Delta x} \tag{3.75}$$

The amplification factor is now a complex function $G_m(\theta_m) = x(\theta_m) + iy(\theta_m)$ where

$$x = 1 - 2r + 2r \cos \theta_m, \quad y = -c \sin \theta_m \tag{3.76}$$

Since $\sin^2 \theta_m + \cos^2 \theta_m = 1$, we have

$$\frac{[x - (1 - 2r)]^2}{(2r)^2} + \frac{y^2}{c^2} = 1 \tag{3.77}$$

which represents an ellipse on the complex plane whose center is at $x = 1 - 2r$ with a semi-major axis length of $2r$ and semi-minor axis length of c, as shown in Figure 3.10. Its rightmost point is always at $x = 1$, $y = 0$, which corresponds to $\theta_m = 0$; its lowermost point corresponds to $\theta_m = \pi/2$; its leftmost point corresponds to $\theta_m = \pi$; and its topmost point corresponds to $\theta_m = 3\pi/2$.

In order to have a stable scheme, we require, as usual,

$$|G_m| = \left| \frac{A^{n+1}}{A^n} \right| = |1 - 2r + 2r \cos \theta_m - ic \sin \theta_m| \leq 1 \tag{3.78}$$

That is, the ellipse must lie inside a unit circle (see Figure 3.10), which demands

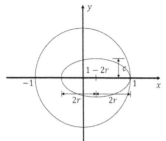

FIGURE 3.10 Amplification factor curve on the complex plane.

$$(1 - 2r) - 2r \geq -1 \rightarrow r \leq \frac{1}{2} \rightarrow \Delta t \leq \frac{Pe\Delta x^2}{2} \qquad (3.79)$$

which is our familiar stability condition due to treating the diffusion term explicitly.

We should also make sure the radius of curvature of the ellipse at $x = 1$ is less than 1, otherwise the ellipse may lie outside the unit circle, as sketched in Figure 3.11.

The radius of curvature (refer to your calculus textbook) of the ellipse is

$$R = \frac{(\dot{x}^2 + \dot{y}^2)^{\frac{3}{2}}}{|\dot{x}\ddot{y} - \ddot{x}\dot{y}|} = \frac{(4r^2 \sin^2 \theta_m + c^2 \cos^2 \theta_m)^{\frac{3}{2}}}{2rc} \qquad (3.80)$$

in which the dots above variables stand for derivatives with respect to θ_m. At the rightmost point, $\theta_m = 0$, and the radius of curvature there is then

$$R = \frac{c^2}{2r} \qquad (3.81)$$

Therefore, another condition for stability is

$$R = \frac{c^2}{2r} \leq 1 \rightarrow c^2 \leq 2r \rightarrow \Delta t \leq \frac{2}{Pe} \qquad (3.82)$$

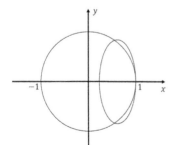

FIGURE 3.11 Amplification factor curve going outside the unit circle.

The stability condition of this scheme is henceforth

$$c^2 \le 2r \le 1 \qquad (3.83)$$

Notice that this condition demands small time step size Δt for both high (see Equation (3.82)) and low (see Equation (3.79)) Pe. This condition can be represented by a region on the r-c diagram, as shown in Figure 3.12. This result is classical and can be found in many textbooks.

Let us do a few numerical experiments to test if this condition is reliable.

Suppose the Péclet number $Pe = 50$ and we use $N = 101$ node points in the range of $-1 \le x \le 0$ so that $\Delta x = 0.01$ (here we use the finite difference method for simplicity). Notice that one boundary condition is $\phi = 0$ at $x = -\infty$ (i.e. no dye far upstream), but obviously we cannot (and do not need to) do calculations on an infinite x-interval. Instead, we simply choose an x-range long enough so that the ϕ value at its left boundary is not affected by the dye injection at the right boundary. For the current $Pe = 50$ case, an x-range of unity length is good enough. We can easily find the time step size constraint established by the stability condition is

$$\Delta t \le \min\left(\frac{Pe\Delta x^2}{2} = 0.0025, \frac{2}{Pe} = 0.04\right) = 0.0025 \qquad (3.84)$$

Two time step sizes $\Delta t = 0.0025$ and $\Delta t = 0.0026$ are tested and in both tests we carry out calculations for 4000 time steps. You are encouraged to write your own code to perform the calculation as it is very straightforward. The results at the finial time moment are shown in Figure 3.13, which agrees with the prediction of our stability analysis. Also you can see from the stable solution that the effects of dye

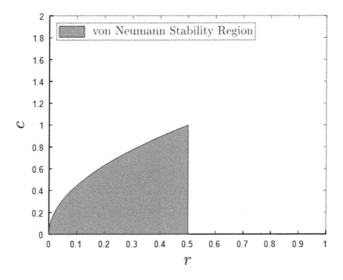

FIGURE 3.12   von Neumann stability region of the FTCS scheme.

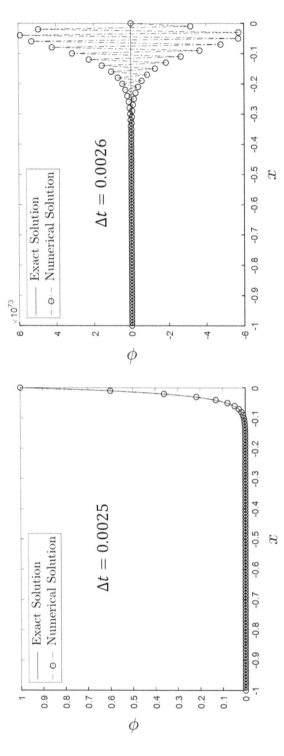

FIGURE 3.13 Numerical tests of the von Neumann stability condition.

injection are confined to a certain distance from the injection spot due to the balance between convection and diffusion.

Let us try another test: $Pe = 38.75$, $N = 21$ (so $\Delta x = 0.05$), and $\Delta t = 0.0775$. These parameters amount to $r = 0.8$ and $c = 1.55$, which is located outside the stability region shown in Figure 3.12. The result after 10,000 steps of calculation is shown in the left graph of Figure 3.14. The result is not very accurate but it is stable; in fact it is even bounded, which directly contradicts the prediction of the classical stability condition.

If we change the parameters just a little bit: $Pe = 38.5$, $N = 21$, and $\Delta t = 0.077$, which correspond to $r = 0.8$ and $c = 1.54$, the result becomes unstable as you can see from the right graph in Figure 3.14.

Let us use the matrix stability analysis method to investigate the stability condition again. The matrix form of the FTCS scheme is

$$
\begin{pmatrix} \phi_1^n \\ \phi_2^n \\ \phi_3^n \\ \vdots \\ \phi_N^n \\ \phi_{N+1}^n \end{pmatrix} = \begin{bmatrix} 1 & 0 & 0 & \cdots & 0 & 0 \\ r+\frac{c}{2} & 1-2r & r-\frac{c}{2} & \cdots & 0 & 0 \\ 0 & r+\frac{c}{2} & 1-2r & r-\frac{c}{2} & 0 & 0 \\ 0 & 0 & \ddots & \ddots & \ddots & 0 \\ 0 & 0 & \ddots & \ddots & \ddots & 0 \\ 0 & 0 & \cdots & r+\frac{c}{2} & 1-2r & r-\frac{c}{2} \\ 0 & 0 & \cdots & 0 & 0 & 1 \end{bmatrix} \times \begin{pmatrix} \phi_1^{n-1} \\ \phi_2^{n-1} \\ \phi_3^{n-1} \\ \vdots \\ \phi_N^{n-1} \\ \phi_{N+1}^{n-1} \end{pmatrix} \quad (3.85)
$$

The eigenvalues of this matrix (see Section 3.1.4) are 1, 1, and

$$
\lambda_m = 1 - 2r + 2\sqrt{r^2 - \frac{c^2}{4}} \cos\left(\frac{m\pi}{N}\right); \ m = 1, \ 2, \ \ldots, N-1 \quad (3.86)
$$

The scheme is only stable when $\max |\lambda_m| \leq 1$, which entails

$$
2\sqrt{\max(0, \ 2r - 1)} \leq c \leq 2\sqrt{r}; \ r \leq 1 \quad (3.87)
$$

The stability region on the r-c diagram representing this condition is shown in Figure 3.15.

According to this condition, the case $r = 0.8$, $c = 1.55$ is stable since $2\sqrt{2 \times 0.8 - 1} = 1.549 \leq 1.55 \leq 2\sqrt{0.8} = 1.79$; and the case $r = 0.8$, $c = 1.54$ should be unstable, which agrees with the test results of our numerical experiments.

The problem of the von Neumann analysis (as presented) is that its presumed standing wave (a wave that only grows/decays in amplitude but not moving) solution does not reflect the physics of the unsteady convection–diffusion equation, whose solution is traveling waves (more precisely, wave packets). The first sign of this fact is the complex amplification factor we encounter for the first time in von Neumann analyses (see Equation (3.73)). Such complex amplification factors make waves move. Since the waves are moving, they will not cause instability even when

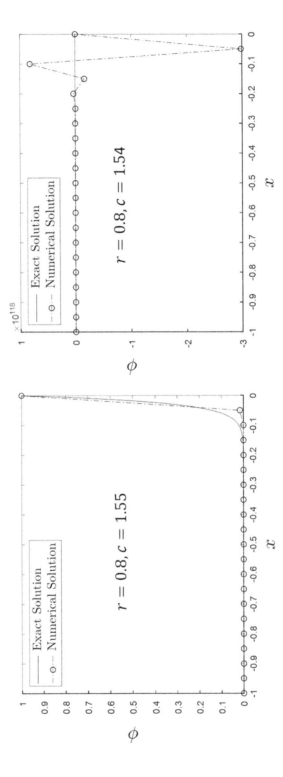

FIGURE 3.14 Numerical results contradicting the von Neumann stability condition.

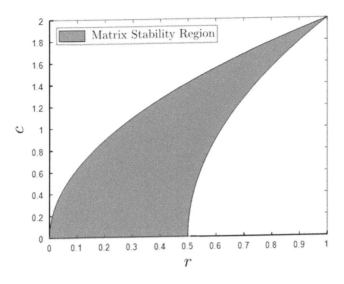

FIGURE 3.15   Matrix stability region of the FTCS scheme.

the amplification factor magnitude is greater than one, because the waves leave the computational domain before they become infinite. The only way to blow up the solution is having a wave that keeps growing and at the same time is being reflected back and forth between the computational domain boundaries. This requires two growing waves with exactly the same wave number and frequency but propagating in opposite directions are both solutions to the finite difference equations, and are both supported by the boundary conditions.

Equipped with this new physical picture, we can have the correct stability condition by using the von Neumann analysis. We need to allow both the wave frequency ω_m (see Equation (3.88) below) and wave number ξ_m to be complex. The frequency is complex because the wave should move; the wave number is complex because the steady-state solution, Equation (3.91), which will be introduced shortly, indicates that there are such wave solutions whose amplitude varies spatially. The wave mode is then

$$\phi_m^n = e^{i\left(\xi_m x - \omega_m t\right)} \tag{3.88}$$

Substituting this form of wave mode into the finite difference equation, we will end up with a relationship between ω_m and ξ_m, say $\omega_m = f\left(\xi_m\right)$. The scheme is unstable if we find even one single point on this f curve at which two group velocities of opposite signs coexist with a positive imaginary part of ω_m. Such an analysis is presented in Appendix B[4].

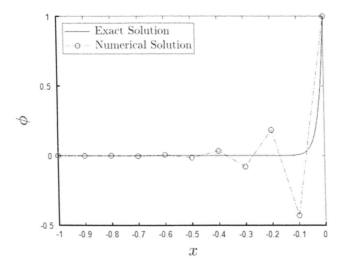

FIGURE 3.16   Spurious sawtooth solution at the steady state.

Let us move on to another issue. If we use 11 points ($\Delta x = 0.1$), $\Delta t = 0.01$ to solve the $Pe = 50$ case ($r = 0.02$, $c = 0.1$) with the FTCS scheme, the steady-state solution is shown in Figure 3.16.

Obviously the result is not satisfactory. A spurious sawtooth pattern exists close to the right boundary. Notice that this sawtooth wave is different from what we observed in Section 3.1, which are oscillations in time (see Figure 3.8); this pattern, on the contrary, is wiggles spreading in space. Oscillations in time may not violate the maximum principle as long as they do not exceed the initial extrema; but the existence of any wiggles i.e. local extrema at the steady state, no matter how small, is a breach of the maximum principle (if there are no source terms in the governing convection–diffusion equation, see Section 1.2.2, Chapter 1).

Therefore, although these wiggles remain finite, thanks to stability, they are unbounded (no matter how small they are).

Since this is a phenomenon of unboundedness, it should disappear as the scheme becomes bounded (positive). The positiveness condition of FTCS will be given in Equation (3.102). We have to, however, expound on this issue a little more, because we will find that some nonpositive schemes can produce wiggle-free steady-state solutions as well, and we will want to know why.

Let us see why such a sawtooth wave arises in the first place. Noticing it persists and reaches a steady state, it must be a solution to the FTCS finite difference equation

$$\frac{\phi_j^{n+1} - \phi_j^n}{\Delta t} + \frac{\phi_{j+1}^n - \phi_{j-1}^n}{2\Delta x} - \frac{1}{Pe}\frac{\phi_{j-1}^n - 2\phi_j^n + \phi_{j+1}^n}{\Delta x^2} = 0 \qquad (3.89)$$

at the steady state. Since at the steady state ϕ stops changing, i.e. $\phi^{n+1} = \phi^n$, we may drop the unsteady term and directly solve

$$\frac{\phi_{j+1} - \phi_{j-1}}{2\Delta x} - \frac{1}{Pe} \frac{\phi_{j-1} - 2\phi_j + \phi_{j+1}}{\Delta x^2} = 0 \qquad (3.90)$$

We substitute the ansatz

$$\phi_j = q^j \qquad (3.91)$$

into Equation (3.90), then we have

$$q^{j+1} - q^{j-1} - R(q^{j-1} - 2 + q^{j+1}) = 0 \qquad (3.92)$$

in which

$$R = \frac{2}{Pe\Delta x} \qquad (3.93)$$

This equation reduces to

$$(1 - R)q^2 + 2Rq - (1 + R) = 0 \qquad (3.94)$$

whose two solutions are

$$q_1 = 1; \; q_2 = -\frac{1 + R}{1 - R} \qquad (3.95)$$

For the current case, $\Delta x = 0.1$, $Pe = 50$, so

$$R = \frac{2}{Pe\Delta x} = 0.4 \qquad (3.96)$$

And

$$q_2 = -\frac{1 + R}{1 - R} = -2.333 \qquad (3.97)$$

Therefore, a particular solution[5] to the finite difference equation is

$$\phi_j = (-2.333)^j = (-1)^j (2.333)^j \qquad (3.98)$$

which is a sawtooth wave that changes sign and grows by 2.333 times in magnitude per every node moving to the right (i.e. everytime when j increases by one). These conclusions are in agreement with the observation of Figure 3.16. The sawtooth wave, therefore, can be avoided if $q_2 > 0$ i.e. $R > 1$, that is,

$$\Delta x \le \frac{2}{Pe} \rightarrow c \le 2r \tag{3.99}$$

We can define a so-called mesh Péclet number

$$Pe_\Delta = Pe\Delta x \tag{3.100}$$

then the condition requires that the mesh Péclet number should not exceed 2.

Therefore, to have a wiggle-free steady-state solution to the 1-D convection–diffusion equation with the FTCS scheme, we have to meet the following condition:

$$2\sqrt{\max(0, 2r - 1)} \le c \le 2r \le 2 \tag{3.101}$$

The region corresponding to this condition on the r-c diagram is shown in the left graph of Figure 3.17.

If we want bounded solutions not only at the final steady state, but at every time moment with the FTCS scheme, we have to require the scheme to be positive, which is the case if

$$c \le 2r; \ r \le 0.5 \tag{3.102}$$

This conclusion can be readily made by moving all terms at the nth moment in Equation (3.89) to the right side and requiring all the coefficients to be positive. This condition is shown in the right graph of Figure 3.17.

If we want solutions not only bounded all the time, but also wiggle-free all the time (notice that in an unsteady process wiggles do not violate the maximum principle i.e. they are bounded, as long as they do not exceed the initial extrema of the solution), we have to require $0 \le \lambda_m \le 1$ where λ_m are the eigenvalues of the scheme matrix, see Equation (3.86). And you may find this requirement leads to the following condition:

$$\sqrt{\max(0, 4r - 1)} \le c \le 2r; \ r \le 0.5 \tag{3.103}$$

The corresponding region on the r-c diagram is shown in Figure 3.18.

Now I believe it is clear to you that the all-time wiggle-free condition is the most stringent one and the stability condition is the most relaxed one. If we use r_{awf}, r_b, r_{swf} and r_s to denote the region on the r-c diagram corresponding to the all-time wiggle-free condition, the boundedness condition, the steady-state wiggle-free condition and the stability condition, we have

$$r_{awf} \in r_b \in r_{swf} \in r_s \tag{3.104}$$

We may also identify the effects of numerical treatments on the performance of the whole scheme. For example, in the boundedness condition, one restriction ($c \le 2r$)

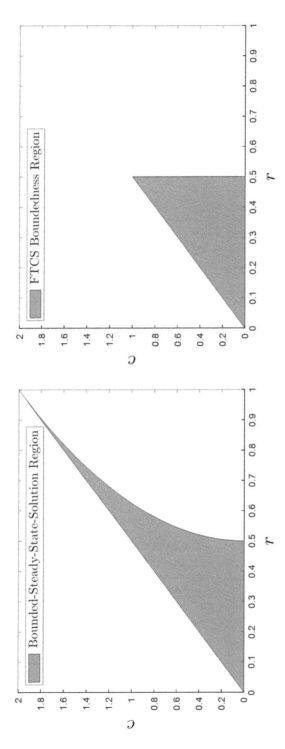

FIGURE 3.17 Bounded steady state and all-time bounded regions of the FTCS scheme.

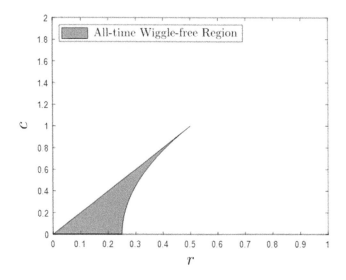

FIGURE 3.18   All-time wiggle-free region of the FTCS scheme.

is due to the central difference used for the convection term while the other ($r \leq 0.5$) is because of the explicit treatment of the diffusion term.

The boundedness condition $c \leq 2r$ is a very stringent one for grid-spacing Δx, especially at a high Pe. For example, if $Pe = 1000$, we have to use a $\Delta x \leq 0.002$. This, in turn, demands a very small Δt due to the second restriction $r \leq 0.5$. The situation becomes even worse in multi-dimensional flow calculations; a point has been already discussed in Section 3.1.4. Such small time step sizes may lead to unacceptably long calculation time. The time step size constraint can be relieved by treating the diffusion terms implicitly, but the grid-spacing limitation is still a concern. It is therefore desirable to find out ways to avoid the sawtooth wiggles without using too small mesh sizes.

3.2.3   LOCAL MESH REFINEMENT

As the wiggles appear when $\Delta x > 2/Pe$ and they are concentrated in the region where ϕ changes significantly in a short distance (see Figure 3.16) (such a region is called a boundary layer in the current context), what happens if we only use a mesh fine enough ($\Delta x < 2/Pe$) in this short region?

To implement this idea we have to first estimate the thickness of this region. Outside the boundary layer convection is much stronger than diffusion if Pe is high; inside the boundary layer, however, diffusion is of the same order as convection, so that

$$\frac{\Delta \phi}{\delta} \sim \frac{1}{Pe} \frac{\Delta \phi}{\delta^2} \rightarrow \delta = \mathcal{O}\left(\frac{1}{Pe}\right) \tag{3.105}$$

in which the first two terms are order-of-magnitude estimates of the convection and diffusion terms. $\Delta\phi$ is the change in ϕ across the boundary layer and δ is the thickness of this layer.

By inspecting the steady-state solution in Figure 3.16, we find for our 1-D example

$$\delta \approx \frac{6}{Pe} \tag{3.106}$$

We have to make sure $\Delta x \le 2/Pe$ within this layer. So let us put 4 nodes inside the boundary layer and 4 nodes outside of it. That is

$$\Delta x = \begin{cases} \frac{\delta}{4} = \frac{1.5}{Pe} & -\delta < x < 0 \\ \frac{1-\delta}{4} & -1 < x < -\delta \end{cases} \tag{3.107}$$

For the $Pe = 50$ case, $\Delta x = 0.03$ inside the boundary layer and 0.22 out of it. Notice that $0.22 > 2/Pe = 0.04$. The mesh is shown in Figure 3.19. Two virtual nodes are added to the two ends of the x-range.

Since we are only interested in the steady-state solution, we will solve the steady 1-D convection–diffusion equation

$$\frac{d\phi}{dx} - \frac{1}{Pe}\frac{d^2\phi}{dx^2} = 0 \tag{3.108}$$

This time, we will use the finite volume method.

Integrating the differential equation over the ith control volume, we have

$$\phi_e - \phi_w - \frac{1}{Pe}\left[\left(\frac{d\phi}{dx}\right)_e - \left(\frac{d\phi}{dx}\right)_w\right] = 0 \tag{3.109}$$

We may approximate the face derivatives as

$$\left(\frac{d\phi}{dx}\right)_e \approx \frac{\phi_{i+1} - \phi_i}{\delta x_i}; \quad \left(\frac{d\phi}{dx}\right)_w \approx \frac{\phi_i - \phi_{i-1}}{\delta x_{i-1}} \tag{3.110}$$

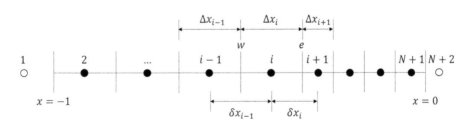

FIGURE 3.19 Local mesh refinement.

We can use linear interpolation to evaluate ϕ values at the control volume faces:

$$\phi_e \approx \frac{\Delta x_i \phi_{i+1} + \Delta x_{i+1} \phi_i}{\Delta x_i + \Delta x_{i+1}}; \quad \phi_w \approx \frac{\Delta x_{i-1} \phi_i + \Delta x_i \phi_{i-1}}{\Delta x_i + \Delta x_{i-1}} \tag{3.111}$$

The finite difference formula is then

$$-\frac{1}{\delta x_{i-1}}\left(\frac{\Delta x_i}{2} + \frac{1}{Pe}\right)\phi_{i-1} + \left(\frac{\Delta x_{i+1}}{2\delta x_i} + \frac{1}{Pe\delta x_i} - \frac{\Delta x_{i-1}}{2\delta x_{i-1}} + \frac{1}{Pe\delta x_{i-1}}\right)\phi_i$$
$$-\frac{1}{\delta x_i}\left(-\frac{\Delta x_i}{2} + \frac{1}{Pe}\right)\phi_{i+1} = 0 \tag{3.112}$$

Well, it is quite complicated. Let us use the average of the two neighboring node values to evaluate the face ϕ values instead:

$$\phi_e \approx \frac{\phi_i + \phi_{i+1}}{2}; \quad \phi_w \approx \frac{\phi_{i-1} + \phi_i}{2} \tag{3.113}$$

Then the finite difference equation becomes much simpler:

$$-\left(1 + \frac{2}{Pe\delta x_{i-1}}\right)\phi_{i-1} + \frac{2}{Pe}\left(\frac{1}{\delta x_i} + \frac{1}{\delta x_{i-1}}\right)\phi_i - \left(-1 + \frac{2}{Pe\delta x_i}\right)\phi_{i+1} = 0 \tag{3.114}$$

This formula is valid at all real nodes. At the two virtual nodes, we apply the boundary conditions:

$$\phi_1 + \phi_2 = 0; \quad \phi_{N+1} + \phi_{N+2} = 2 \tag{3.115}$$

All these finite difference equations form a tridiagonal system and can be solved by using the MATLAB backslash or the TDMA algorithm. The result for the $Pe = 50$ case is shown in Figure 3.20.

The result is pretty good compared with the solution obtained by using a uniform mesh and more node points (see Figure 3.16). There are no wiggles in the boundary layer, which agrees with our expectation because $\Delta x < 2/Pe$ in this region. Since $\Delta x > 2/Pe$ outside the boundary layer, we should see wiggles there. In fact wiggles do exist outside the boundary layer, but they are very small. The reason is because the fine mesh in the boundary layer does its work and the numerical result drops to the correct small value at the interface between the two sets of mesh. This small ϕ value then performs as a boundary condition in the coarse-mesh-zone calculations. So although we still have an oscillating solution in this zone, it no longer varies between $\phi = 1$ and $\phi = 0$ but between this small ϕ value and $\phi = 0$. Usually we can simply live with such small wiggles. Or we can bound them with a MATLAB command like phi = max(0,min(phi,1)) in the code after we obtain the ϕ values.

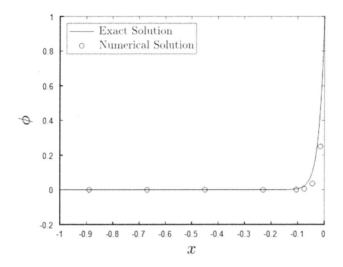

FIGURE 3.20   Numerical solution of the $Pe = 50$ case with local mesh refinement.

The nonuniform mesh and the simple average used for face ϕ value evaluations (see Equation (3.113)) do not affect the accuracy of the scheme, which is still second-order accurate like a common central difference scheme (CDS) although the truncation error is only first-order (Wesseling, 2001). You can verify this conclusion with numerical experiments. See problem 13 in exercises at the end of this chapter.

With the local mesh refinement technique we might reach Pe-independent accuracy and Pe-independent computation cost. For example, we still use 8 real nodes in total and put 4 of them inside the boundary layer but now we calculate a $Pe = 500$ case. The result is shown in Figure 3.21 and Table 3.5, which confirms the Péclet-number independency of both accuracy and cost (notice that we used the same number of nodes and solved the same number of equations in all cases. So, the computation costs are the same). Physically the Péclet number represents the ratio of the strength of convection to diffusion effects. As Pe is large, convection dominates and dye can only diffuse to the upstream for a very short distance, as shown in Figure 3.21.

If we use linear interpolation (see Equation (3.111)) instead of simple average (see Equation (3.113)) to evaluate the face ϕ values, the result for the $Pe = 500$ case is shown in Figure 3.22.

Surprisingly the result is not as good as the one obtained using simple average. This is because the coefficient before ϕ_i in Equation (3.112) is

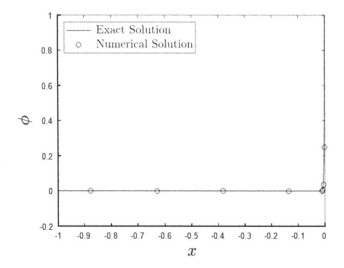

FIGURE 3.21 Numerical solution of the $Pe = 500$ case with local mesh refinement.

TABLE 3.5
Maximum Global Errors at Different Péclet Numbers

Pe	10	50	500
Maximum global error	0.2223	0.2224	0.2224

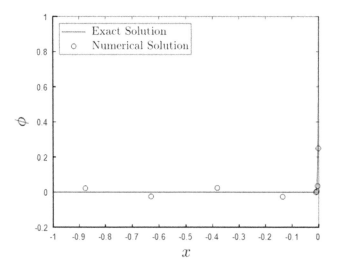

FIGURE 3.22 Numerical solution of the $Pe = 500$ case with linear interpolation.

$$\frac{\Delta x_{i+1}}{2\delta x_i} + \frac{1}{Pe\delta x_i} - \frac{\Delta x_{i-1}}{2\delta x_{i-1}} + \frac{1}{Pe\delta x_{i-1}} = \frac{\Delta x_i(\Delta x_{i+1} - \Delta x_{i-1})}{4\delta x_i\delta x_{i-1}} + \frac{1}{Pe\delta x_i}$$
$$+ \frac{1}{Pe\delta x_{i-1}} \tag{3.116}$$

which may become small or even negative as Pe is large, since $\Delta x_{i+1} < \Delta x_{i-1}$ due to the mesh refinement. This makes the scheme farther away from a positive i.e. bounded scheme.

Linear interpolation, however, works better if the mesh is coarsened along the flow direction, which happens for example when we simulate the wake region behind a solid body immersed in a flow. One therefore has to choose the proper face ϕ value evaluation methods based on the specific physical problem in hand.

Although the local mesh refinement technique works great, it has a major drawback: we have to know where the flow variables change quickly so a local mesh refinement is needed. Unfortunately, such information usually is not available *a priori*.

Since the wiggles at the steady state are caused by using the central difference for the convection term, it is hence natural to explore the other schemes for the convection term, which may perform better.

3.3 SCHEMES FOR CONVECTION TERM

3.3.1 First-Order Upwind Scheme

As Pe becomes large, convection dominates and the transport of any physical property is strongly affected by the upstream. Part of the reason why CDS becomes unbounded at high Pe number is that it does not represent this physics and gives the same weight to the downstream node value as the upstream node value in evaluating face values (see Equation (3.113)). If we follow this lead, we may come up with a fix by only using the upstream (upwind) node value to evaluate the face value:

$$\phi_e = \begin{cases} \phi_i, & u_e > 0 \\ \phi_{i+1}, & u_e < 0 \end{cases} \tag{3.117}$$

and

$$\phi_w = \begin{cases} \phi_{i-1}, & u_w > 0 \\ \phi_i, & u_w < 0 \end{cases} \tag{3.118}$$

See Figure 3.23. This treatment, when applied to the steady 1-D convection–diffusion equation results in the first-order upwind scheme (FUS):

$$\begin{cases} \dfrac{\phi_i - \phi_{i-1}}{\Delta x} - \dfrac{1}{Pe}\dfrac{\phi_{i-1} - 2\phi_i + \phi_{i+1}}{\Delta x^2} = 0, & Pe > 0 \\ \dfrac{\phi_{i+1} - \phi_i}{\Delta x} - \dfrac{1}{Pe}\dfrac{\phi_{i-1} - 2\phi_i + \phi_{i+1}}{\Delta x^2} = 0, & Pe < 0 \end{cases} \tag{3.119}$$

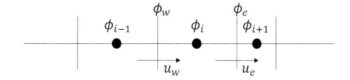

FIGURE 3.23   First-order upwind scheme.

or

$$
\begin{cases}
-\left(1 + \frac{1}{Pe\Delta x}\right)\phi_{i-1} + \left(1 + \frac{2}{Pe\Delta x}\right)\phi_i - \left(\frac{1}{Pe\Delta x}\right)\phi_{i+1} = 0, & Pe > 0 \\
\left(\frac{1}{Pe\Delta x}\right)\phi_{i-1} + \left(1 - \frac{2}{Pe\Delta x}\right)\phi_i - \left(1 - \frac{1}{Pe\Delta x}\right)\phi_{i+1} = 0, & Pe < 0
\end{cases}
\tag{3.120}
$$

The Péclet number is negative when the fluid flows in the negative x-direction.

It is clearly a positive scheme so it does not generate wiggles. The numerical result for the $Pe = 50$ case using 10 uniform control volumes and FUS is shown in Figure 3.24.

If we expand ϕ_{i-1} and ϕ_{i+1} in the FUS formula into Taylor's series, we find that the equation FUS really solves is

$$
\left(\frac{d\phi}{dx}\right)_i - \left(\frac{\Delta x}{2} + \frac{1}{Pe}\right)\left(\frac{d^2\phi}{dx^2}\right)_i + \mathcal{O}(\Delta x^2) = 0
\tag{3.121}
$$

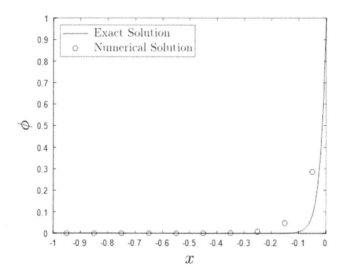

FIGURE 3.24   Numerical solution with the first-order upwind scheme.

An additional numerical or artificial diffusion term is introduced into this equation. It is this term that prevents the generation of wiggles since diffusion smooths out differences in a flow field including wiggles. In fact, all even-order derivatives behave like diffusion terms and have similar wiggle-suppressing features. A scheme that has such an artificial diffusion term as the leading term in its truncation error is called a diffusive or dissipative scheme. If a scheme, on the other hand, has an odd-order derivative as the leading truncation error term, it is said to be dispersive. One example of dispersive schemes is the central difference for the convection term:

$$\frac{\phi_{i+1} - \phi_{i-1}}{2\Delta x} = \left(\frac{d\phi}{dx}\right)_i + \frac{\Delta x^2}{6}\left(\frac{d^3\phi}{dx^3}\right)_i + \mathcal{O}(\Delta x^4) = 0 \qquad (3.122)$$

A dispersive scheme tends to promote wiggles near the locations where the solution changes quickly.

So the artificial diffusion is good in this regard. The problem is that this term in the FUS is proportional to Δx, which is grossly large and makes the scheme only first-order accurate.

Can we use a second-order accurate upwind scheme to approximate the convection term, so that it is both second-order accurate and bounded? Well, before we try this idea, let us see one theorem.

3.3.2   GODUNOV THEOREM

The Godunov theorem (Wesseling, 2001) states that linear numerical schemes for the convection term that have the property of not generating new extrema (wiggles) can be at most first-order accurate.

This theorem seems to destroy our hope of finding a high-order accurate scheme for the convection term that does not generate wiggles. But the situation is not really that pessimistic. There are at least three ways to circumvent the Godunov prophecy.

First, as we have seen, local mesh refinement can help suppress wiggles generated by an unbounded high-order scheme.

Second, a high-order scheme may be unbounded but does not always produce wiggles.

And most importantly, Godunov theorem does not exclude the possibility of nonlinear high-order bounded schemes.

3.3.3   SECOND- AND HIGHER-ORDER UPWIND SCHEMES

Let us first try using the second-order upwind difference to approximate the convection term. Referring to Table 2.1, we can construct the following second-order upwind scheme (SUS) for the steady 1-D convection–diffusion equation:

$$\begin{cases} \dfrac{3\phi_i - 4\phi_{i-1} + \phi_{i-2}}{2\Delta x} - \dfrac{1}{Pe}\dfrac{\phi_{i-1} - 2\phi_i + \phi_{i+1}}{\Delta x^2} = 0, & Pe > 0 \\[3mm] \dfrac{-3\phi_i + 4\phi_{i+1} - \phi_{i+2}}{2\Delta x} - \dfrac{1}{Pe}\dfrac{\phi_{i-1} - 2\phi_i + \phi_{i+1}}{\Delta x^2} = 0, & Pe < 0 \end{cases} \tag{3.123}$$

which reduces to

$$\begin{cases} \phi_{i-2} - \left(4 + \dfrac{2}{Pe\Delta x}\right)\phi_{i-1} + \left(3 + \dfrac{4}{Pe\Delta x}\right)\phi_i - \left(\dfrac{2}{Pe\Delta x}\right)\phi_{i+1} = 0, & Pe > 0 \\[3mm] \left(\dfrac{2}{Pe\Delta x}\right)\phi_{i-1} + \left(3 - \dfrac{4}{Pe\Delta x}\right)\phi_i - \left(4 - \dfrac{2}{Pe\Delta x}\right)\phi_{i+1} + \phi_{i+2} = 0, & Pe < 0 \end{cases} \tag{3.124}$$

on a uniform mesh.

If we want to derive this scheme by using the finite volume method, we only need to evaluate the face ϕ values with a linear interpolation of two upwind nodal values (Figure 3.25).

That is

$$\phi_e = \begin{cases} \phi_i + \left(\dfrac{\phi_i - \phi_{i-1}}{\Delta x}\right)\dfrac{\Delta x}{2} = \dfrac{3\phi_i - \phi_{i-1}}{2}, & u_e > 0 \\[3mm] \phi_{i+1} - \left(\dfrac{\phi_{i+2} - \phi_{i+1}}{\Delta x}\right)\dfrac{\Delta x}{2} = \dfrac{3\phi_{i+1} - \phi_{i+2}}{2}, & u_e < 0 \end{cases} \tag{3.125}$$

By substituting this formula and a similar one for ϕ_w into Equation (3.109), we will obtain the scheme.

As Godunov predicted, this second-order accurate linear scheme is not positive i.e. unbounded.

Will this scheme generate wiggles? The results using this scheme with a 10-node uniform mesh are shown in Figure 3.26.

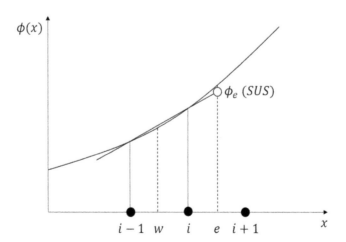

FIGURE 3.25 Second-order upwind scheme.

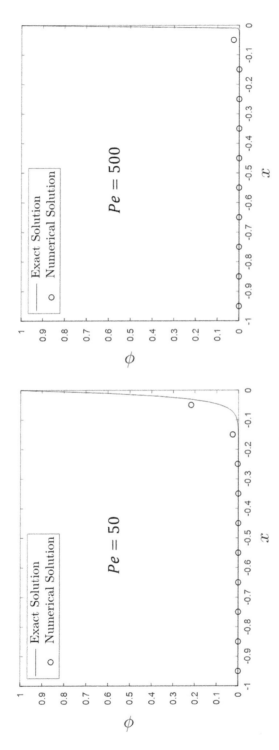

FIGURE 3.26 Numerical solutions with the second-order upwind scheme.

The solutions contain no wiggles no matter how large Pe is and are more accurate than the FUS results. Why does it perform so well? Well, we can substitute the ansatz $\phi_i = q^i$ into the SUS formula and check if a steady-state solution with $q < 0$ is allowed. We will only consider the $Pe > 0$ situation. The substitution gives

$$1 - \left(4 + \frac{2}{Pe\Delta x}\right)q + \left(3 + \frac{4}{Pe\Delta x}\right)q^2 - \left(\frac{2}{Pe\Delta x}\right)q^3 = 0 \qquad (3.126)$$

which is

$$(1 - q)\left[\frac{2}{Pe\Delta x}q^2 - \left(3 + \frac{2}{Pe\Delta x}\right)q + 1\right] = 0 \qquad (3.127)$$

You may easily find that one solution is $q_1 = 1$ and the other two solutions q_2 and q_3 are both positive because

$$q_2 + q_3 = \frac{3Pe\Delta x + 2}{2} > 0; \; q_2 q_3 = \frac{Pe\Delta x}{2} > 0 \qquad (3.128)$$

A sawtooth type of steady-state solution is therefore impossible. Notice that although SUS does not support the sawtooth wiggles, it may still create other types of unbounded solutions (e.g. smooth overshoot/undershoot) because it is not a positive scheme (recall that the boundedness condition is more stringent than the steady-state wiggle-free condition).

Another popular upwind-biased linear scheme is the third-order accurate quadratic upstream interpolation for convective kinematics (QUICK) scheme (Leonard, 1979) which shows less artificial diffusion than SUS and better boundedness than CDS. The idea is shown in Figure 3.27 from which one may find that the

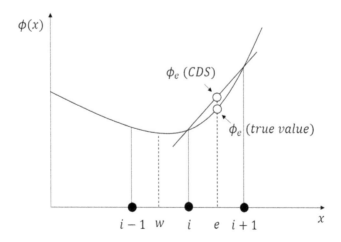

FIGURE 3.27   QUICK scheme.

discrepancy between the estimate of the interfacial values (ϕ_e, ϕ_w, etc.) given by CDS and their true values is due to the curvature of the ϕ function. A curvature correction to CDS was thus proposed as follows

$$\phi_e = \frac{\phi_i + \phi_{i+1}}{2} - \frac{C}{8} \tag{3.129}$$

where the curvature correction is

$$C = \begin{cases} \phi_{i+1} + \phi_{i-1} - 2\phi_i & \text{if } u_e > 0 \\ \phi_i + \phi_{i+2} - 2\phi_{i+1} & \text{if } u_e < 0 \end{cases} \tag{3.130}$$

The formula for ϕ_w can be written down similarly.

The solutions to the steady 1-D convection–diffusion equation using the QUICK scheme with a 10-node mesh are shown in Figure 3.28.

As we can see, for the small Pe case the QUICK scheme result is very good, yet it has wiggles as Pe is large. This is because QUICK is not a positive scheme and unlike SUS, it may produce sawtooth wiggles. Using the q^j ansatz you may find that the (steady state) wiggle-free condition of the QUICK scheme is

$$\Delta x \leq \frac{8}{3Pe} \to Pe_\Delta = Pe\Delta x \leq \frac{8}{3} \tag{3.131}$$

There are many other second- or higher-order upwind schemes. They can be written in a unified form (assume $u_e > 0$):

$$\phi_e = \phi_i + \frac{1 - \kappa}{4}\left(\phi_i - \phi_{i-1}\right) + \frac{1 + \kappa}{4}\left(\phi_{i+1} - \phi_i\right) \tag{3.132}$$

If $\kappa = -1$, it is SUS; if $\kappa = 0$, it is called Fromm's scheme; if $\kappa = 1/3$, it is the third-order upwind scheme (TUS), if $\kappa = 1/2$, it is QUICK; and finally if $\kappa = 1$, it is CDS. We may rewrite Equation (3.132) in a more instructive form:

$$\phi_e = \frac{\phi_i + \phi_{i+1}}{2} - \frac{1 - \kappa}{4}\left(\phi_{i-1} - 2\phi_i + \phi_{i+1}\right) \tag{3.133}$$

That is, these upwind schemes try to reduce the dispersive nature of CDS by subtracting a dispersion term from it.

3.3.4 Deferred-Correction Approach

An issue of applying second- or higher-order schemes like SUS and QUICK is that it involves ϕ values at many nodes in its finite difference formula. For example, SUS results in a 5-point (pentadiagonal) scheme for 1-D problems, and a 9-point

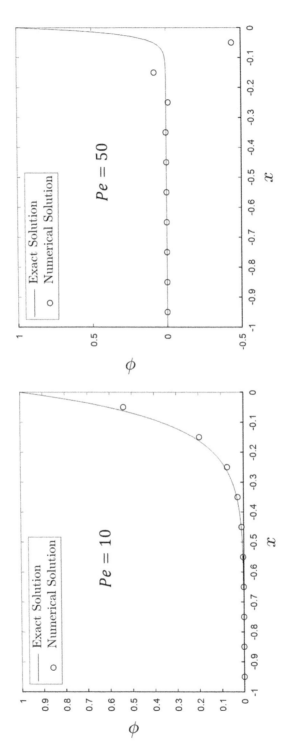

FIGURE 3.28 Numerical solutions with the QUICK scheme.

scheme for 2-D problems. They are more difficult to solve than the familiar tridiagonal systems. One common way to avoid this difficulty is using the deferred-correction approach (Khosla & Rubin, 1974), which expresses a higher-order numerical scheme symbolically as

$$L_\Delta^H \phi_i = L_\Delta^L \phi_i - (L_\Delta^L - L_\Delta^H) \phi_i = RHS_i \tag{3.134}$$

in which L_Δ^L is a lower-order scheme, say FUS. In the deferred-correction approach, the lower-order scheme is solved implicitly i.e. it will be put on the left side of the equation; the difference between the higher and lower order schemes $(L_\Delta^L - L_\Delta^H) \phi_i$ is treated explicitly. That is, it is moved to and join the right-hand-side (RHS_i) of the equation:

$$L_\Delta^L \phi_i^n = (L_\Delta^L - L_\Delta^H) \phi_i^{n-1} + RHS_i^{n-1} \tag{3.135}$$

Since FUS only involves three ϕ values (for 1-D problems) in each finite difference equation, Equations like (3.135) form a tridiagonal system and can be solved easily. Then we update the right side of these equations with the latest ϕ values, and solve them again. This process continues until convergence is reached.

3.3.5 HYBRID SCHEMES

It is possible to construct hybrid schemes that switch between different schemes according to the local mesh Péclet number (see Equation (3.100)) to exploit the best of these schemes e.g. a blending of FUS and CDS (Spalding, 1972):

$$\phi_e = \begin{cases} \left[\left(1 + \frac{Pe_\Delta}{2}\right) \frac{\phi_i}{2} + \left(1 - \frac{Pe_\Delta}{2}\right) \frac{\phi_{i+1}}{2} \right] & |Pe_\Delta| < 2 \\ \phi_i & Pe_\Delta \geq 2 \\ \phi_{i+1} & Pe_\Delta \leq -2 \end{cases} \tag{3.136}$$

This scheme tends toward FUS as Pe_Δ is large; and becomes CDS as $Pe_\Delta = 0$. It is bounded but of less than second-order accuracy.

Another example of such hybrid schemes is the stability-guaranteed second-order difference scheme (SGSD) (Li & Tao, 2002):

$$\phi_e = \beta \phi_e^{CDS} + (1 - \beta) \phi_e^{SUS}; \quad \beta = \frac{2}{2 + |Pe_\Delta|} \tag{3.137}$$

which combines CDS and SUS. This scheme has at least second-order accuracy but it is not a positive scheme (due to SUS).

3.3.6   BOUNDED SECOND-ORDER SCHEMES

According to Godunov's theorem, bounded second- or higher-order accurate schemes for the convection term can only be nonlinear. Such schemes are termed total-variation-diminishing (TVD, see Section 1.2.2 in Chapter 1) schemes (Harten, 1983).

The total variation of a solution is defined as

$$TV(\phi) = \sum_i \left| \phi_i - \phi_{i-1} \right| \tag{3.138}$$

and a TVD scheme guarantees that the total variation does not increase i.e. they will never create new extrema in the computation process when used to solve a 1-D unsteady convection equation

$$\frac{\partial \phi}{\partial t} + u \frac{\partial \phi}{\partial x} = 0 \tag{3.139}$$

as long as the Courant number (CFL number) (see Equation (3.75)) is less than a certain critical value. If we use a TVD scheme to solve the steady 1-D convection–diffusion equation, it should produce bounded solutions no matter how coarse a mesh we use.

One example of such schemes is the van Leer scheme (van Leer, 1977), which calculates the face values as follows (assume $Pe > 0$)

$$\phi_e = \begin{cases} \phi_i + \dfrac{2\left(\phi_i - \phi_{i-1}\right)}{\phi_{i+1} - \phi_{i-1}} \left(\dfrac{\phi_i + \phi_{i+1}}{2} - \phi_i \right) & if \ 0 \le \dfrac{\phi_i - \phi_{i-1}}{\phi_{i+1} - \phi_{i-1}} \le 1 \\ \phi_i & otherwise \end{cases} \tag{3.140}$$

This scheme is upwind-biased, unconditionally bounded, and is second-order accurate except at local extrema of the solution, where accuracy drops to the first-order.

We may understand the behavior of such schemes better if we rewrite the above formula as

$$\phi_e = \phi_i + \Psi(r_i)\left(\frac{\phi_i + \phi_{i+1}}{2} - \phi_i \right) = \phi_i + \Psi(r_i)\left(\frac{\phi_{i+1} - \phi_i}{2} \right) \tag{3.141}$$

where

$$r_i = \frac{\phi_i - \phi_{i-1}}{\phi_{i+1} - \phi_i} \tag{3.142}$$

You may find that Equation (3.141) may represent many schemes. For example, it is

FUS if $\Psi(r_i) = 0$; it is SUS if $\Psi(r_i) = r_i$; it is QUICK if $\Psi(r_i) = 3/4 + r_i/4$; it is TUS if $\Psi(r_i) = 2/3 + r_i/3$; it is Fromm's scheme if $\Psi(r_i) = 1/2 + r_i/2$; it is CDS if $\Psi(r_i) = 1$; if $\Psi(r_i) = 2$, it will be a first-order downwind scheme. The $\Psi(r_i)$ functions of these linear schemes are linear.

For the van Leer scheme,

$$\Psi(r_i) = \frac{r_i + |r_i|}{1 + r_i} \tag{3.143}$$

which is a nonlinear function of r_i. When we look at the definition of r_i, we can see it measures the smoothness of the solution in the vicinity of point i. If $r_i = 1$, the solution is linear, or perfectly smooth; if $r_i < 0$, there is a local extremum at point i. So to keep the solution bounded but still of the second-order accuracy generally, we will want to use different schemes to evaluate ϕ_e according to r_i, say using FUS when $r_i < 0$, but using CDS when $r_i = 1$, etc. So such nonlinear schemes are also hybrid schemes, but not schemes that just combine two schemes, instead they use a spectrum of schemes according to the local smoothness of the solution.

From a different perspective, Equation (3.141) can be understood as an interpolation formula and the second term on its right side is a slope times the distance between the east face and point i. So the function Ψ is used to limit this slope to avoid overshoot/undershoot based on r_i, which can also be viewed as the ratio of two slopes. For this reason, Ψ is also called a limiter (a slope limiter here, but if we decide to evaluate the whole flux $u\phi$ in the convection term $d(u\phi)/dx$ with such schemes, Ψ becomes a flux limiter).

From one more point of view, these schemes can be written as

$$\phi_e = \frac{\phi_i + \phi_{i+1}}{2} - [1 - \Psi(r_i)]\left(\frac{\phi_{i+1} - \phi_i}{2}\right) \tag{3.144}$$

That is, these schemes add some artificial diffusion to CDS to make it bounded, yet at the same time do not affect its accuracy.

You can see the variable r_i has lots of information in it and the TVD schemes try to make good use of such information to reach both high accuracy and boundedness. Researches (Sweby, 1984; Leonard, 1988; Gaskell & Lau, 1988; Wei, Yu, Tao, Kawaguchi, & Wang, 2003, among many others) revealed that a second-order accurate scheme is TVD if its Ψ function falls in a certain region on the r_i-Ψ or Sweby diagram, as shown in Figure 3.29.

This region is

$$\begin{cases} \Psi(r_i) = 0 & r_i < 0 \\ r_i \leq \Psi(r_i) \leq 2r_i & 0 \leq r_i \leq 0.5 \\ r_i \leq \Psi(r_i) \leq 1 & 0.5 \leq r_i \leq 1 \\ 1 \leq \Psi(r_i) \leq r_i & 1 \leq r_i \leq 2 \\ 1 \leq \Psi(r_i) \leq 2 & r_i > 2 \end{cases} \tag{3.145}$$

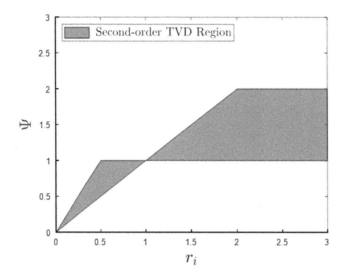

FIGURE 3.29   Second-order TVD region.

TABLE 3.6
Limiters of Several TVD Schemes

Scheme	$\Psi(r_i)$
Van Leer	$\frac{r_i + \lvert r_i \rvert}{1 + r_i}$
Minmod	$\max[0,\ \min(1,\ r_i)]$
MUSCL	$\max\left[0,\ \min\left(2r_i,\ \frac{r_i+1}{2},\ 2\right)\right]$
Superbee	$\max[0,\ \min(2r_i,\ 1),\ \min(r_i,\ 2)]$
Sweby	$\max[0,\ \min(\beta r_i,\ 1),\ \min(r_i,\ \beta)];\ 1 \le \beta \le 2$

Some second-order TVD schemes are listed in Table 3.6.

A problem of the TVD schemes is that the artificial diffusion they introduce to suppress wiggles is still a little excessive, which renders them inappropriate for certain applications in which physical wiggles do exist and should be sustained e.g. in large eddy simulation of turbulent flows[6].

3.3.7   ENO AND WENO SCHEMES

The TVD schemes are generally only second-order accurate. Harten et al. (Harten, Engquist, & Osher, 1987) proposed the essentially non-oscillatory (ENO) schemes, which can be any-order accurate and still suppress oscillations. It is based on a repeated "choose-construct-move" procedure as follows.

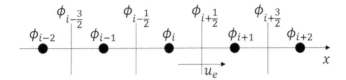

FIGURE 3.30 Face value evaluation with the ENO scheme.

Since different schemes for the convective term differ only in the way to evaluate face values, let us focus on the east face of the ith control volume, as shown in Figure 3.30.

Imagine we are standing on this face, face $(i + 1/2)$. We look around and see two points: point i and point $(i + 1)$. We first choose one of them based on the face velocity u_e: if $u_e > 0$, we choose the left point, otherwise we choose the right point, i.e. we choose the upwind point. Suppose we choose the left point, then we may construct a first-order estimate of ϕ_e: $\phi_e = \phi_i$; then we move to this point i and look around and see two faces: face $(i - 1/2)$ and face $(i + 1/2)$. We will choose one of them based on which one has a smaller absolute value of the so-called first divided difference

$$D^1_{i-\frac{1}{2}} = \frac{\phi_i - \phi_{i-1}}{\Delta x}; \quad D^1_{i+\frac{1}{2}} = \frac{\phi_{i+1} - \phi_i}{\Delta x} \qquad (3.146)$$

Suppose $|D^1_{i-1/2}| < |D^1_{i+1/2}|$, we then use the former to construct a second-order estimate of ϕ_e: $\phi_e = \phi_i + D^1_{i-1/2}(x_e - x_i)$. If $|D^1_{i-1/2}| > |D^1_{i+1/2}|$, then $\phi_e = \phi_i + D^1_{i+1/2}(x_e - x_i)$. Why? Because the absolute value of such divided differences measures how much ϕ varies between neighboring nodes (see Equation (3.138)), and I want to avoid evaluating a derivative based on an abrupt change in ϕ. Now I move to, say face $(i - 1/2)$ if this is the one I just chose. Then I look around, I see points $(i - 1)$ and i. I choose one of them based on which has a smaller absolute value of the second divided difference:

$$D^2_{i-1} = \frac{D^1_{i-\frac{1}{2}} - D^1_{i-\frac{3}{2}}}{2\Delta x}; \quad D^2_i = \frac{D^1_{i+\frac{1}{2}} - D^1_{i-\frac{1}{2}}}{2\Delta x} \qquad (3.147)$$

Suppose this time point i wins, then we can construct a third-order accurate, bounded estimate of ϕ_e: $\phi_e = \phi_i + D^1_{i-1/2}(x_e - x_i) + D^2_i(x_e - x_i)(x_e - x_{i-1})$. Notice that I am still standing on face $(i - 1/2)$, that is why I subtract x_i and x_{i-1} from x_e: because these are the two points around face $(i - 1/2)$. This is the procedure to construct ENO schemes. Just remember that an nth-order divided difference is the difference between two $(n - 1)^{th}$-order divided differences, then divided by $n\Delta x$; also when you add a new term into the ϕ_e formula, those x values you have to subtract from x_e come from the point you are standing on (if you are standing on a point) and the symmetrical points around it.

This procedure requires calculating lots of divided differences in each step yet eventually only one of them is chosen to construct the scheme. To make better use of these divided differences, one can use their weighted average to form numerical schemes. Such schemes are called weighted essentially non-oscillatory (WENO) schemes (Liu, Osher, & Chan, 1994). More details of such schemes can be found in the cited literature.

Now let us use the TVD schemes listed in Table 3.6 and the third-order ENO scheme to solve the steady 1-D convection–diffusion equation. We use a 20-node uniform mesh to calculate a $Pe = 50$ case. The deferred-correction approach is used to implement such high-order schemes. The results are shown in Figure 3.31. The performance of these schemes are comparable. They all give bounded solutions. The third-order ENO scheme is more accurate than the second-order TVD schemes.

3.3.8 HARTEN'S LEMMA

I have intentionally used steady 1-D convection–diffusion equations to introduce you to the TVD schemes. The reason is because once we use such schemes for unsteady flows, a Courant or CFL condition has to be satisfied to ensure the TVD property of such schemes. For example, if we use a TVD scheme to solve a 1-D unsteady convection–diffusion equation

$$\frac{\partial \phi}{\partial t} + \frac{\partial \phi}{\partial x} = \frac{1}{Pe} \frac{\partial^2 \phi}{\partial x^2} \tag{3.148}$$

and if we adopt the Euler explicit scheme for time advancing (and a central difference for the diffusion term, of course), you may find the solution has wiggles or unstable if we use a too large time step size. How to determine the time step size limit? This question can be answered by Harten's lemma.

Harten (Harten, 1983) proved the following lemma: if the finite difference equation associated to a scheme can be written in the following form

$$\phi_i^{n+1} = \phi_i^n + C_{i+1/2}^n \left(\phi_{i+1}^n - \phi_i^n \right) - D_{i-1/2}^n \left(\phi_i^n - \phi_{i-1}^n \right) \tag{3.149}$$

then it is TVD if

$$C_{i+1/2}^n \geq 0; \quad D_{i+1/2}^n \geq 0; \quad C_{i+1/2}^n + D_{i+1/2}^n \leq 1 \tag{3.150}$$

Notice that this condition is a sufficient condition for TVD of both linear and nonlinear schemes and, therefore, it is a little bit different from the positiveness condition that is both sufficient and necessary for TVD of linear schemes.

If we use the Euler explicit scheme for time marching, a TVD scheme for the convection term, CDS for the diffusion term in solving the unsteady 1-D convection–diffusion equation, we have

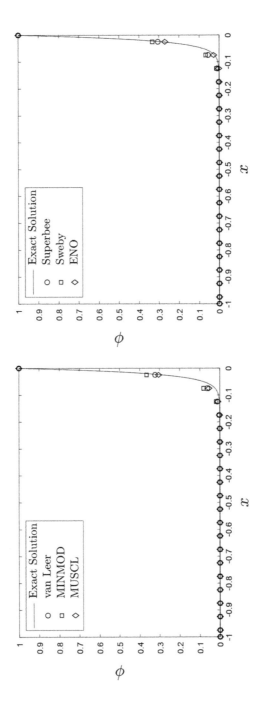

FIGURE 3.31 Numerical results of TVD and ENO schemes.

$$\phi_i^{n+1} = \phi_i^n - c\left\{\left[\phi_i^n + \frac{\Psi(r_i)}{2}\left(\phi_{i+1}^n - \phi_i^n\right)\right] - \left[\phi_{i-1}^n + \frac{\Psi(r_{i-1})}{2}\left(\phi_i^n - \phi_{i-1}^n\right)\right]\right\}$$
$$+ r\left(\phi_{i-1}^n - 2\phi_i^n + \phi_{i+1}^n\right) \tag{3.151}$$

where c and r are the Courant number and diffusion number, respectively. It can be written in the form of Equation (3.149):

$$\phi_i^{n+1} = \phi_i^n + \left[r - c\frac{\Psi(r_i)}{2}\right]\left(\phi_{i+1}^n - \phi_i^n\right) - \left[r + c - c\frac{\Psi(r_{i-1})}{2}\right]\left(\phi_i^n - \phi_{i-1}^n\right) \tag{3.152}$$

To keep the TVD property, in virtue of Harten's lemma, we need

$$\begin{cases} r - c\frac{\Psi(r_i)}{2} \geq 0 \\ r + c - c\frac{\Psi(r_i)}{2} \geq 0 \\ 2r + c - c\Psi(r_i) \leq 1 \end{cases} \tag{3.153}$$

Since for all TVD schemes $0 \leq \Psi(r_i) \leq 2$, we find

$$c \leq \min(r, 1 - 2r) \tag{3.154}$$

This is a sufficient condition i.e. it may be more stringent than necessary for TVD. To push the condition a little closer to the really necessary one, we may rewrite Equation (3.152) as

$$\phi_i^{n+1} = \phi_i^n + r\left(\phi_{i+1}^n - \phi_i^n\right) - \left[r + c + \frac{c}{2}\left(\frac{\Psi(r_i)}{r_i} - \Psi(r_{i-1})\right)\right]\left(\phi_i^n - \phi_{i-1}^n\right) \tag{3.155}$$

which is again in the form of Equation (3.149). And in order for it to be TVD, we have to require:

$$\begin{cases} r + c\left[1 + \frac{1}{2}\left(\frac{\Psi(r_{i+1})}{r_{i+1}} - \Psi(r_i)\right)\right] \geq 0 \\ 2r + c\left[1 + \frac{1}{2}\left(\frac{\Psi(r_{i+1})}{r_{i+1}} - \Psi(r_i)\right)\right] \leq 1 \end{cases} \tag{3.156}$$

The first condition is always satisfied by the TVD schemes. We only need to focus on the second condition, which is,

$$c \le \frac{1 - 2r}{1 + \frac{1}{2} \max\left(\frac{\Psi(r_{i+1})}{r_{i+1}} - \Psi(r_i)\right)} \tag{3.157}$$

As an example, if the van Leer scheme is used, the CFL condition to ensure TVD is then

$$c \le \frac{1 - 2r}{1 + \frac{1}{2} \max\left[\frac{r_{i+1} + |r_{i+1}|}{r_{i+1}(1 + r_{i+1})} - \frac{r_i + |r_i|}{1 + r_i}\right]} = \frac{1}{2} - r \tag{3.158}$$

The maximum of the denominator is obtained as $r_{i+1} \to 0$, $r_i \to 0$. This condition is again usually more stringent than necessary. Therefore, we may use

$$c \le \max(\min(r, 1 - 2r), 0.5 - r) \tag{3.159}$$

as the CFL condition. Obviously $r < 0.5$ also has to be met. Especially, when a TVD scheme (together with the Euler explicit scheme) is used to solve the 1-D unsteady convection equation, i.e. the pure convection/advection equation, Equation (3.139), the Courant number should be less than or equal to 0.5. You may readily verify this condition with numerical experiments.

When we choose numerical schemes to solve a fluid flow/transport phenomenon problem, we have to consider all the tradeoffs between different scheme performance criteria (e.g. efficiency and accuracy) due to different options (e.g. explicit/implicit, first-order/higher-order, bounded/unbounded schemes). The decision has to be made according to the problem in hand, as well as the resources available (e.g. software, hardware, time, and even programming skills). The bottom line is that we have to always use consistent and stable schemes in our calculations. A bounded scheme usually has to be used if discontinuity is present in the flow field (e.g. the abrupt change in physical properties across the phase interface in a multiphase flow), or if a pure convection equation is to be solved.

3.4   PROPER BOUNDARY CONDITIONS

The 1-D convection–diffusion equations we solved hitherto all come with two Dirichlet boundary conditions. Can we use other types of boundary conditions?

For example, by inspecting the exact solution (see Figure 3.31) of the steady 1-D convection–diffusion equation, it is obvious that ϕ is almost a constant near the left boundary, which indicates $d\phi/dx \approx 0$ there. So it seems permissible to use a Neumann condition at the inlet. To show my point, I will use a really small number ε ("eps" in MATLAB, which is 2^{-52}) instead of zero as the ϕ derivative at the inlet:

$$\begin{cases} \frac{d\phi}{dx} = \varepsilon & x = -1 \\ \phi = 1 & x = 0 \end{cases} \tag{3.160}$$

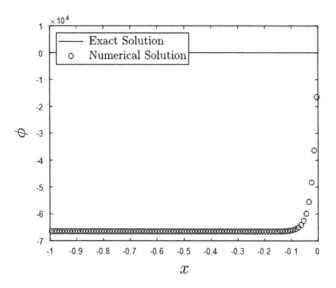

FIGURE 3.32 Numerical solution with Neumann condition at the flow inlet.

The steady-state solution for a $Pe = 50$ case with using 100 nodes is shown in Figure 3.32.

The result is totally wrong. We can understand this phenomenon by examining the exact solution to the steady 1-D convection–diffusion equation. The exact solution subject to the boundary conditions given in Equation (3.160) is

$$\phi = 1 + \frac{\epsilon}{Pe}e^{Pe}(e^{Pe \cdot x} - 1) \tag{3.161}$$

As you can see, if we simply set the ϕ derivative at the inlet to zero, we should have a solution $\phi = 1$ everywhere. But because the exponential function of the Péclet number is huge, even an extremely small deviation from the perfect inlet condition results in a significant departure from the desired solution.

Recall that the Lax equivalence theorem (see Section 3.1.3) tells us we can only have converged solutions to a mathematically well-posed problem. The Neumann condition at the inlet creates an ill-posed problem, which prevents us from obtaining a converged solution even with a consistent and stable numerical scheme. We should therefore avoid using the Neumann condition at an inflow boundary.

Now let us switch the inflow and outflow boundary condition types to

$$\begin{cases} \phi = 0 & x = -1 \\ \frac{d\phi}{dx} = b & x = 0 \end{cases} \tag{3.162}$$

which employs a Dirichlet condition at the inlet and a Neumann condition at the outlet. The result for a $Pe = 50$ case with $b = 50$ is shown in Figure 3.33.

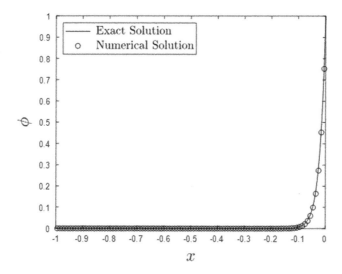

FIGURE 3.33 Numerical solution with Neumann condition at the flow outlet.

The result is very good. The Neumann condition works well at an outlet. This again can be understood by studying the exact solution to the steady 1-D convection–diffusion equation under the given conditions:

$$\phi = b\frac{e^{-Pe\cdot(-x)} - e^{-Pe}}{Pe} \qquad (3.163)$$

Notice that $x \leq 0$ in our computational domain, so the exponential functions in this formula are very small. As a consequence, any perturbation in b only causes a miniature change in solution ϕ, so the problem is "stable" i.e. well-posed.

Although such mathematical explanations make sense, a little physics is usually a better guide when we need it. The transport of physical properties is from upstream to downstream. We hence need the upstream information if we want to determine what is going to happen downstream. The inlet is upstream of the whole computational domain, so we need definite information specified there i.e. a Dirichlet condition; on the contrary, at the outlet, which is downstream of the whole computational domain, we do not need very specific information there, so we can give the properties some flexibility and let them adjust themselves with a Neumann condition. Of course we can use a Dirichlet condition at the outlet as well, but in the real world, flow properties at the outlet are rarely known. In fact they are usually something we need to solve for. For this reason Dirichlet condition is seldomly used at an outlet. A common practice is to use a long enough computational domain so that the outlet is far from the region where properties change significantly. Then we can reasonably assume properties no longer change along the flow direction at the outlet i.e.

$$\frac{\partial \phi}{\partial x} = 0 \tag{3.164}$$

where the x-direction is assumed to be the flow direction.

Another possible choice to set up the outflow condition is solving an advection equation

$$\frac{\partial \phi}{\partial t} + U \frac{\partial \phi}{\partial x} = 0 \tag{3.165}$$

at the outlet. Here U is the average flow velocity at the outlet. The assumption underlying this condition is that ϕ is being advected out of the computational domain, no matter how it distributes close to the outlet. For example, we may use this condition when $\partial \phi / \partial x$ may be not zero.

There is a noticeable exception to these rules of proper boundary conditions, which is pressure. It is common to use a Neumann condition for pressure at the flow inlet while applying a Dirichlet condition to the flow outlet. It is because pressure variations do not transport like the other properties, e.g. mass, momentum, and energy, which are carried by the flow from the upstream to the downstream at a finite speed. Pressure variations, on the other hand travel at an infinite[7] speed in all directions in an incompressible subsonic flow. It simply does not care where the inflow or outflow is.

Exercises

1. If we use central differences for both the unsteady and diffusion terms in the 1-D unsteady diffusion equation, what is the stability condition of such a scheme?
2. Use the matrix stability analysis method to show that the Euler implicit scheme is unconditionally stable for the 1-D unsteady diffusion equation.
3. Reproduce the numerical results shown in Table 3.2 with the Euler implicit scheme.
4. Find the order of the truncation error of the Crank–Nicolson scheme.
5. Show that the Crank–Nicolson scheme is unconditionally stable with using the von Neumann stability analysis method.
6. What is the all-time wiggle-free condition of the Euler explicit scheme when it is used to solve the 1-D unsteady diffusion equation?
7. Solve the start-up of Couette flow example with the second-order Runge–Kutta method. Reproduce the results in Table 3.4.
8. Prove that the third-order Runge–Kutta method presented in this chapter is stable as $r \le 0.6282$ and wiggle-free as $r \le 0.399$ when applied to the unsteady 1-D diffusion equation.
9. Show that the Adams–Bashforth scheme is stable when used to solve the 1-D unsteady diffusion equation if $r \le 0.25$.
10. A so-called ω-scheme can be used to solve the 1-D unsteady diffusion equation

$$\frac{\partial \phi}{\partial t} = \frac{\partial^2 \phi}{\partial y^2} \tag{3.166}$$

The finite different formula is

$$\frac{\phi_j^{n+1} - \phi_j^n}{\Delta t} = \omega \left(\frac{\phi_{j-1}^{n+1} - 2\phi_j^{n+1} + \phi_{j+1}^{n+1}}{\Delta y^2} \right) + (1 - \omega) \left(\frac{\phi_{j-1}^n - 2\phi_j^n + \phi_{j+1}^n}{\Delta y^2} \right) \tag{3.167}$$

Under what conditions is the ω-scheme stable? Verify your conclusion by solving the start-up of Couette flow with this scheme.

11. If we use the Euler explicit scheme for time advancing, the FUS for the convection term and the CDS for the diffusion term to solve the 1-D unsteady convection–diffusion equation, what is its stability condition? What is the steady-state wiggle-free condition? What is the boundedness condition? What is the all-time wiggle-free condition? Do these conditions relate with each other as expected? Write a program to verify these conditions.

12. If we use use the Euler explicit scheme for time advancing, the first-order downwind scheme for the convection term ($\phi_e = \phi_{i+1}$; $\phi_w = \phi_i$ for flow velocity $u > 0$); and the CDS for the diffusion term in solving the 1-D unsteady convection–diffusion equation, what is its stability condition? What is the steady-state wiggle-free condition? What is the boundedness condition? What is the all-time wiggle-free condition? Do these conditions relate with each other as expected? Write a program to verify these conditions.

13. Reproduce the steady 1-D convection–diffusion equation solutions shown in Section 3.2.3. Using the same mesh setup as described in the section to solve the $Pe = 50$ case. Then double the number of control volumes inside and outside the boundary layer. Do we have second-order accuracy with using simple average to evaluate face ϕ values?

14. Show that QUICK can produce wiggle-free solution to the 1-D steady convection–diffusion equation if $\Delta x \leq 8/(3Pe)$.

15. Use a 20-node uniform mesh and one of the TVD schemes listed in Table 3.6 to solve the steady-state 1-D convection–diffusion equation at $Pe = 50$.

16. Use the third-order ENO scheme to solve the steady-state solution to the 1-D convection–diffusion equation at $Pe = 50$. Use a 20-node uniform mesh.

17. If we use FTCS to discretize the 2-D unsteady convection–diffusion equation

$$\frac{\partial \phi}{\partial t} + u\frac{\partial \phi}{\partial x} + v\frac{\partial \phi}{\partial y} = \frac{1}{Pe} \left(\frac{\partial^2 \phi}{\partial x^2} + \frac{\partial^2 \phi}{\partial y^2} \right) \tag{3.168}$$

what is the boundedness condition?

NOTES

1 Here we assume that the solution of the physical problem is always finite.
2 A scheme is still consistent if the truncation error is finite at a boundary, as long as the global error present there approaches zero as the mesh is refined.
3 In fact, even when we use $r \leq 1$, there may still be very small wiggles in the solution in the first few steps. To totally remove such wiggles, a more stringent criterion applies: $r \leq 0.5$. See Section 3.1.7 for an analysis of this type.
4 In case you are worried, the von Neumann analyses presented in Section 3.1 are still valid as the amplification factors are all real in those examples. The traditional way of von Neumann analysis is also valid when dealing with an infinite computational domain. In such cases the traveling waves can still grow to infinity since they have an infinite domain to play with.
5 Notice that this is only a particular solution (one component in the complete solution). The complete solution is $\phi_j = C_1 q_1^j + C_2 q_2^j$ where C_1 and C_1 are constants that can be determined by using boundary conditions.
6 Their artificial diffusion may be taken advantage of by the so-called implicit large eddy simulation of turbulent flows though.
7 You can replace this word with "very high" if you feel uneasy because of Einstein.

4 Numerical Algorithms

4.1 INTRODUCTION

The purpose of this chapter is to cover some iterative numerical algorithms commonly used in computational fluid dynamics (CFD). The finite difference method usually results in a large system of linear equations, and how to solve such equations efficiently is the main concern with regard to numerical algorithms. The numerical algorithms, or numerical solution methods, can be divided into two categories, namely the direct methods and iterative methods. A direct method gives the exact solution to the finite difference equations (note that it is not the exact solution to the original differential equation. They differ by the global error) in one go. For example the tridiagonal matrix algorithm (TDMA; also known as Thomas algorithm) is a direct method. On the other hand, an iterative method approaches the exact solution of finite difference equations by generating successive approximations to it. And it takes infinite iterations to really reach the exact solution. Of course we do not need to wait that long because typically we only want a sufficiently accurate estimate of the exact solution. The difference between the estimate solution given by an iterative method and the exact solution to the finite difference equations is called iterative error.

The sources of error are shown in Figure 4.1. The total error, which is the difference between the numerical solution and the true physical values, can include the modeling error due to unsound assumptions we make to model the physical phenomena, the global (or discretization) error because of replacing differential equations with finite difference equations, and the solution error comprising the iterative and round-off errors.

Direct methods are free from iterative errors. They seem, therefore, superior to the iterative methods, which is, however, not true. This is because to solve a sparse system of N linear equations, direct methods are usually not very efficient. For example the Gauss elimination method, which is a direct method, needs $\mathcal{O}(N^3)$ operations (Tannehill, Anderson, & Pletcher, 1997) to solve such a system, while the LU decomposition method (another direct method) requires $\mathcal{O}(N^2)$ operations (Wesseling, 2001). On the contrary, some iterative methods may only need $\mathcal{O}(N)$ operations to solve the same system. Therefore iterative methods are usually more preferable in CFD.

We will study the performance of different iterative methods by using them to calculate an example, namely the steady fully developed laminar duct flow.

As Figure 4.2 shows, an incompressible fluid flows steadily through a square duct. The duct size $H = 1\ m$. The flow is laminar and fully developed. The flow is in the z-direction.

In this case the governing equation of the flow velocity is

DOI: 10.1201/9781003138822-4

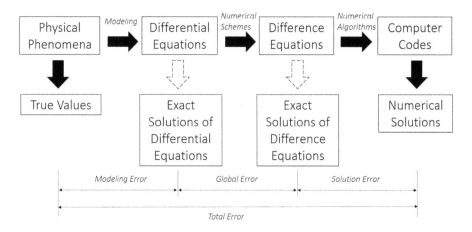

FIGURE 4.1   Sources of error.

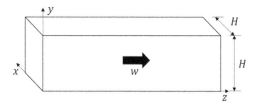

FIGURE 4.2   Fully developed laminar duct flow example setup.

$$-\left(\frac{\partial^2 w}{\partial x^2} + \frac{\partial^2 w}{\partial y^2}\right) = -\frac{1}{\mu}\frac{dp}{dz} = f \qquad (4.1)$$

This is a Poisson equation. At the duct walls the no-slip condition, i.e. $w = 0$ applies. We will simulate a case with $f = 1\,(m \cdot s)^{-1}$. We want to find out the velocity distribution in the duct, $w(x, y)$.

This is a two-dimensional problem since velocity is a function of both x and y coordinates. So we set up a uniform 2-D mesh over the duct cross-section, with $N+1$ nodes along each direction, as shown in Figure 4.3.

The mesh size in both directions is

$$\Delta x = \Delta y = h = \frac{1}{N} \qquad (4.2)$$

We will simply replace the derivatives in Equation (4.1) with central differences (see Table 2.1). For interior nodes $2 \le i,\ j \le N$ we have

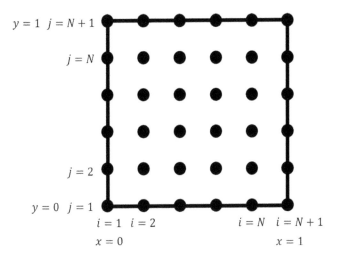

FIGURE 4.3   Finite difference mesh for the duct flow.

$$\frac{-w_{i-1,j} + 2w_{i,j} - w_{i+1,j}}{\Delta x^2} + \frac{-w_{i,j-1} + 2w_{i,j} - w_{i,j+1}}{\Delta y^2} = f = 1 \qquad (4.3)$$

Since $\Delta x = \Delta y = h$ we have

$$- w_{i-1,j} - w_{i,j-1} + 4w_{i,j} - w_{i+1,j} - w_{i,j+1} = h^2 \qquad (4.4)$$

At the duct surfaces, fluid velocity is zero, which is the no-slip condition

$$w_{1,j} = w_{N+1,j} = w_{i,1} = w_{i,N+1} = 0 \qquad (4.5)$$

This system of finite difference equations (4.4) and (4.5), if written in matrix form, is no longer a tridiagonal system since now each $w_{i,j}$ is affected by four neighboring nodes. So now we have a pentadiagonal system of equations, as shown in Figure 4.4, in which the five diagonals P, N, E, W, S denote the coefficients before ws of the present node, and its north, east, west, and south neighboring nodes in the finite difference equations, respectively.

We will use various iterative algorithms to solve this problem. To test the efficiency of different numerical algorithms, we will use a pretty large 129 × 129 mesh within the flow domain ($N = 128$). Let us begin with some basic iterative methods (BIMs).

4.2   BASIC ITERATIVE METHODS

For a system of linear equations

$$A\phi = b \qquad (4.6)$$

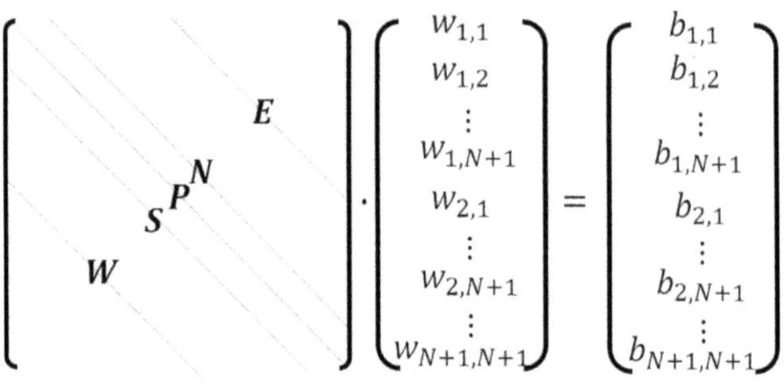

$$\begin{bmatrix} & & & & \\ & & & E & \\ & & S^P \!\!^N & & \\ W & & & & \end{bmatrix} \cdot \begin{pmatrix} w_{1,1} \\ w_{1,2} \\ \vdots \\ w_{1,N+1} \\ w_{2,1} \\ \vdots \\ w_{2,N+1} \\ \vdots \\ w_{N+1,N+1} \end{pmatrix} = \begin{pmatrix} b_{1,1} \\ b_{1,2} \\ \vdots \\ b_{1,N+1} \\ b_{2,1} \\ \vdots \\ b_{2,N+1} \\ \vdots \\ b_{N+1,N+1} \end{pmatrix}$$

FIGURE 4.4   Matrix form of the finite difference equations.

where A is the coefficient matrix, ϕ the unknown solution vector, and b the constant vector, we can solve it with the following iterative method

$$M\phi^k = N\phi^{k-1} + b \tag{4.7}$$

where the superscripts k and $k - 1$ denote the present iteration (suppose it is the kth iteration) and the last, $(k - 1)^{th}$ iteration. Obviously we should have

$$M - N = A \tag{4.8}$$

We call

$$r^k = b - A\phi^k \tag{4.9}$$

the residual after k iterations. You may find that Equation (4.7) is equivalent to

$$M\delta\phi^k = r^{k-1}; \quad \delta\phi^k = \phi^k - \phi^{k-1} \tag{4.10}$$

Any iterative method of the form of Equation (4.7) or (4.10) is called a BIM. The simplest BIMs are the Jacobi and Gauss–Seidel iteration methods.

4.2.1   JACOBI AND GAUSS–SEIDEL ITERATION METHODS

We can directly use Equation (4.4) to generate new estimate of w^k based on the old approximation w^{k-1} as follows

$$w_{i,j}^k = \frac{h^2 + w_{i-1,j}^{k-1} + w_{i,j-1}^{k-1} + w_{i,j+1}^{k-1} + w_{i+1,j}^{k-1}}{4} \tag{4.11}$$

This is the Jacobi iteration method. If written in the "delta" form, it is

$$w_{i,j}^k = w_{i,j}^{k-1} + \delta w_{i,j}^k \tag{4.12}$$

And (cf. Equation (4.4))

$$\delta w_{i,j}^k = \frac{r_{i,j}^{k-1}}{4} = \frac{h^2 + w_{i-1,j}^{k-1} + w_{i,j-1}^{k-1} + w_{i,j+1}^{k-1} + w_{i+1,j}^{k-1} - 4w_{i,j}^{k-1}}{4} \tag{4.13}$$

After all the interior nodes are updated, we go ahead to the next $(k + 1)^{th}$ iteration. We repeat iterating until a sufficiently accurate solution is obtained. To determine if a solution is sufficiently accurate, one may require certain forms of norm of the residual, for example the maximum of the absolute value of r^k to be less than a prescribed small tolerance ε

$$\|r^k\| = \max(|r^k|) < \varepsilon \tag{4.14}$$

If we use $w_{i,j}^0 = 0$ as initial values, and set $\varepsilon = 10^{-8}$, the Jacobi iteration method MATLAB code takes about 4 s to reach the desired accuracy on my PC with a 6-core 2.7 GHz CPU.

The velocity distribution is shown in a contour graph, Figure 4.5, in which the exact solution and numerical solution are both plotted but they match each other so

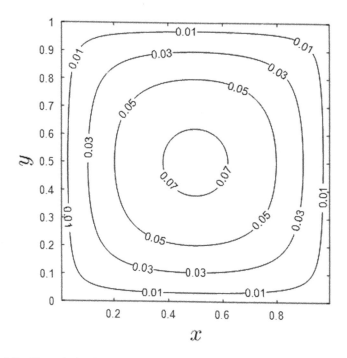

FIGURE 4.5   Numerical solution of the laminar duct flow.

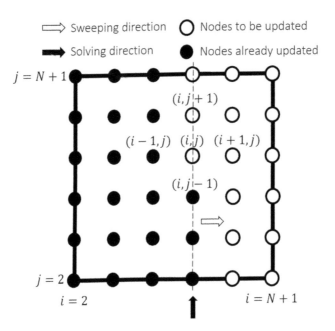

FIGURE 4.6   Gauss–Seidel method.

perfectly that it looks like there is only one set of curves. The exact solution can be found by using finite Fourier transform (Deen, 1998), which is

$$w(x, y) = \sum_{n=1,3,...}^{\infty} \frac{4}{(n\pi)^3} \left\{ 1 - \frac{\sinh(n\pi y) + \sinh[n\pi(1-y)]}{\sinh(n\pi)} \right\} \sin(n\pi x) \quad (4.15)$$

If we always use the most recently updated values in calculating $w_{i,j}^k$, we have the Gauss–Seidel iteration method. Say if we adopt the order of solving equations as shown in Figure 4.6, the Gauss–Seidel iteration formula is then

$$w_{i,j}^k = \frac{h^2 + w_{i-1,j}^k + w_{i,j-1}^k + w_{i,j+1}^{k-1} + w_{i+1,j}^{k-1}}{4} \quad (4.16)$$

Notice that all terms on the right hand of this equation have been calculated. By using Gauss–Seidel method for our duct flow, the time needed for the maximum residual to drop to 10^{-8} is reduced to about 1.7 s on my PC.

4.2.2   SUCCESSIVE OVER-RELAXATION (SOR) METHOD

We observe that the iterative methods generate a series of approximations that approach the exact solution. Why should we wait for the iterations to slowly converge to the solution instead of extrapolating the existing approximations to predict the exact solution? This is the idea underlying the successive over-relaxation (SOR) method.

Suppose the iteration formula is

$$\phi^k = B\phi^{k-1} + d \tag{4.17}$$

We may modify it to

$$\phi^k = \omega(B\phi^{k-1} + d) + (1 - \omega)\phi^{k-1} \tag{4.18}$$

where ω is called relaxation factor. If $\omega < 1$, or as under-relaxed, ϕ^k is an inter-polation, or weighted average of the new value and old value; if $\omega > 1$, or as over-relaxed, ϕ^k becomes an extrapolation based on the new value and the old value. It can be shown that ω must be less than 2, otherwise the iterations diverge.

For our duct flow example, if we choose $\omega = 1.93$ and using the Gauss–Seidel method, the time needed for the maximum residual to drop to 10^{-8} is only 0.17 s, which is ten times faster than the regular Gauss–Seidel method. This technique is therefore very useful, especially for solving the current Poisson type of equations. As we will see later, we usually need to solve a Poisson equation for pressure in calculating incompressible flow fields. SOR can greatly accelerate such solving processes.

4.2.3 ALTERNATING DIRECTION IMPLICIT (ADI) METHOD

As we have seen in the last chapter, we can solve a steady-state problem by advancing a related unsteady problem for prolonged time. For example we can add a pseudo transient term to Equation (4.1)

$$\frac{\partial w}{\partial t} - \left(\frac{\partial^2 w}{\partial x^2} + \frac{\partial^2 w}{\partial y^2}\right) = f \tag{4.19}$$

The steady-state solution to this equation is the desired solution to Equation (4.1).

Equation (4.19) is discretized as follows

$$\frac{w_{i,j}^n - w_{i,j}^{n-1}}{\Delta t} - (L_{\Delta x} w_{i,j} + L_{\Delta y} w_{i,j}) = f \tag{4.20}$$

where $L_{\Delta x}$ and $L_{\Delta y}$ are the central difference operators. The idea of alternating di-rection implicit (ADI) method is using implicit scheme in alternating spatial direc-tions through iterations. It proceeds with the following sequence in each time step:

$$\frac{w_{i,j}^{n-\frac{1}{2}} - w_{i,j}^{n-1}}{\Delta t/2} - \left(L_{\Delta x} w_{i,j}^{n-\frac{1}{2}} + L_{\Delta y} w_{i,j}^{n-1}\right) = f^{n-1} \tag{4.21}$$

$$\frac{w_{i,j}^n - w_{i,j}^{n-\frac{1}{2}}}{\Delta t/2} - \left(L_{\Delta x} w_{i,j}^{n-\frac{1}{2}} + L_{\Delta y} w_{i,j}^n \right) = f^{n-\frac{1}{2}} \qquad (4.22)$$

So in the first half of a time step the central difference in the x-direction is rendered in an implicit form and in the second half of the time step the central difference in the y-direction is treated implicitly. In each half of a time step the equations are tridiagonal and can be solved efficiently with the TDMA algorithm. For example, in Equation (4.21), we may set $j = 2$, then the unknowns are $w_{i,2}$, $w_{i-1,2}$, and $w_{i+1,2}$ at the $(n - 1/2)$th moment. Once we solve these unknowns, we move on to $j = 3$ and so on. It can be proved by using the von Neumann or matrix stability analysis methods that ADI is unconditionally stable.

The time cost for solving the current duct flow example with ADI is about 0.28 s if an optimal Δt is used. A similar method by Strang (Strang, 1968) will be introduced in Chapter 7.

4.2.4 Strongly Implicit Procedure (SIP) Method

SIP method was proposed by Stone (Stone, 1968). It is an incomplete lower-upper (ILU) decomposition method.

One of the direct methods that can be used to solve a system of linear equations $A\phi = b$ is the LU decomposition method, which decomposes the coefficient matrix A into a lower triangular matrix L and an upper triangular matrix U

$$A = LU \qquad (4.23)$$

Then ϕ is obtained by forward and backward substitutions:

$$Lx = b; \quad U\phi = x \qquad (4.24)$$

In MATLAB, one line is enough to realize Equation (4.24): `phi = U\L\b`.

One problem of the LU decomposition method is that the decomposition of A matrix is pretty time-consuming, although the forward and backward substitutions are typically quite efficient. So, the LU method may not be a good choice if we have to do the decomposition repeatedly e.g. when the A matrix varies through iterations.

To fix this problem, we may try to find an approximate but fast, instead of an exact but slow decomposition procedure. To this end, we require both L and U contain only three nonzero diagonals and the main-diagonal elements of U are all ones, as shown in Figure 4.7.

Although L and U are pretty simple matrices, their product M is not a pentadiagonal matrix, instead it is a heptadiagonal matrix. It has two extra diagonals corresponding to the coefficients before the northwestern (NW) and southeastern (SE) neighboring node values, as shown in Figure 4.7. So $M = LU$ is not exactly the same as A. However, it is possible to form an iterative method based on such an approximate decomposition:

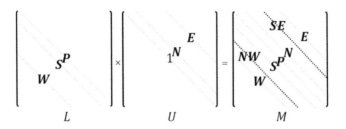

FIGURE 4.7   Approximate LU decomposition.

$$M\phi^k = N\phi^{k-1} + b \tag{4.25}$$

where

$$N = M - A \tag{4.26}$$

As iterations converge the solution is the exact solution of $A\phi = b$.

Since matrix N is the difference between heptadiagonal matrix M and original pentadiagonal matrix A, it should at least contain the two extra diagonals of M (M_{NW} and M_{SE}). Obviously the closer M is to A, the faster should the iterative method be. That is, we should require $N\phi$ to be as close to a zero vector as possible. This can be done by allowing N also being a heptadiagonal matrix, so that

$$N_P \cdot \phi_P + N_W \cdot \phi_W + N_E \cdot \phi_E + N_N \cdot \phi_N + N_S \cdot \phi_S$$
$$+ M_{NW} \cdot \phi_{NW} + M_{SE} \cdot \phi_{SE} \approx 0 \tag{4.27}$$

How can we make sure this equation is almost zero for an arbitrary ϕ? Stone's trick is assuming

$$\phi_{NW} = \alpha (\phi_N + \phi_W - \phi_P) \tag{4.28}$$

and

$$\phi_{SE} = \alpha (\phi_S + \phi_E - \phi_P) \tag{4.29}$$

where α is a certain constant and its optimal value is about 0.93. Now by collecting ϕ terms and requiring their coefficients to vanish, we find

$$N_N = N_W = -\alpha M_{NW}; \ N_S = N_E = -\alpha M_{SE}; \ N_P = \alpha (M_{NW} + M_{SE}) \tag{4.30}$$

Then using Equation (4.26) we may express the M matrix diagonals as linear combinations of A matrix diagonals, which enables us to determine the L and U matrices as follows

$$
\left\{
\begin{aligned}
L_W(i, j) &= \frac{A_W(i,j)}{1 + \alpha U_N(i-1,j)} \\
L_S(i, j) &= \frac{A_S(i,j)}{1 + \alpha U_E(i,j-1)} \\
L_P(i, j) &= A_P(i, j) + \alpha\,[L_W(i, j)\,U_N(i-1, j) + L_S(i, j)\,U_E(i, j-1)] \quad (4.31) \\
&\quad - L_W(i, j)\,U_E(i-1, j) - L_S(i, j)\,U_N(i, j-1) \\
U_N(i, j) &= [A_N(i, j) - \alpha L_W(i, j)\,U_N(i-1, j)]/L_P(i, j) \\
U_E(i, j) &= [A_E(i, j) - \alpha L_S(i, j)\,U_E(i, j-1)]/L_P(i, j)
\end{aligned}
\right.
$$

Here (i, j) are the indices of the meshing nodes.

After we have L and U, we can solve for ϕ with the following steps

1. Calculate the residual vector

$$
r^k = b - A\phi^{k-1} \tag{4.32}
$$

2. Find correction $\delta\phi^k$ by solving

$$
LU\delta\phi^k = r^k \tag{4.33}
$$

which is done by using the forward and backward substitutions.

3. Update ϕ

$$
\phi^k = \phi^{k-1} + \delta\phi^k \tag{4.34}
$$

4. Repeat steps (1) through (3) until a sufficiently accurate solution is obtained.

A MATLAB function that realizes this algorithm is found in Appendix A.3 Stone's Strongly Implicit Procedure (SIP). The time required to solve the duct flow example using this function is about 0.7 s. Based on numerical experiments, SIP indeed can be a lot faster than the other methods discussed in this chapter when solving non-Poisson type of equations e.g. the convection–diffusion equation, especially in 3-D.

4.3   KRYLOV SUBSPACE METHODS

From the discussion of Section 4.2 we can see that typically faster convergence can be obtained as more direct-solving processes are introduced to the iterative procedure. The Krylov subspace methods incorporate significant direct-solving components to achieve high efficiency.

4.3.1   CONJUGATE GRADIENT METHOD

Conjugate gradient (CG) method can be used to solve a system of linear equations $A\phi = b$ when A is a symmetrical positive definite matrix, that is

$$A^T = A; \; x^T A x > 0 \tag{4.35}$$

where A^T is the transpose of an $n \times n$ matrix A and x is any nonzero column vector with n real elements. The coefficient matrix in our steady fully developed laminar duct flow problem is such a matrix.

Solving $A\phi = b$ is equivalent to searching the minimum of

$$f = \frac{1}{2}\phi^T A \phi - \phi^T b \tag{4.36}$$

since f reaches its minimum as its gradient $\nabla f = A\phi - b$ vanishes. Therefore, one may solve $A\phi = b$ by finding new ϕ^k values along the negative gradient $-\nabla f(\phi^{k-1})$ direction, which is the direction that f drops most quickly. Notice $-\nabla f(\phi^{k-1}) = b - A\phi^{k-1} = r^{k-1}$ is the residual at the $(k-1)^{th}$ iteration. That is

$$\phi^k = \phi^{k-1} + a_k p_k; \; p_k = r^{k-1} \tag{4.37}$$

Here a_k is a constant to be determined. This method converges rather slowly. One way to improve the convergence is to require the searching direction in each iteration to be conjugate to the searching directions in previous iterations.

Two column vectors u and v are said to be conjugate with respect to matrix A if

$$u^T A v = 0 \tag{4.38}$$

Notice that $u^T A v$ is the same as $v^T A u$ since for a symmetrical positive definite A we have

$$u^T A v = u^T A^T v = (Au)^T v = v^T A u \tag{4.39}$$

The last step in the derivation is due to the fact that $u^T v = v^T u$ for two column vectors u and v.

We may therefore define an inner product with respect to A as

$$\langle u, v \rangle_A = u^T A v = v^T A u = \langle v, u \rangle_A \tag{4.40}$$

Equation (4.38) can then be written as

$$\langle u, v \rangle_A = \langle v, u \rangle_A = 0 \tag{4.41}$$

So if u is conjugate to v, v is also conjugate to u: they are mutually conjugate, that is mutually "orthogonal" with respect to A. As we know from linear algebra, any real vector with n elements can be represented as a linear combination of n mutually orthogonal base vectors $p_1, p_2, \ldots, p_n$, each of which has n real elements:

$$v = \sum_{i=1}^{n} a_i p_i \tag{4.42}$$

In fact, n mutually conjugate vectors can serve as such base vectors as well.

Now suppose the exact solution to $A\phi = b$ is ϕ^{∞} and we begin our iteration with an initial guess ϕ^0, both of which have n real elements. Then of course their difference is a vector with n real elements and can be expressed as

$$\phi^{\infty} - \phi^0 = \sum_{i=1}^{n} a_i p_i \tag{4.43}$$

That is,

$$\phi^{\infty} = \phi^0 + a_1 p_1 + a_2 p_2 + \ldots + a_n p_n \tag{4.44}$$

which implies that the accurate solution can be obtained after at most n steps if we iterate like

$$\phi^1 = \phi^0 + a_1 p_1; \ \phi^2 = \phi^1 + a_2 p_2; \ldots; \phi^n = \phi^{n-1} + a_n p_n \tag{4.45}$$

That is why we require the searching direction at each step to be conjugate to the directions of previous iterations.

Based on what we discussed above, we may change Equation (4.37) to

$$\phi^k = \phi^{k-1} + a_k p_k; \ p_k = r^{k-1} + b_{k-1} p_{k-1} \tag{4.46}$$

To assure mutual conjugation, we require

$$\langle p_{k-1}, p_k \rangle_A = \langle p_{k-1}, r^{k-1} \rangle_A + b_{k-1} \langle p_{k-1}, p_{k-1} \rangle_A = 0 \tag{4.47}$$

That is

$$b_{k-1} = -\frac{\langle p_{k-1}, r^{k-1} \rangle_A}{\langle p_{k-1}, p_{k-1} \rangle_A} \tag{4.48}$$

Then we may determine a_k in Equation (4.46) by requiring $A\phi^k = b$ (it is nice to hope for good), so that

$$A\phi^k = A\phi^{k-1} + a_k A p_k = b \tag{4.49}$$

That is

$$(b - r^{k-1}) + a_k A p_k = b \qquad (4.50)$$

Therefore

$$a_k A p_k = r^{k-1} \qquad (4.51)$$

which entails

$$a_k (r^{k-1})^T A p_k = (r^{k-1})^T r^{k-1} \qquad (4.52)$$

Eventually we have

$$a_k = \frac{(r^{k-1})^T r^{k-1}}{\langle r^{k-1}, p_k \rangle_A} \qquad (4.53)$$

The residual at the kth iteration is

$$r^k = r^{k-1} - a_k A p_k \qquad (4.54)$$

as one can easily derive from Equation (4.46). Equation (4.51) is our hope, which only occurs when $r^k = 0$, that is when we get the exact solution at the current iteration. You might immediately find out that with the choice of Equation (4.53), we have

$$(r^{k-1})^T r^k = 0 \qquad (4.55)$$

which can be used to obtain alternative formulas for a_k and b_k like the following two

$$a_k = \frac{(r^{k-1})^T r^{k-1}}{\langle p_k, p_k \rangle_A} \qquad (4.56)$$

$$b_k = \frac{(r^k)^T r^k}{(r^{k-1})^T r^{k-1}} \qquad (4.57)$$

The steps to realize this algorithm are as follows:

1. Give an initial guess of solution ϕ^0.
2. Calculate the initial residual $r^0 = b - A\phi^0$. Set $p_1 = r^0$ and $k = 1$.
3. Calculate a_k by using Equation (4.53) or (4.56).
4. Update the estimate of the solution using $\phi^k = \phi^{k-1} + a_k p_k$.
5. Calculate the new residual using $r^k = r^{k-1} - a_k A p_k$, i.e. Equation (4.54).
6. Calculate b_k by using Equation (4.48) or (4.57).
7. Find the new searching vector $p_{k+1} = r^k + b_k p_k$. Set $k = k + 1$.

8. Repeat steps (3) through (7) until convergence is reached, e.g. $\max|r^k|$ is small enough.

Using this method, it only takes about 0.02 s to calculate our duct flow field.

To apply the CG method, one has to form the coefficient matrix A which is a $(n \times m) \times (n \times m)$ matrix and the constant vector b which is a $(n \times m) \times 1$ vector for a $n \times m$ mesh. Notice that with such a mesh, we have $n \times m$ unknowns, and the same number of finite difference equations. Since each equation corresponds to one row of the coefficient matrix, the matrix has $n \times m$ rows (see Figure 4.4). Based on our finite difference equation, we can create five $n \times m$ matrices to store the coefficients before the five unknowns in each equation. Then we may assemble these five matrices to form the big $(n \times m) \times (n \times m)$ coefficient matrix shown in Figure 4.4. A function can be found in Appendix A.4 Assemble Diagonal Matrices to Form Coefficient Matrix for this purpose.

The indices of the unknowns, that is the indices of the mesh nodes, which are also the indices of elements in the aforementioned five matrices are shown in Figure 4.8.

Notice that the bottom row (the south boundary) of node points corresponds to elements on the first column (the left or west "boundary") of the matrices. That is, when you look at such a matrix, if you want to make a connection between the numbers in the matrix to the node points, you will have to rotate your computer screen 90 degrees anticlockwise (rotate your head in the opposite direction also works). Typically, we do not need to worry about this detail until we have to plot the results out of such matrices.

4.3.2 CONDITION NUMBER AND PRECONDITIONED CONJUGATE GRADIENT METHOD

The condition number of a matrix A is

$$cond(A) = \frac{\max(|\lambda(A)|)}{\min(|\lambda(A)|)} \tag{4.58}$$

where $\lambda(A)$ denotes the eigenvalues of matrix A. In solving $A\phi = b$, the condition number gives an upper limit of the ratio of the relative error in solution ϕ to the

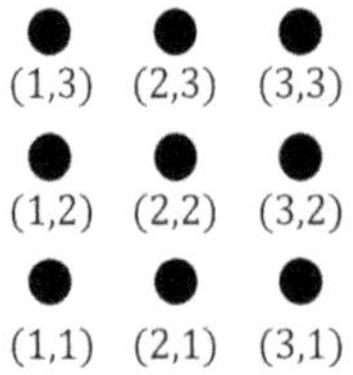

$$
\begin{array}{ccc}
\bullet & \bullet & \bullet \\
(1,3) & (2,3) & (3,3) \\
\bullet & \bullet & \bullet \\
(1,2) & (2,2) & (3,2) \\
\bullet & \bullet & \bullet \\
(1,1) & (2,1) & (3,1)
\end{array}
\qquad
\begin{bmatrix}
(1,1) & (1,2) & (1,3) \\
(2,1) & (2,2) & (2,3) \\
(3,1) & (3,2) & (3,3)
\end{bmatrix}
$$

FIGURE 4.8   Indices of mesh nodes and corresponding elements in matrices.

relative error in the vector b. Suppose an error ϵ_b is introduced to b, which results in an error ϵ_ϕ in ϕ, that is

$$A(\phi + \epsilon_\phi) = b + \epsilon_b \tag{4.59}$$

Easily we have

$$A\epsilon_\phi = \epsilon_b \tag{4.60}$$

Let v_i be the eigenvector corresponding to the eigenvalue λ_i, that is

$$Av_i = \lambda_i v_i \tag{4.61}$$

Such eigenvectors are mutually orthogonal if A is symmetrical positive definite. In such a case the solution vector ϕ can be expressed as

$$\phi = \sum_{i=1}^{n} a_i v_i \tag{4.62}$$

And in virtue of Equation (4.61), we have

$$b = \sum_{i=1}^{n} a_i \lambda_i v_i \tag{4.63}$$

So that

$$|\lambda|_{min} \le \frac{\|b\|}{\|\phi\|} = \frac{\|\sum_{i=1}^{n} a_i \lambda_i v_i\|}{\|\sum_{i=1}^{n} a_i v_i\|} \le |\lambda|_{max} \tag{4.64}$$

Similarly

$$|\lambda|_{min} \le \frac{\|\epsilon_b\|}{\|\epsilon_\phi\|} \le |\lambda|_{max} \tag{4.65}$$

Therefore

$$\frac{\frac{\|b\|}{\|\phi\|}}{\frac{\|\epsilon_b\|}{\|\epsilon_\phi\|}} = \frac{\frac{\|\epsilon_\phi\|}{\|\phi\|}}{\frac{\|\epsilon_b\|}{\|b\|}} \le \frac{|\lambda|_{max}}{|\lambda|_{min}} = cond(A) \tag{4.66}$$

It can be shown (Golub & Van Loan, 1989) that the number of iterations required to solve $A\phi = b$ by using the CG method is proportional to the square root of the condition number of matrix A. As one may readily verify by using the built-in function "cond" in MATLAB, usually the condition number of the coefficient matrix A is proportional to the size of A. Therefore, for a CFD simulation involving a large number of nodes and in turn a large-size coefficient matrix, the CG method becomes less efficient.

One technique that can fix this problem is called preconditioning. The idea is to multiply the governing equation by a preconditioning matrix M^{-1} so that the new coefficient matrix $M^{-1}A$ has a much lower condition number than that of the original matrix A and thence entails a much faster method:

$$M^{-1}A\phi = M^{-1}b \tag{4.67}$$

One way to form the preconditioning matrix M is using the incomplete Cholesky factorization method. One version of this method is shown in Figure 4.9.

Here the diagonals N, E in the upper triangular matrix U, and W, S in the lower triangular matrix L are directly inherited from A. The main diagonal elements of L and U are the same, and equal to the inverse of the corresponding elements of D, thus we have

$$L_{i,i} = U_{i,i} = \frac{1}{D_{i,i}} = d_i = A_{i,i} - \frac{A_{i-1,i}A_{i,i-1}}{d_{i-1}} - \frac{A_{i-n,i}A_{i,i-n}}{d_{i-n}} \tag{4.68}$$

You might check that the condition number of $M^{-1}A$ indeed is much less than the condition number of A.

Once we construct L, D, and U, we can apply the preconditioning matrix M to the CG method, whose formulas are

$$\phi^k = \phi^{k-1} + a_k p_k; \ p_k = z^{k-1} + b_{k-1}p_{k-1}; \ p_1 = z^0 \tag{4.69}$$

$$a_k = \frac{(z^{k-1})^T r^{k-1}}{\langle p_k, p_k \rangle_A} \tag{4.70}$$

$$b_k = \frac{(z^k)^T r^k}{(z^{k-1})^T r^{k-1}} \tag{4.71}$$

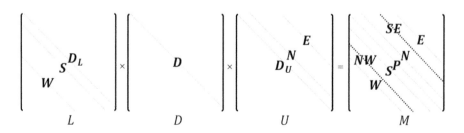

FIGURE 4.9   Incomplete Cholesky factorization.

$$r^k = r^{k-1} - a_k A p_k \qquad (4.72)$$

where

$$M z^k = r^k \qquad (4.73)$$

Notice that the only extra cost we have to pay is to calculate the residual of the preconditioned system $z^k = M^{-1}(b - A\phi^k) = M^{-1}r^k$, which can be easily found by using MATLAB: $z = U\backslash D\backslash L\backslash r$.

A program that realizes this method can be found in Appendix A.5 Incomplete Cholesky Conjugate Gradient (ICCG) Method. It takes about 0.25 s to calculate the duct flow field. Although ICCG does not accelerate solving our 2-D duct flow, it significantly increases convergence speed for more practical 3-D problems (Wesseling, 2001).

A modified incomplete Cholesky conjugate gradient (MICCG) method replaces Equation (4.68) with

$$d_i = A_{i,i} - \frac{(A_{i-1,i} + A_{i-1,i+n-1})A_{i,i-1}}{d_{i-1}} - \frac{(A_{i-n,i} + A_{i-n,i-n+1})A_{i,i-n}}{d_{i-n}} \qquad (4.74)$$

which shows better convergence speed than ICCG in many flow simulations (Tao, 2000).

MATLAB has a built-in function pcg that solves a system of linear equations with the preconditioned CG method. Using this built-in function the duct flow example can be solved in 0.044 s.

4.3.3   GMRES, BiCGSTAB, AND '\'

The CG method only works for equations with symmetrical coefficient matrices. The coefficient matrix resulted from discretization of a convection-diffusion equation is usually, however, not symmetrical. The requirement of symmetrical coefficient matrices can be relaxed for the preconditioned CG methods as long as the coefficient matrix A is such that $A_{i,j} \leq 0$ as $i \neq j$ (Tao, 2000), which is the case when we use a positive scheme to discretize governing equations.

Two popular methods that solve asymmetrical system of equations are the generalized minimal residual (GMRES) method and the bi-conjugate gradient stabilized (BiCGSTAB) method (Saad, 2003). I will not dig into details of these methods except reminding you that MATLAB has two built-in functions, gmres and bicgstab designed to realize these two methods.

Another method to solve $A\phi = b$ is using the MATLAB left division operator '\'. That is $\phi = A\backslash b$. You might be surprised to see how fast this method is for a small to medium size of system of equations. For instance it only takes 0.05 s to solve our duct flow example.

Notice that the comparison of the execution time of different methods listed above might not be totally fair because programs we write have to be first

translated into executable codes before they really do their work, while the
MATLAB built-in functions or operators like '\' may use some shortcuts.
When solving momentum equations like those discussed in Chapter 5, the '\'
operator is sometimes slower than the SIP method for 2-D cases, and usually
much slower for 3-D cases. Also one can easily verify that the TDMA procedure is much faster than '\' when solving a large tridiagonal system of linear
equations.

One should understand that the efficiency of a program is not solely determined
by the adopted numerical algorithm, but also by your programming style. Details
matter. For example if a constant matrix is used in iterations, you should construct
this matrix outside rather than inside the iteration loops for better efficiency.
If such a matrix is a coefficient matrix, we will want to do the time-consuming
part like inversion or LU decomposition, etc. outside the loop, so that it is done
only once.

4.4 FFT METHOD

Many methods introduced in this chapter are able to solve all kinds of equations. If
we restrict ourselves to the Poisson equation, the champion solver is probably the
fast Fourier transform (FFT) method.

For a vector

$$v = [v_1, v_2, ..., v_N] \tag{4.75}$$

its discrete Fourier transform (DFT) is

$$\hat{v} = [\hat{v}_1, \hat{v}_2, ..., \hat{v}_N] \tag{4.76}$$

where

$$\hat{v}_n = \sum_{j=1}^{N} v_j e^{-\frac{(2\pi i)(n-1)(j-1)}{N}} \tag{4.77}$$

The vector v can be recovered by using the inverse DFT (IDFT):

$$v_j = \frac{1}{N} \sum_{n=1}^{N} \hat{v}_n e^{-\frac{(2\pi i)(n-1)(j-1)}{N}} \tag{4.78}$$

Although apparently $\mathcal{O}(N^2)$ flops are required to perform a DFT or IDFT of a vector
with N elements, a very efficient and elegant algorithm called the FFT can reduce the
cost to $\mathcal{O}(N \log N)$ flops, which is close to the ideal cost limit $\mathcal{O}(N)$. In MATLAB,
the DFT and IDFT are calculated with the built-in functions fft and ifft,
using the FFT algorithm.

A very simple property of DFT and IDFT is that

$$if \; v = f + g, \; then \; \hat{v} = \hat{f} + \hat{g} \tag{4.79}$$

Let us define the left-shifted v-vectors as

$$v^l = [v_2, \ldots, v_N, v_1] \tag{4.80}$$

and the right-shifted v-vector

$$v^r = [v_N, v_1, v_2, \ldots, v_{N-1}] \tag{4.81}$$

By making use of the definition of DFT, we may find that

$$\hat{v}^l_n = e^{\frac{(2\pi i)(n-1)}{N}} \hat{v}_n \tag{4.82}$$

and

$$\hat{v}^r_n = e^{-\frac{(2\pi i)(n-1)}{N}} \hat{v}_n \tag{4.83}$$

Now consider a 1-D diffusion equation

$$\frac{d^2u}{dx^2} = f(x) \tag{4.84}$$

If we discretize the equation with central difference on a uniform mesh (see Figure 4.10), it becomes

$$\frac{u_{i-1} - 2u_i + u_{i+1}}{h^2} = f(x_i) = f_i \tag{4.85}$$

in which h is the grid spacing. Of course boundary conditions are applied at the two boundary nodes. Let us suppose both conditions are Dirichlet:

$$u_1 = a; \; u_{N+1} = b \tag{4.86}$$

To solve these finite difference equations with DFT, we construct two vectors

$$U = [0, u_2, u_3, \ldots, u_{N-1}, u_N, 0, -u_N, -u_{N-1}, \ldots, -u_3, -u_2] \tag{4.87}$$

and

FIGURE 4.10 Uniform mesh for the 1-D diffusion example.

$$F = \left[0, f_2 - \frac{a}{h^2}, f_3, \dots, f_{N-1}, f_N - \frac{b}{h^2}, 0, -\left(f_N - \frac{b}{h^2}\right), -f_{N-1}, \dots, -f_3, \right.$$

$$\left. -\left(f_2 - \frac{a}{h^2}\right) \right] \tag{4.88}$$

You may find that

$$\frac{U^r - 2U + U^l}{h^2} = F \tag{4.89}$$

where U^r and U^l are the right-shifted and left-shifted U vectors. Taking DFT of the vectors on both sides of Equation (4.89), we have

$$\frac{1}{h^2}\left[e^{-\frac{(2\pi i)(n-1)}{m}} - 2 + e^{\frac{(2\pi i)(n-1)}{m}} \right] \hat{U}_n = \hat{F}_n; \; n = 1, \; 2, \; \dots, \; m \tag{4.90}$$

where $m = 2N$ is the total number of elements of the U or F vector; $\hat{U}_n$ and $\hat{F}_n$ are the elements of the $\hat{U}$ and $\hat{F}$ vectors (the DFT of U and F). Clearly

$$\hat{U}_n = \frac{\hat{F}_n}{\mu_n}; \; \mu_n = -\frac{4}{h^2} \sin^2 \frac{\pi(n-1)}{m} \tag{4.91}$$

Since $f(x)$ is a known function, $\hat{F}$ is a vector that can be readily evaluated by using the MATLAB built-in function fft. Once we have $\hat{F}$, the DFT of U, $\hat{U}$ can be calculated immediately from Equation (4.91); then the U vector, which is what we really want, can be computed very efficiently by using the MATLAB built-in inverse FFT function ifft. Due to a math subtlety, we have to take the real part of the U vector obtained with the inverse FFT. Notice that the solution $u_2, u_3, \dots, u_N$ is located in the first half of the U vector.

Let us solve an example:

$$\frac{d^2 u}{dx^2} - u = -5 \cos(2x) + 10 \sin(3x + 1) \tag{4.92}$$

subject to boundary conditions

$$u(0) = 1 - \sin(1); \; u(1) = \cos(2) - \sin(4) \qquad (4.93)$$

The exact solution to this problem is

$$u(x) = \cos(2x) - \sin(3x + 1) \qquad (4.94)$$

This is an example I created by using the so-called manufactured solution method (Salari & Knupp, 2000). That is, instead of trying to find the exact solution to a differential equation, we can simply substitute a function (e.g. Equation (4.94)) into a differential operator (e.g. the left side of Equation (4.92)) to produce a differential equation (e.g. Equation (4.92)) and the initial/boundary conditions (e.g. Equation (4.93)). Then we can test our code by using it to solve this equation and check if the numerical solution can match the solution we manufactured.

Although the left side of Equation (4.92) has a $-u$ term, it does not add any difficulty to the procedure. We only need to change μ_n a little

$$\mu_n = -\frac{4}{h^2} \sin^2 \frac{\pi(n-1)}{m} - 1 \qquad (4.95)$$

If we use a nine-node uniform mesh ($N = 8$), the solution is listed in Table 4.1.

You probably wonder why we put so many redundant elements like $-u_N, \; -u_{N-1}, \; \ldots, \; -u_2$ in the U vector. The reason is because we want to make up something like a complete sine wave with these extra elements, so that the FFT can give us more accurate results. If we have different boundary conditions, we will set up the vectors differently, as summarized in Table 4.2.

Notice that the F vector has to be set up accordingly. The rule is to replace every u_i in the U vector with f_i; and if a Dirichlet boundary condition is applied, say $u_1 = a$, or $u_1 + u_2 = a$, then f_2 has to be changed to $f_2 - a/h^2$.

Now let us consider the 2-D Poisson equation

$$\frac{\partial^2 u}{\partial x^2} + \frac{\partial^2 u}{\partial y^2} = f(x, y) \qquad (4.96)$$

over a rectangular computational domain, subject to Dirichlet conditions at all four boundaries

$$u_{1,j} = a_j; \; u_{N+1,j} = b_j; \; u_{i,1} = c_i; \; u_{i,M+1} = d_i \qquad (4.97)$$

We use a mesh like the one shown in Figure 4.3, except that there are $N + 1$ nodes along the x-direction while $M + 1$ nodes along the y-direction, so that the grid spacing is h_x in the x-direction but h_y in the y-direction. The finite difference equation is

TABLE 4.1

Solution to the Second-Order Differential Equation Using the FFT Method

Node	x	Numerical Solution	Exact Solution
2	0.125	-0.0150	-0.0120
3	0.250	-0.1112	-0.1064
4	0.375	-0.1240	-0.1186
5	0.500	-0.0629	-0.0582
6	0.625	0.0484	0.0519
7	0.750	0.1770	0.1789
8	0.875	0.2859	0.2866

TABLE 4.2

U Vectors at Different Boundary Conditions in the FFT Method

Boundary Condition Type	Boundary Conditions	U Vector
Periodic	$u_1 = u_{N+1}$	$[u_1, u_2, ..., u_N]$
Periodic	$\begin{cases} u_1 = u_N \\ u_2 = u_{N+1} \end{cases}$	$[u_2, ..., u_N]$
Dirichlet-Dirichlet	$\begin{cases} u_1 = a \\ u_{N+1} = b \end{cases}$	$[0, u_2, ..., u_N, 0, -u_N, ..., -u_2]$
Dirichlet–Dirichlet	$\begin{cases} u_1 + u_2 = a \\ u_N + u_{N+1} = b \end{cases}$	$[u_2, ..., u_N, -u_N, ..., -u_2]$
Dirichlet–Neumann	$\begin{cases} u_1 = a \\ u_{N+1} = u_N \end{cases}$	$[0, u_2, ..., u_N, u_N, ..., u_2, -u_2, ..., -u_N, -u_N, ..., -u_2]$
Dirichlet–Neumann	$\begin{cases} u_1 + u_2 = a \\ u_{N+1} = u_N \end{cases}$	$[u_2, ..., u_N, u_N, ..., u_2, -u_2, ..., -u_N, -u_N, ..., -u_2]$
Neumann–Dirichlet	$\begin{cases} u_1 = u_2 \\ u_N + u_{N+1} = b \end{cases}$	$[u_2, ..., u_N, -u_N, ..., -u_2, -u_2, ..., -u_N, u_N, ..., u_2]$
Neumann–Neumann	$\begin{cases} u_1 = u_2 \\ u_{N+1} = u_N \end{cases}$	$[u_2, ..., u_N, u_N, ..., u_2]$

$$\frac{u_{i-1,j} - 2u_{i,j} + u_{i+1,j}}{h_x^2} + \frac{u_{i,j-1} - 2u_{i,j} + u_{i,j+1}}{h_y^2} = f(x_i, y_j) = f_{i,j} \quad (4.98)$$

If we focus on a specific j, we may define two column vectors

$$U_j = \left[0;\ u_{2,j};\ ...;u_{N,j};\ 0;\ -u_{N,j};\ ...;-u_{2,j} \right] \qquad (4.99)$$

and

$$F_j = \left[0;\ f_{2,j} - \frac{a_j}{h_x^2};\ ...;f_{N,j} - \frac{b_j}{h_x^2};\ 0;\ -\left(f_{N,j} - \frac{b_j}{h_x^2}\right);\ ...;-\left(f_{2,j} - \frac{a_j}{h^2}\right) \right] \quad (4.100)$$

Notice that if $j = 2$, you will also have to subtract c_i/h_y^2 from $f_{i,j}$; and if $j = M$, subtract d_i/h_y^2 from $f_{i,j}$.

Then by taking DFT of both sides of Equation (4.98), we have

$$\mu_n \hat{U}_{n,j} + \frac{1}{h_y^2}\left(\hat{U}_{n,j-1} - 2\hat{U}_{n,j} + \hat{U}_{n,j+1} \right) = \hat{F}_{n,j};\ \mu_n = -\frac{4}{h_x^2}\sin^2\frac{\pi(n-1)}{2N} \quad (4.101)$$

At this point, we have two choices. One choice is to solve these $\hat{U}$ values with the TDMA algorithm since they form tridiagonal systems of equations. In fact we have to take this route if the mesh is nonuniform in the y-direction; another choice is to use DFT again, which is as follows.

We store these $\hat{U}_{n,j}$ (j fixed) as a column vector, which has $2N$ rows in total. Such vectors for $j = 2,\ 3,\ ...,\ N$ form a matrix, which has $2N$ rows and $N - 1$ columns. Now if we focus on one row of this matrix, say the ith row, it would be $\left[\hat{U}_{i,2},\ \hat{U}_{i,3}, ...,\ \hat{U}_{i,N} \right]$. If we extend it to a longer row vector

$$\hat{V}_i = \left[0,\ \hat{U}_{i,2}, ...,\ \hat{U}_{i,N},\ 0,\ -\hat{U}_{i,N}, ...,\ -\hat{U}_{i,2} \right] \qquad (4.102)$$

and also extend the ith row of the $\hat{F}$ matrix to

$$\hat{G}_i = \left[0,\ \hat{F}_{i,2}, ...,\ \hat{F}_{i,N},\ 0,\ -\hat{F}_{i,N}, ...,\ -\hat{F}_{i,2} \right] \qquad (4.103)$$

You may find that Equation (4.101) is indeed

$$\mu_n \hat{V}_n + \frac{1}{h_y^2}\left(\hat{V}_n^r - 2\hat{V}_n + \hat{V}_n^l \right) = \hat{G}_n \qquad (4.104)$$

So if we take DFT on both sides of this equation again, we end up with

$$\mu_n \hat{\hat{V}}_{n,p} + \nu_p \hat{\hat{V}}_{n,p} = \hat{\hat{G}}_{n,p}; \; \nu_p = -\frac{4}{h_y^2} \sin^2 \frac{\pi(p-1)}{2M} \tag{4.105}$$

Eventually we find that the double DFT of the unknowns (the $u_{i,j}$ matrix) is a matrix whose elements are

$$\hat{\hat{V}}_{n,p} = \frac{\hat{\hat{G}}_{n,p}}{\mu_n + \nu_p} \tag{4.106}$$

where $\hat{\hat{G}}_{n,p}$ are the elements of the double DFT of the (extended) F_{ij} matrix. The u values can be found by performing double inverse DFT to $\hat{\hat{V}}$. Notice that we have to set $\mu_1 + \nu_1$ to an arbitrary nonzero value. Otherwise we will run into a 0/0 error.

I guess you probably have been lost in this Fourier jungle. A MATLAB program can be found in Appendix A.6 2-D Poisson Solver, which uses this method to solve the Poisson equation

$$\frac{\partial^2 u}{\partial x^2} + \frac{\partial^2 u}{\partial y^2} = 10 \tag{4.107}$$

subject to Dirichlet conditions at all four boundaries of a square domain

$$u(0, y) = 3y^2; \; u(1, y) = 3y^2 + 2; \; u(x, 0) = 2x^2; \; u(x, 1) = 2x^2 + 3 \tag{4.108}$$

The exact solution to this problem is

$$u(x, y) = 2x^2 + 3y^2 \tag{4.109}$$

which again was created by using the manufactured solution method.

If we use this method to solve our duct flow example, it accomplishes the calculation in 0.004 s, which is amazing.

Another method which can compete with the FFT method is the multigrid method. For more information about this method refer to e.g. (Wesseling, 1991).

Exercises
1. Reproduce the flow field of the duct flow example using the Gauss–Seidel method, the ADI method, and the SIP method.
2. Now try a line Gauss–Seidel method to solve the duct flow example. Notice that there are five unknowns in each finite difference equation. We can move two of them to the right side and treat them explicitly i.e. using their most recent values; and solve the other three unknowns, which should be all on the same row or the same column:

either w_N, w_P, and w_S or w_W, w_P, and w_E. Obviously all these unknowns have to be solved simultaneously, with e.g. the TDMA algorithm.

3. Then let us use the SOR method to accelerate the line Gauss–Seidel method. Using a relaxation factor $\omega = 1.93$.

4. Follow the algorithm shown in Section 4.3.1 to compose a function realizing the CG method.

5. Reproduce the duct flow field using the CG method and the MATLAB left division.

6. Reproduce the results shown in Table 4.1 with the FFT method. Compare its performance with the TDMA algorithm.

7. Solve Equation (4.92) with the FFT method again, but this time change the boundary conditions to $u(0) = 1 - \sin(1)$ and $du/dx = 0$ at $x = 0.331$.

8. Reproduce the duct flow field using the FFT method.

9. Let us solve the following Poisson equation

$$\frac{\partial^2 u}{\partial x^2} + \frac{\partial^2 u}{\partial y^2} = -2\cos(x)\cos(y) \qquad (4.110)$$

on the square domain $0 \leq x,\ y \leq \pi$. At each boundary a vanishing gradient condition applies. Using a 256×256 uniform mesh and the FFT method to carry out the calculation. Compare the result with the exact solution $u(x, y) = \cos(x)\cos(y)$.

10. Solve the duct flow field example again. This time use the FFT method along the x-direction but the TDMA algorithm along the y-direction.

5 Navier–Stokes Solution Methods

5.1 ODD–EVEN DECOUPLING

5.1.1 EXAMPLE: 1-D FLOW THROUGH FILTER

Until now, we have been always considering transport phenomena like mass transfer within a known flow field. What if we must solve for the flow field itself? In such a situation, we have to solve the Navier–Stokes equations, which are the convection–diffusion equations governing the transport of fluid mass and momentum.

Let us begin with a simple example, namely the 1-D flow through a filter. Liquid water is forced to flow through a horizontal filter, as shown in Figure 5.1.

The filtration material inside the filter exerts a friction force on the water, which has to be compensated by the pressure drop applied to the inlet and outlet, so that a constant water flow can be sustained. The governing equations are

$$\frac{du}{dx} = 0 \tag{5.1}$$

$$0 = -\frac{dp}{dx} - cu^2 \tag{5.2}$$

So, in this case the fluid experiences a body force—the friction force, which is assumed to be proportional to the fluid kinetic energy, whence the u^2 term. Because of this nonlinear term, the solving process of these equations has to be iterative.

Due to the multiple layers of different filtration material in the filter, the parameter c in Equation (5.2) is not a constant, but is supposed to follow the formula

$$c = 50 - 30x \tag{5.3}$$

The pressure at the filter inlet is $p_{in} = 20$ and $p_{out} = 0$ at the outlet. We assume all these given parameters are with correct and consistent units. We want to find out the water speed as well as the pressure distribution throughout the filter.

We first split the x-range $0 \le x \le 1$ into three equal-sized control volumes, and place a node at the center of each control volume. Two virtual nodes are added beyond the two ends. We index the nodes, and the corresponding control volumes as shown in Figure 5.2.

If we integrate the continuity equation, Equation (5.1), with e.g. the third control volume, we have

DOI: 10.1201/9781003138822-5

FIGURE 5.1   1-D flow through a filter example setup.

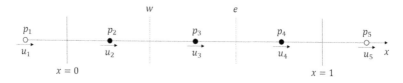

FIGURE 5.2   Finite volume mesh for the 1-D flow through a filter example.

$$u_e - u_w = 0 \tag{5.4}$$

The subscripts e and w denote the east and west surfaces of the control volume. We use the average of the two neighboring node velocities to evaluate each face velocity:

$$0.5(u_4 + u_3) - 0.5(u_2 + u_3) = 0 \rightarrow u_4 - u_2 = 0 \tag{5.5}$$

Here is a tricky point. Are we using the central difference for this convection term of the continuity equation? Well, yes and no. Yes, because $\partial u/\partial x$ is discretized to a central difference $(u_4 - u_2)/2\Delta x$ ($2\Delta x$ is cancelled); no, because when we talk about using a numerical scheme for a convection term, say $\partial(u\phi)/\partial x$, we usually mean a specific method used to evaluate ϕ, not u (although many schemes do evaluate the flux $u\phi$ together, especially in the aerodynamics context). In the continuity equation $\phi = 1$, so we do not need, and did not use any schemes for ϕ here.

   A similar formula exists for every interior (real) control volume. How about the two virtual nodes? We will apply the two Dirichlet pressure boundary conditions there:

$$p_1 + p_2 = 2p_{in}; \quad p_4 + p_5 = 2p_{out} \tag{5.6}$$

You might be a little surprised that I complete a group of equations which apparently are only about velocities with two boundary conditions all about pressures. The point is that, the continuity equation, although does not involve pressure, is in a certain sense the governing equation of pressure for an incompressible flow. That is, pressure has to adjust itself to make sure the continuity equation is satisfied. This point will become clear when we discuss more complex examples.

Now let us move on to the momentum equation. Integrating Equation (5.2) over the third control volume leads to

$$0 = -(p_e - p_w) - \int_w^e cu^2 dx \tag{5.7}$$

Since pressure does not distinguish upstream or downstream, we may use the central difference scheme (CDS) for the pressure term i.e. using the average of nodal values as the face value:

$$p_e \approx \frac{p_4 + p_3}{2}; \ p_w \approx \frac{p_3 + p_2}{2} \rightarrow p_e - p_w \approx \frac{p_4 - p_2}{2} \tag{5.8}$$

It is tempted to move the u^2 term out of the integration symbol and integrate c to find the second term on the right side of Equation (5.7). This method, however, does not work in more general situations, in which the velocity u may be not a constant. Also notice that Equation (5.7) in fact is a simplified momentum equation i.e. an equation for velocity, not for pressure, and we are supposed to solve the velocity values from such equations. Taking into account all these considerations, we will approximate the integral in Equation (5.7) as follows:

$$\int_w^e cu^2 dx \approx c_3 u_3^2 \Delta x \approx c_3 u_3^o u_3 \Delta x \tag{5.9}$$

where u_3^o is the old velocity value obtained in the previous iteration. This nonlinear term is linearized by using the so-called Picard linearization: $u^2 \approx u^o u$. Another way to linearize this term is using the Newton linearization: $u^2 \approx (u^o)^2 + 2u^o(u - u^o) = 2u^o u - (u^o)^2$. The Newton linearization is more accurate, but also more complex than the Picard linearization.

The finite difference equation is thence

$$u_3 = \frac{1}{c_3 u_3^o \Delta x} \frac{p_2 - p_4}{2} = d_3 \frac{p_2 - p_4}{2} \tag{5.10}$$

Notice that this coefficient d_3 is not a constant, it changes through iterations.

At the inlet we apply the same momentum equation:

$$u_{in} \approx \frac{u_1 + u_2}{2} = \frac{1}{c_{in} u_{in}^o \Delta x}(p_1 - p_2) = d_{in}(p_1 - p_2) \tag{5.11}$$

Notice that $c_{in} = 50$. At the outlet we apply a Neumann condition:

$$\left(\frac{\partial u}{\partial x}\right)_{out} = 0 \rightarrow u_5 - u_4 = 0 \tag{5.12}$$

Now we have five equations from discretization of the continuity equation and five equations from discretization of the momentum equation. And we have ten unknowns: five velocity values and five pressure values. So we have just enough number of equations to solve for these unknowns.

If we write the first five ($i = 1, 2, \ldots, 5$) equations in the form of

$$a_{i1}u_1 + a_{i2}u_2 + \ldots + a_{i5}u_5 + b_{i1}p_1 + b_{i2}p_2 + \ldots + b_{i5}p_5 = r_i \qquad (5.13)$$

and similarly write the last five ($i = 1, 2, \ldots, 5$) equations in the form of

$$l_{i1}u_1 + l_{i2}u_2 + \ldots + l_{i5}u_5 + m_{i1}p_1 + m_{i2}p_2 + \ldots + m_{i5}p_5 = q_i \qquad (5.14)$$

These coefficients form four matrices A, B, L, M, and two vectors R and Q, whose elements are a_{ij}, b_{ij}, l_{ij}, m_{ij}, r_i and q_i, respectively:

$$A = \begin{bmatrix} 0 & 0 & 0 & 0 & 0 \\ -1 & 0 & 1 & 0 & 0 \\ 0 & -1 & 0 & 1 & 0 \\ 0 & 0 & -1 & 0 & 1 \\ 0 & 0 & 0 & 0 & 0 \end{bmatrix}; B = \begin{bmatrix} 1 & 1 & 0 & 0 & 0 \\ 0 & 0 & 0 & 0 & 0 \\ 0 & 0 & 0 & 0 & 0 \\ 0 & 0 & 0 & 0 & 0 \\ 0 & 0 & 0 & 1 & 1 \end{bmatrix}; R = \begin{bmatrix} 2p_{in} \\ 0 \\ 0 \\ 0 \\ 2p_{out} \end{bmatrix} \qquad (5.15)$$

and

$$L = \begin{bmatrix} 0.5 & 0.5 & 0 & 0 & 0 \\ 0 & 1 & 0 & 0 & 0 \\ 0 & 0 & 1 & 0 & 0 \\ 0 & 0 & 0 & 1 & 0 \\ 0 & 0 & 0 & -1 & 1 \end{bmatrix}; M = \begin{bmatrix} -d_{in} & d_{in} & 0 & 0 & 0 \\ -\frac{d_2}{2} & 0 & \frac{d_2}{2} & 0 & 0 \\ 0 & -\frac{d_3}{2} & 0 & \frac{d_3}{2} & 0 \\ 0 & 0 & -\frac{d_4}{2} & 0 & \frac{d_4}{2} \\ 0 & 0 & 0 & 0 & 0 \end{bmatrix}; Q = \begin{bmatrix} 0 \\ 0 \\ 0 \\ 0 \\ 0 \end{bmatrix} \qquad (5.16)$$

so that the ten finite difference equations can be written in a matrix form

$$\begin{bmatrix} A & B \\ L & M \end{bmatrix} \begin{bmatrix} U \\ P \end{bmatrix} = \begin{bmatrix} R \\ Q \end{bmatrix} \qquad (5.17)$$

where U and P are the column vectors that contain all the unknown u values and p values, respectively. You should be able to create such matrices in MATLAB, and assemble them using a statement like Mat = [A, B; L, M]. So the solution procedure is

1. Give initial values to the U vector and P vector.
2. Calculate coefficients d_{in}, d_2 etc.

3. Form matrices and vectors involved in Equations (5.15) through (5.17). Assemble them.
4. Compute the new U and P vectors by using e.g. MATLAB left division.
5. Repeat steps (2) through (4) until convergence is reached.

If you follow this procedure, you may find the velocity values jumping back and forth between two values through iterations. This is because the momentum equation we are solving is nonlinear. Numerical methods might not be able to converge to the real solution of a nonlinear equation even with a consistent and stable numerical scheme if the solution changes a lot across iterations. A technique called under-relaxation has to be used to slow down the solution variations. That is, at the end of step (4), we add

$$U = \alpha U^{new} + (1 - \alpha)U^o \tag{5.18}$$

where $0 < \alpha < 1$ is called under-relaxation factor, U^{new} is the calculated new velocity vector and U^o is the old velocity vector obtained in the previous iteration.

The numerical results are shown in Figure 5.3.

The velocity result is excellent. It matches the exact solution $u = \sqrt{2\Delta p/70} = 0.7559$ perfectly. On the contrary, the pressure result is bad: the familiar sawtooth wiggles show up. You may think it is because our mesh is very coarse, and a mesh refinement should fix all the problems. Well, you can do the experiment. No matter how fine a mesh we use, such wiggles persist.

These wiggles are due to the so-called odd–even decoupling. Look at Equation (5.10), you can see u_3 is only related to two pressures with even indices (p_2 and p_4). Similarly, a velocity with an even index, say u_2 is only related to pressures with odd indices (p_1 and p_3). So we essentially did calculations on two separate sets of variables: one set only involves the odd-indexed velocities and even-indexed pressures and the other only involves the even-indexed velocities and odd-indexed pressures. These two groups are totally decoupled except at the boundaries via boundary conditions. We eventually end up with two sets of separate solutions.

There are several ways to resolve this problem. The first way is using a one-sided instead of central difference for the pressure term. For example, using a forward difference:

$$p_e \approx p_{i+1}; \; p_w \approx p_i \rightarrow p_e - p_w \approx p_{i+1} - p_i \tag{5.19}$$

Then the discretized momentum equation becomes

$$u_i = \frac{1}{c_i u_i^o \Delta x}(p_i - p_{i+1}) \tag{5.20}$$

Now we have both odd and even pressure indices. If you try this method, you may find the wiggles disappear. The problem of this method is its accuracy, since a one-sided difference is only first-order accurate.

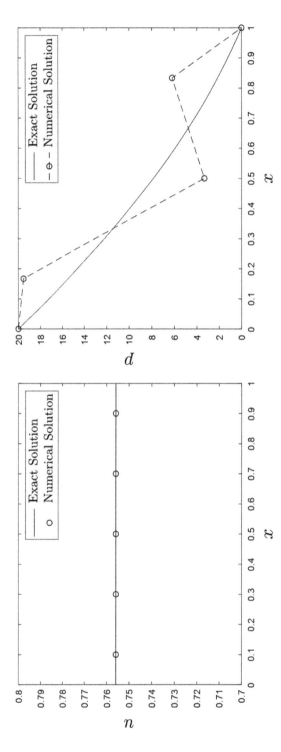

FIGURE 5.3 Numerical results of the 1-D flow through a filter example.

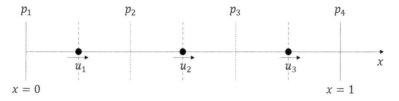

FIGURE 5.4   Staggered mesh.

The second way to fix the odd–even decoupling issue is using a staggered mesh (Harlow & Welch, 1965). That is, we store velocities and pressures at different positions. In fact we have already used this technique when we apply the momentum equation to the inlet (refer to Equation (5.11)). You can see we defined an inlet velocity, which was very naturally coupled with the two pressure values straddling the inlet. There are no any odd–even decoupling issue there.

5.1.2   STAGGERED MESH

If we use a staggered mesh to solve our filter example, we can shift the pressure points by one half of Δx, as shown in Figure 5.4.

You can see I also delete the two virtual nodes since they are really unnecessary now: we have two pressure nodes right at the two boundaries, so we do not need virtual nodes to enforce the pressure boundary conditions. Also we do not need virtual nodes to implement the velocity boundary conditions because, well, we simply do not have any such conditions. Notice that the velocities are still at the center of their original control volumes, but the pressures now have their own control volumes, whose faces coincide with the velocity nodes.

If we integrate the continuity equation over the ith pressure control volume, we have

$$u_i - u_{i-1} = 0 \tag{5.21}$$

This is valid for all interior pressure nodes i.e. $i = 2,\ 3$. For the two boundary pressure nodes, we apply their Dirichlet conditions:

$$p_1 = p_{in};\ p_4 = p_{out} \tag{5.22}$$

Integration of the momentum equation over the jth velocity control volume results in

$$u_j = \frac{1}{c_j u_j^o \Delta x}(p_j - p_{j+1}) = d_j(p_j - p_{j+1});\ j = 1,\ 2,\ 3 \tag{5.23}$$

No odd–even decoupling exists at all. Notice that although the pressure term in Equation (5.23) looks like a one-sided difference, it is indeed a central difference.

You should be able to formulate and solve these finite difference equations by following the same procedure outlined in the previous subsection. The numerical results are now very good, as shown in Figure 5.5.

The exact velocity value is 0.7559 and pressure is

$$p_{exact} = 20 - 0.7559^2(50x - 15x^2) \qquad (5.24)$$

There are a few more methods we can use to circumvent the odd–even decoupling problem, but the discussion of them will be postponed. We will make the most of the good performance of the staggered mesh, and see what methods we can use to solve the Navier–Stokes equations.

5.2   NAVIER–STOKES SOLUTION METHODS

5.2.1   Coupled vs. Segregated Methods

We have solved the filter example by calculating the velocity and pressure in a coupled manner. That is, they are solved simultaneously (see Equation (5.17)). I call such a method a "coupled method." On the contrary, if we solve the momentum and continuity equations separately, one after another, the method then belongs to the so-called segregated methods. Generally speaking the coupled methods usually are more robust and efficient than the segregated methods; the segregated methods, on the other hand, typically demands less computer memory than the coupled methods.

One of the most widely used segregated Navier–Stokes solution methods is the semi-implicit method for pressure-linked equations (SIMPLE) (Patankar & Spalding, 1972). Well, I should say it is not that simple. Let us see how it works.

5.2.2   SIMPLE Method

The SIMPLE method is a segregated method. I will take the filter flow as an example to explain the SIMPLE procedure. The same staggered mesh illustrated in Figure 5.4 will be used.

We begin with the momentum equation

$$u_i^* = \frac{1}{c_i u_i^o \Delta x}(p_i^* - p_{i+1}^*) = d_i(p_i^* - p_{i+1}^*) \qquad (5.25)$$

where u^* and p^* are the intermediate flow field which will be corrected later; u^o is the old velocity field obtained in the previous iteration. To proceed, we simply use p^o from the last iteration as p^*. So quickly we can calculate all u_i^* values.

Then we go ahead to handle the continuity equation

$$u_i - u_{i-1} = 0 \qquad (5.26)$$

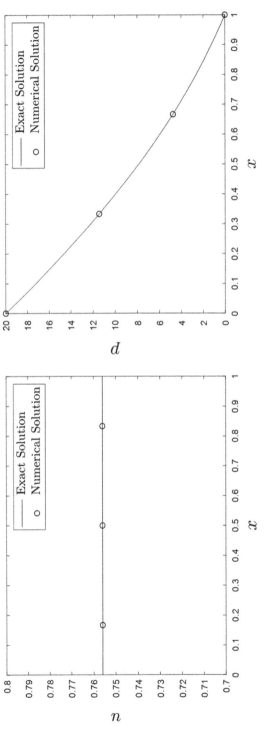

FIGURE 5.5 Numerical results with the staggered mesh.

When we use a coupled method, we solve the momentum and continuity equations simultaneously so that the solution always satisfies these two equations at the same time. Now with a segregated method like SIMPLE, we solve these two equations separately. So the u_i^* values we obtained may not satisfy the continuity equation. That is, if we plug them into the left side of Equation (5.26), the result $(u_i^* - u_{i-1}^*)$ usually is not zero. Therefore, we have to make corrections to the u_i^* values so that they may satisfy the continuity equation. But the problem is, the corrected velocity values then may no longer satisfy the momentum equation. The key to break this circle is to correct pressure and velocity at the same time:

$$p = p^* + p' \qquad (5.27)$$

and

$$u = u^* + u' \qquad (5.28)$$

so that the corrected flow field satisfies both equations:

$$u_i^* + u_i' = d_i[(p_i^* + p_i') - (p_{i+1}^* + p_{i+1}')] \qquad (5.29)$$

and

$$(u_i^* + u_i') - (u_{i-1}^* + u_{i-1}') = 0 \qquad (5.30)$$

Then we find that there must exist a relationship between the velocity and pressure corrections:

$$u_i' = d_i(p_i' - p_{i+1}') \qquad (5.31)$$

by virtue of Equations (5.25) and (5.29). Substituting this relationship into Equation (5.30), we end up with

$$d_i(p_i' - p_{i+1}') - d_{i-1}(p_{i-1}' - p_i') = -(u_i^* - u_{i-1}^*) \qquad (5.32)$$

i.e.

$$d_{i-1}p_{i-1}' - (d_{i-1} + d_i)p_i' + d_i p_{i+1}' = u_i^* - u_{i-1}^* \qquad (5.33)$$

This is a pressure Poisson equation (PPE) with variable coefficients. Notice that the right side of this equation can be readily evaluated and usually is not zero; also such equations form a tridiagonal system, which can be easily solved by the TDMA algorithm. At the two boundaries, we are given the pressure values (Dirichlet conditions), so we do not need to, and should not correct the pressure values there:

$$p_1' = 0; \; p_4' = 0 \tag{5.34}$$

which serves as the boundary conditions for the tridiagonal system of p' equations.

Once we have solved p', the velocity corrections u' can be calculated using Equation (5.31). And once we have p' and u', we immediately have the new flow field by updating $p*$ and $u*$ according to Equations (5.27) and (5.28).

This is the logic underlying the SIMPLE procedure. In practice, we may follow the steps below:

1. Give initial values to u and p.
2. Solve the momentum equation based on the old u and p values to find an intermediate velocity field $u*$.
3. Solve the pressure-correction Equation (5.33) for p'.
4. Calculate the velocity corrections u' according to Equation (5.31).
5. Update the flow field according to Equations (5.27) and (5.28).
6. Repeat steps (2) through (5) until convergence is reached.

Again, under-relaxation is needed to prevent the solutions from varying too quickly. The numerical results obtained with the SIMPLE method are as good as those shown in Figure 5.5.

5.2.3 PROJECTION METHOD

Another segregated Navier–Stokes solution method, which is also extremely popular, is the projection method (Chorin, 1968). To apply this method to our filter example, we should first add a pseudo unsteady term to the momentum equation (if we were solving an unsteady flow problem, then we had already had the unsteady term):

$$\frac{\partial u}{\partial t} = -\frac{dp}{dx} - cu^2 \tag{5.35}$$

The solution begins with an initial guess of the flow field. The desired solution is obtained when marching the equation to the steady state.

We first semi-discretize the momentum equation:

$$\frac{u^{n+1} - u^n}{\Delta t} = -\frac{dp^{n+1}}{dx} - cu^n u^{n+1} \tag{5.36}$$

Then we solve it in two fractional time steps:
Step 1:

Solve an intermediate velocity field $u*$ from

$$\frac{u* - u^n}{\Delta t} + cu^n u* = -\frac{dp^n}{dx} \tag{5.37}$$

Here we treat the friction force term (semi)-implicitly to accelerate convergence.
   Step 2:

$$\frac{u^{n+1} - u*}{\Delta t} = -\frac{dp'}{dx} \tag{5.38}$$

where

$$p' = p^{n+1} - p^n \tag{5.39}$$

is the pressure correction. To obtain an equation for p', we take the x-derivative of both sides of Equation (5.38), and require the velocity at the $(n + 1)^{th}$ moment to satisfy the continuity equation i.e. its x-derivative vanishes:

$$\frac{0 - du*/dx}{\Delta t} = -\frac{d^2 p'}{dx^2} \tag{5.40}$$

This is again a PPE. It can be easily discretized using the CDS

$$p'_{i-1} - 2p'_i + p'_{i+1} = \frac{\Delta x}{\Delta t}(u_i^* - u_{i-1}^*) \tag{5.41}$$

Such a system of finite difference equations of course is very easy to solve. Once we have p', the new velocity field can be calculated from Equation (5.38):

$$u_i^{n+1} = u_i^* - \frac{\Delta t}{\Delta x}\left(p'_{i+1} - p'_i\right) \tag{5.42}$$

The new pressure field of course is

$$p_i^{n+1} = p_i^n + p'_i \tag{5.43}$$

In summary, the projection method procedure is:

1. Solve discretized Equation (5.37) for $u*$.
2. Solve Equation (5.41) for p'.
3. Update the flow field by using Equations (5.42) and (5.43).

4. Repeat steps (1) through (3) until the steady-state or desired time moment is reached.

Another version of the projection method is totally dropping the pressure gradient term in step one, then the solution of the PPE will be pressure, instead of pressure correction.

Step 1:

Solve an intermediate velocity field $u*$ from

$$\frac{u* - u^n}{\Delta t} + cu^n u* = 0 \tag{5.44}$$

Step 2:

$$\frac{u^{n+1} - u*}{\Delta t} = -\frac{dp^{n+1}}{dx} \tag{5.45}$$

Take the x-derivative of both sides of this equation, and require the velocity at the $(n + 1)^{th}$ moment to satisfy the continuity equation:

$$\frac{0 - du*/dx}{\Delta t} = -\frac{d^2 p^{n+1}}{dx^2} \tag{5.46}$$

This is the PPE, which is discretized as

$$p_{i-1}^{n+1} - 2p_i^{n+1} + p_{i+1}^{n+1} = \frac{\Delta x}{\Delta t}(u_i^* - u_{i-1}^*) \tag{5.47}$$

The new pressure field then can be solved. Once we have p^{n+1}, the new velocity field can be calculated from Equation (5.45):

$$u_i^{n+1} = u_i^* - \frac{\Delta t}{\Delta x}\left(p_{i-1}^{n+1} - p_i^{n+1}\right) \tag{5.48}$$

A good feature of the projection method is that, unlike the PPE of the SIMPLE method (see Equation (5.33)), the coefficients in the PPE of the projection method are constants. This enables us to use very efficient Poisson equation solvers. For example, if you want to use the LU decomposition method to solve $Ap' = b$ (the matrix form of the PPE), you only need to do the time-consuming LU decomposition of matrix A once, then p' may be solved quite efficiently by using the same L and U matrices in every time step.

5.2.4 CO-LOCATED MESH AND MOMENTUM INTERPOLATION METHOD

Although the staggered mesh provides wonderful coupling between velocity and pressure, sometimes we may want to use a co-located mesh i.e. a mesh on which velocity and pressure values are stored at the same locations. Then of course we will have to find a way to address the odd–even decoupling issue. A very effective approach for this purpose is the momentum interpolation method (MIM) (Rhie & Chow, 1983). It is probably the easiest to show how it works in the framework of the projection method.

Suppose we still use the co-located mesh displayed in Figure 5.2 for the filter example. We first solve the preliminary velocity field from the momentum equation (5.37). If we use central difference for the pressure term, we will have

$$\frac{u^* - u^n}{\Delta t} + cu^n u^* = -\frac{dp^n}{dx} \rightarrow u_i^* = a_i + d_i \frac{p_{i-1}^n - p_{i+1}^n}{2} \tag{5.49}$$

where

$$a_i = \frac{u_i^n}{1 + c_i \Delta t u_i^n}; \; d_i = \frac{\Delta t}{\Delta x (1 + c_i \Delta t u_i^n)} \tag{5.50}$$

The odd–even decoupling occurs at this step since the pressure indices, $i - 1$ and $i + 1$, are either both odd, or both even. Well, since this is an incorrect coupling, let us drop the pressures to delete it:

$$u_i^{**} = a_i \tag{5.51}$$

Then we go ahead to solve the PPE, Equation (5.40). If we integrate this equation over the ith control volume, we will have

$$\frac{p_{i-1}' - 2p_i' + p_{i+1}'}{\Delta x^2} = \frac{1}{\Delta t} \frac{u_e^* - u_w^*}{\Delta x} \tag{5.52}$$

That is, we need the surface velocities. Notice that the right side of this equation comes from the continuity equation, so we are free to use the average of neighboring nodal velocities to evaluate these surface velocities. But the problem is, we only have u^{**}, the velocity with the pressure gradient dropped. It is not likely an accurate velocity field because of such a brutal operation. Then how should we evaluate the face velocities? Now it is time to play the trick. We add the pressure gradient back to the face u_e^{**}, but this time with the correct coupling:

$$u_e^* = \frac{u_i^{**} + u_{i+1}^{**}}{2} + \frac{d_i + d_{i+1}}{2}\left(p_i^n - p_{i+1}^n\right) \tag{5.53}$$

i and $i + 1$ cannot be both odd, or both even, and we are free of the odd–even decoupling issue.

After obtaining pressure corrections, we may update the flow field:

$$u_e^{n+1} = u_e^* + \frac{d_i + d_{i+1}}{2}\left(p_i' - p_{i+1}'\right); \ p_i^{n+1} = p_i^n + p_i' \tag{5.54}$$

Then we can interpolate the face velocity values to find the nodal velocities.

The procedure using the projection method and MIM to solve the filter flow example is:

1. Solve Equation (5.49) for an intermediate velocity field u^*.
2. Calculate the pseudo velocity field u^{**} by dropping the pressure terms.
3. Form face velocities according to Equation (5.53).
4. Solve Equation (5.52) for pressure corrections p'.
5. Update the flow field according to (5.54).
6. Repeat steps (1) through (5) until the steady-state or desired time moment is reached.

Obviously, we can also use another version of the projection method in the procedure above. That is, we directly drop the pressure term from the momentum equation in the first step. Then we can skip step 2 and directly go to step 3, and the PPE is then an equation of pressure, instead of pressure correction.

The same technique can be applied to other solution methods e.g. SIMPLE. For the current example, we may first solve the preliminary velocity field

$$u_i^* = \frac{1}{2c_i u_i^o \Delta x}(p_{i-1}^* - p_{i+1}^*) = d_i \frac{p_{i-1}^* - p_{i+1}^*}{2} \tag{5.55}$$

Then we go ahead to use the continuity equation to find the pressure corrections. At this step we use both the face velocity and the MIM technique:

$$u_e^* = \frac{1}{2}(u_i^* + u_{i+1}^*) - \frac{1}{2}\left(d_i \frac{p_{i-1}^* - p_{i+1}^*}{2} + d_{i+1}\frac{p_i^* - p_{i+2}^*}{2}\right)$$
$$+ \frac{d_i + d_{i+1}}{2}(p_i^* - p_{i+1}^*) \tag{5.56}$$

Then we assume

$$u_e' \approx \frac{d_i + d_{i+1}}{2}(p_i' - p_{i+1}') \tag{5.57}$$

We require that the corrected face velocities should satisfy continuity:

$$(u'_e + u^*_e) - (u'_w + u^*_w) = 0 \tag{5.58}$$

which results in the PPE for pressure corrections:

$$-\frac{d_{i-1} + d_i}{2}p'_{i-1} + \frac{d_{i-1} + 2d_i + d_{i+1}}{2}p'_i - \frac{d_i + d_{i+1}}{2}p'_{i+1} = -(u^*_e - u^*_w) \tag{5.59}$$

After the pressure corrections are solved, we can use them to correct the flow field:

$$p_i = p^*_i + p'_i; \ u_e = u^*_e + \frac{d_i + d_{i+1}}{2}(p'_i - p'_{i+1}) \tag{5.60}$$

Again, you can form nodal velocities by interpolation.

Let us solve some classical flow examples with these Navier–Stokes solution methods.

5.3 EXAMPLE: LID-DRIVEN CAVITY FLOW

5.3.1 PROBLEM STATEMENT, MESH, AND FORMULAS

As Figure 5.6 shows, a square cavity is filled with an incompressible fluid with constant density and viscosity. The top boundary (lid) of the cavity moves at a constant speed U, which causes the fluid to flow. Our purpose is to calculate the steady-state flow field in this cavity.

By using the cavity side length H, the lid velocity U as the reference length, and velocity scales, the Navier–Stokes equations can be normalized as

$$\frac{\partial u}{\partial x} + \frac{\partial v}{\partial y} = 0 \tag{5.61}$$

$$\frac{\partial u}{\partial t} + \frac{\partial(u \cdot u)}{\partial x} + \frac{\partial(v \cdot u)}{\partial y} - \frac{1}{Re}\left(\frac{\partial^2 u}{\partial x^2} + \frac{\partial^2 u}{\partial y^2}\right) = -\frac{\partial p}{\partial x} \tag{5.62}$$

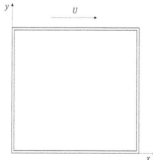

FIGURE 5.6 Lid-driven cavity flow example setup.

$$\frac{\partial v}{\partial t} + \frac{\partial (u \cdot v)}{\partial x} + \frac{\partial (v \cdot v)}{\partial y} - \frac{1}{Re}\left(\frac{\partial^2 v}{\partial x^2} + \frac{\partial^2 v}{\partial y^2}\right) = -\frac{\partial p}{\partial y} \tag{5.63}$$

where Reynolds number

$$Re = \frac{\rho U H}{\mu} \tag{5.64}$$

We will simulate the $Re = 100$ case with the SIMPLE method on a staggered mesh.

The staggered mesh is shown in Figure 5.7. u and v are stored at the faces of the p control volumes, and the u control volumes are staggered in x-direction while v control volumes are staggered in y-direction with respect to the p control volumes.

The x momentum equation (5.62) is integrated over the (i, j)th u control volume, as shown in Figure 5.8.

$$\int_s^n \int_w^e \left[\frac{\partial u}{\partial t} + \frac{\partial E}{\partial x} + \frac{\partial F}{\partial y} = -\frac{\partial p}{\partial x}\right] dxdy \tag{5.65}$$

where w, e, s, and n denote the west, east, south, and north faces of the $u_{i,j}$ control volume and

$$E = u \cdot u - \frac{1}{Re}\left(\frac{\partial u}{\partial x}\right) \tag{5.66}$$

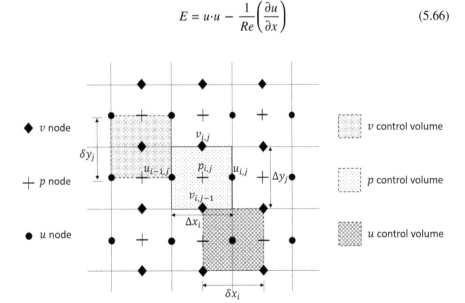

FIGURE 5.7 Staggered mesh for the lid-driven cavity flow.

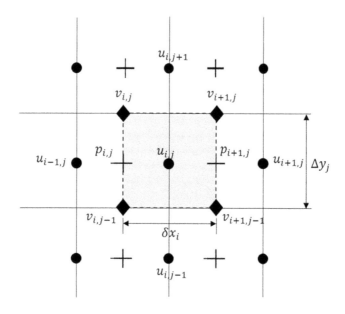

FIGURE 5.8   U control volumes.

$$F = v \cdot u - \frac{1}{Re}\left(\frac{\partial u}{\partial y}\right) \tag{5.67}$$

are the momentum fluxes in the x and y-directions.

With common finite volume method assumptions, Equation (5.65) becomes

$$\frac{(u_{i,j}^{n+1} - u_{i,j}^{n})\delta x_i \Delta y_j}{\Delta t} + (E_{i+1,j} - E_{i,j})\Delta y_j + (F_{i,j} - F_{i,j-1})\delta x_i$$
$$= -\Delta y_j (p_{i+1,j} - p_{i,j}) \tag{5.68}$$

We will follow the SIMPLE procedure: we first obtain a preliminary velocity field by solving the momentum equations, then using the continuity equation to correct the flow field. From now on, we will use u^*, v^*, p^* for the preliminary flow field and u, v, p the corrected flow field; u^o, v^o and $p^o = p^*$ denote the (corrected) flow field calculated from the last iteration. After we obtain the flow field at the nth moment, we will keep iterating to solve the Navier–Stokes equations until convergence is reached. The converged solution is the solution at the $(n + 1)$th moment.

Using CDS, the fluxes can be approximated as

$$E_{i+1,j} \approx \frac{1}{4}(u_{i+1,j}^o + u_{i,j}^o)(u_{i+1,j}^* + u_{i,j}^*) - \frac{1}{Re}\frac{u_{i+1,j}^* - u_{i,j}^*}{\Delta x_{i+1}} \tag{5.69}$$

$$E_{i,j} \approx \frac{1}{4}\left(u^{o}_{i-1,j} + u^{o}_{i,j}\right)\left(u^{*}_{i-1,j} + u^{*}_{i,j}\right) - \frac{1}{Re}\frac{u^{*}_{i,j} - u^{*}_{i-1,j}}{\Delta x_i} \tag{5.70}$$

$$F_{i,j} \approx \frac{1}{4}\left(v^{o}_{i+1,j} + v^{o}_{i,j}\right)\left(u^{*}_{i,j+1} + u^{*}_{i,j}\right) - \frac{1}{Re}\frac{u^{*}_{i,j+1} - u^{*}_{i,j}}{\delta y_j} \tag{5.71}$$

$$F_{i,j-1} \approx \frac{1}{4}\left(v^{o}_{i+1,j-1} + v^{o}_{i,j-1}\right)\left(u^{*}_{i,j-1} + u^{*}_{i,j}\right) - \frac{1}{Re}\frac{u^{*}_{i,j} - u^{*}_{i,j-1}}{\delta y_{j-1}} \tag{5.72}$$

Notice we linearized the nonlinear terms like $u*u*$ as $u^{o}u*$ i.e. with the Picard linearization method. Here the first u, u^{o} is the transporting flow field, and we can evaluate it with any accurate methods (my choice is the central difference); the second u, $u*$, however, is a property being transported by u^{o}, which has to be evaluated by a proper numerical scheme for convection term, as discussed in Chapter 3. Also notice that the implicit scheme is used to improve stability and to remove some of the stringent time step size restrictions imposed on an explicit scheme (see Section 3.2.2 in Chapter 3). By using the implicit scheme, we can use an infinite time step size since we are only interested in the steady-state solution. We have to, however, still use a fine enough mesh to avoid unbounded wiggles due to the central difference treatment to the convection terms.

Substituting Equations (5.69), (5.70), (5.71), and (5.72) into Equation (5.68) we obtain the finite difference u-momentum equation

$$a^{u}_{p}u^{*}_{i,j} + a^{u}_{w}u^{*}_{i-1,j} + a^{u}_{e}u^{*}_{i+1,j} + a^{u}_{s}u^{*}_{i,j-1} + a^{u}_{n}u^{*}_{i,j+1} = b^{u} \tag{5.73}$$

where

$$a^{u}_{w} = -\frac{\Delta y_j}{4}\left(u^{o}_{i-1,j} + u^{o}_{i,j}\right) - \frac{\Delta y_j}{Re\Delta x_i} = a^{u}_{wc} - a^{u}_{wd} \tag{5.74}$$

$$a^{u}_{e} = \frac{\Delta y_j}{4}\left(u^{o}_{i+1,j} + u^{o}_{i,j}\right) - \frac{\Delta y_j}{Re\Delta x_{i+1}} = a^{u}_{ec} - a^{u}_{ed} \tag{5.75}$$

$$a^{u}_{s} = -\frac{\delta x_i}{4}\left(v^{o}_{i+1,j-1} + v^{o}_{i,j-1}\right) - \frac{\delta x_i}{Re\delta y_{j-1}} = a^{u}_{sc} - a^{u}_{sd} \tag{5.76}$$

$$a^{u}_{n} = \frac{\delta x_i}{4}\left(v^{o}_{i+1,j} + v^{o}_{i,j}\right) - \frac{\delta x_i}{Re\delta y_j} = a^{u}_{nc} - a^{u}_{nd} \tag{5.77}$$

$$a_t^u = \frac{\delta x_i \Delta y_j}{\Delta t} \tag{5.78}$$

$$a_p^u = a_{wc}^u + a_{ec}^u + a_{sc}^u + a_{nc}^u + a_{wd}^u + a_{ed}^u + a_{sd}^u + a_{nd}^u + a_t^u \tag{5.79}$$

$$b^u = a_t^u u_{i,j}^n + \Delta y_j (p_{i,j}^* - p_{i+1,j}^*) \tag{5.80}$$

Notice that $p^* = p^o$.

Such equations form a pentadiagonal system, which can be solved by methods discussed in Chapter 4.

Integrating the y momentum equation over the (i, j)th v control volume (see Figure 5.9), results in

$$\int_s^n \int_w^e \left[\frac{\partial v}{\partial t} + \frac{\partial E}{\partial x} + \frac{\partial F}{\partial y} = -\frac{\partial p}{\partial y} \right] dxdy \tag{5.81}$$

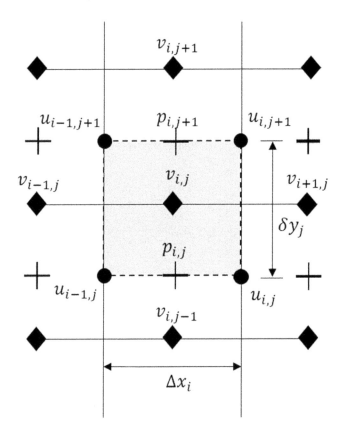

FIGURE 5.9 V control volumes.

where

$$E = u \cdot v - \frac{1}{Re}\left(\frac{\partial v}{\partial x}\right) \tag{5.82}$$

$$F = v \cdot v - \frac{1}{Re}\left(\frac{\partial v}{\partial y}\right) \tag{5.83}$$

are the v-momentum fluxes in the x- and y-directions.

Equation (5.81) is

$$\frac{\left(v_{i,j}^{n+1} - v_{i,j}^{n}\right)\Delta x_i \delta y_j}{\Delta t} + \left(E_{i,j} - E_{i-1,j}\right)\delta y_j + \left(F_{i,j+1} - F_{i,j}\right)\Delta x_i \tag{5.84}$$
$$= -\Delta x_i (p_{i,j+1} - p_{i,j})$$

Using CDS for the fluxes

$$E_{i,j} \approx \frac{1}{4}\left(u_{i,j+1}^{o} + u_{i,j}^{o}\right)\left(v_{i+1,j}^{*} + v_{i,j}^{*}\right) - \frac{1}{Re}\frac{v_{i+1,j}^{*} - v_{i,j}^{*}}{\delta x_i} \tag{5.85}$$

$$E_{i-1,j} \approx \frac{1}{4}\left(u_{i-1,j+1}^{o} + u_{i-1,j}^{o}\right)\left(v_{i-1,j}^{*} + v_{i,j}^{*}\right) - \frac{1}{Re}\frac{v_{i,j}^{*} - v_{i-1,j}^{*}}{\delta x_{i-1}} \tag{5.86}$$

$$F_{i,j+1} \approx \frac{1}{4}\left(v_{i,j+1}^{o} + v_{i,j}^{o}\right)\left(v_{i,j+1}^{*} + v_{i,j}^{*}\right) - \frac{1}{Re}\frac{v_{i,j+1}^{*} - v_{i,j}^{*}}{\Delta y_{j+1}} \tag{5.87}$$

$$F_{i,j} \approx \frac{1}{4}\left(v_{i,j-1}^{o} + v_{i,j}^{o}\right)\left(v_{i,j-1}^{*} + v_{i,j}^{*}\right) - \frac{1}{Re}\frac{v_{i,j}^{*} - v_{i,j-1}^{*}}{\Delta y_j} \tag{5.88}$$

Substituting Equations (5.85), (5.86), (5.87), and (5.88) into Equation (5.84), we have

$$a_p^v v_{i,j}^* + a_w^v v_{i-1,j}^* + a_e^v v_{i+1,j}^* + a_s^v v_{i,j-1}^* + a_n^v v_{i,j+1}^* = b^v \tag{5.89}$$

where

$$a_w^v = -\frac{\delta y_j}{4}\left(u_{i-1,j+1}^{o} + u_{i-1,j}^{o}\right) - \frac{\delta y_j}{Re\delta x_{i-1}} = a_{wc}^v - a_{wd}^v \tag{5.90}$$

$$a_e^v = \frac{\delta y_j}{4}\left(u_{i,j+1}^o + u_{i,j}^o\right) - \frac{\delta y_j}{Re\delta x_i} = a_{ec}^v - a_{ed}^v \tag{5.91}$$

$$a_s^v = -\frac{\Delta x_i}{4}\left(v_{i,j-1}^o + v_{i,j}^o\right) - \frac{\Delta x_i}{Re\Delta y_j} = a_{sc}^v - a_{sd}^v \tag{5.92}$$

$$a_n^v = \frac{\Delta x_i}{4}\left(v_{i,j+1}^o + v_{i,j}^o\right) - \frac{\Delta x_i}{Re\Delta y_{j+1}} = a_{nc}^v - a_{nd}^v \tag{5.93}$$

$$a_t^v = \frac{\Delta x_i \delta y_j}{\Delta t} \tag{5.94}$$

$$a_p^v = a_{wc}^v + a_{ec}^v + a_{sc}^v + a_{nc}^v + a_{wd}^v + a_{ed}^v + a_{sd}^v + a_{nd}^v + a_t^v \tag{5.95}$$

$$b^v = a_t^v v_{i,j}^n + \Delta x_i(p_{i,j}^* - p_{i,j+1}^*) \tag{5.96}$$

Again these equations form a pentadiagonal system and can be solved by using e.g. the SIP method (see Chapter 4).

By integrating continuity Equation (5.61) over the $p_{i,j}$ control volume (see Figure 5.10) we have

$$\Delta y_j(u_{i,j} - u_{i-1,j}) + \Delta x_i(v_{i,j} - v_{i,j-1}) = 0 \tag{5.97}$$

If the velocities u^*, v^* solved from the discretized momentum Equations (5.73) and (5.89) also satisfy Equation (5.97) in each p control volume, we will happily claim success in solving the lid-driven cavity flow field. Unfortunately, usually this is not the case since we have not solved for pressure yet and without the correct pressure field we should not expect the velocity field, which is partly driven by the pressure gradient, to be correct i.e. satisfy the continuity equation.

So how can we find the correct pressure field? We will correct pressure in each p control volume until both the momentum and the continuity equations are satisfied by the corrected pressure and velocity

$$p = p^* + p' \tag{5.98}$$

$$u = u^* + u' \tag{5.99}$$

$$v = v^* + v' \tag{5.100}$$

By substituting the corrected pressure and velocity into the momentum equations we find that the relationships between the velocity and pressure corrections are

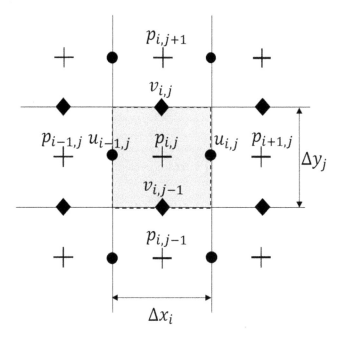

FIGURE 5.10   P control volumes.

$$a_p^u u_{i,j}' + \sum a_{nb}^u u_{nb}' = \Delta y_j (p_{i,j}' - p_{i+1,j}') \tag{5.101}$$

$$a_p^v v_{i,j}' + \sum a_{nb}^v v_{nb}' = \Delta x_i (p_{i,j}' - p_{i,j+1}') \tag{5.102}$$

where the subscript "p" stands for the present (the (i,j)th) node; while the subscript "nb" represents the neighboring nodes. Well, these relationships are pretty complex. To make it really SIMPLE, Patankar and Spalding (Patankar & Spalding, 1972) proposed to neglect the $\sum a_{nb}^u u_{nb}'$ and $\sum a_{nb}^v v_{nb}'$ terms in Equations (5.101) and (5.102), so the velocity corrections are directly proportional to the corresponding pressure corrections:

$$u_{i,j}' \approx d_{i,j}^u (p_{i,j}' - p_{i+1,j}') \tag{5.103}$$

$$v_{i,j}' \approx d_{i,j}^v (p_{i,j}' - p_{i,j+1}') \tag{5.104}$$

where

$$d_{i,j}^u = \Delta y_j / a_{p,\,(i,j)}^u \tag{5.105}$$

$$d_{i,j}^v = \Delta x_i / a_{p,\,(i,j)}^v \tag{5.106}$$

I add an "(i, j)" to the subscripts of a_p^u and a_p^v to remind you that these coefficients are not constants but different at different nodes. Notice that neglecting the $\sum a_{nb}^u u_{nb}'$ and $\sum a_{nb}^v v_{nb}'$ terms will not affect the accuracy of the final converged solution because such terms will vanish (or become very small) as iterations converge i.e. when all velocity corrections are small enough.

By requiring the corrected velocities to satisfy the continuity Equation (5.97), we have

$$
\Delta y_j \left(u_{i,j}' - u_{i-1,j}' \right) + \Delta x_i \left(v_{i,j}' - v_{i,j-1}' \right)
$$
$$
= - \left[\Delta y_j \left(u_{i,j}^* - u_{i-1,j}^* \right) + \Delta x_i \left(v_{i,j}^* - v_{i,j-1}^* \right) \right]
$$

(5.107)

By substituting Equations (5.103) and (5.104) into Equation (5.107), we obtain the p' Poisson equation (PPE)

$$
a_p^p p_{i,j}' + a_w^p p_{i-1,j}' + a_e^p p_{i+1,j}' + a_s^p p_{i,j-1}' + a_n^p p_{i,j+1}' = b^p
$$

(5.108)

where

$$
a_w^p = -\Delta y_j d_{i-1,j}^u = -\Delta y_j^2 / a_{p,(i-1,j)}^u
$$

(5.109)

$$
a_e^p = -\Delta y_j d_{i,j}^u = -\Delta y_j^2 / a_{p,(i,j)}^u
$$

(5.110)

$$
a_s^p = -\Delta x_i d_{i,j-1}^v = -\Delta x_i^2 / a_{p,(i,j-1)}^v
$$

(5.111)

$$
a_n^p = -\Delta x_i d_{i,j}^v = -\Delta x_i^2 / a_{p,(i,j)}^v
$$

(5.112)

$$
a_p^p = -(a_w^p + a_e^p + a_s^p + a_n^p)
$$

(5.113)

$$
b^p = - \left[\Delta y_j \left(u_{i,j}^* - u_{i-1,j}^* \right) + \Delta x_i \left(v_{i,j}^* - v_{i,j-1}^* \right) \right]
$$

(5.114)

These are again pentadiagonal equations. They can be solved by using methods introduced in Chapter 4.

After we obtain pressure corrections, we may correct pressure and velocity according to Equations (5.98), (5.99), (5.100), (5.103), and (5.104). Then the corrected pressures and velocities can be used to calculate the coefficients of the momentum and pressure correction equations in the next iteration of calculation.

5.3.2 UNDER-RELAXATION

The Navier–Stokes equations we are solving are nonlinear, which makes our code prone to diverge. To avoid constantly frustrating divergence (CFD), we may use the under-relaxation technique (see Section 5.1.1) to update our solutions. That is, instead of using

$$u = u^* + u' \tag{5.115}$$

we use

$$u = \alpha(u^* + u') + (1 - \alpha)u^o = \alpha u' + [\alpha u^* + (1 - \alpha)u^o] \tag{5.116}$$

where $0 < \alpha < 1$ is the under-relaxation factor and u^o is the old u value obtained in the previous iteration.

Such under-relaxations can be realized implicitly as follows. u^* is solved from the finite difference equation

$$a_p^u u^* + \sum a_{nb}^u u_{nb}^* = b^u \tag{5.117}$$

which gives

$$u^* = \frac{b^u}{a_p^u} - \sum \frac{a_{nb}^u}{a_p^u} u_{nb}^* \tag{5.118}$$

Therefore

$$\alpha u^* + (1 - \alpha)u^o = \frac{b^u}{a_p^u/\alpha} - \sum \frac{a_{nb}^u}{a_p^u/\alpha} u_{nb}^* + (1 - \alpha)u^o \tag{5.119}$$

If we still call the left side of this equation u^*, then its finite difference equation becomes

$$\frac{a_p^u}{\alpha} u^* + \sum a_{nb}^u u_{nb}^* = b^u + (1 - \alpha)\frac{a_p^u}{\alpha} u^o \tag{5.120}$$

That is, we have to replace a_p^u in Equation (5.117) with a_p^u/α and b^u with $b^u + (1 - \alpha)\frac{a_p^u}{\alpha} u^o$. You may find that if α is small, the first and last terms dominate and $u^* \approx u^o$ while as $\alpha = 1$ the original Equation (5.117) is recovered. The v velocity and pressure can be relaxed similarly. After solving such u^* and v^*, we then obtain the new solution using Equation (5.116), which is now

$$u = u^* + \alpha u' \tag{5.121}$$

You should understand that using under-relaxation does not change the final converged solution, it only changes the convergence speed.

It can be proved that if the relaxation factor for p is α, the optimal relaxation factor for u and v should be close to $1 - \alpha$ (Ferziger & Peric, 2002). For the lid-driven cavity flow, you may use $\alpha = 0.9$ for u and v, and 0.2 for p.

A way to determine convergence is using the deviation (residue) from the continuity equation

$$r = \sum_{i,j} \left| \Delta y_j \left(u^*_{i+1,j} - u^*_{i,j} \right) + \Delta x_i \left(v^*_{i,j+1} - v^*_{i,j} \right) \right| \qquad (5.122)$$

Convergence is reached if this residual is less than a prescribed small number. Now let us see how to apply boundary conditions.

5.3.3 BOUNDARY CONDITIONS IMPLEMENTATION

We will use virtual control volumes and nodes to implement boundary conditions, as shown in Figure 5.11.

The shaded region stands for the flow domain, which is filled with p control volumes. We assume the number of internal p control volumes in x-direction is N and in y-direction is M. One layer of virtual p control volumes with corresponding

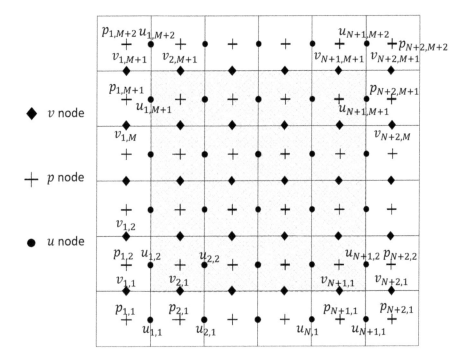

FIGURE 5.11 Indices of u, v, and p control volumes.

virtual nodes are added around the flow domain. The size of the virtual control volume is the same as their neighboring internal control volume.

For our lid-driven cavity flow, $u = v = 0$ at the west, east, and south boundaries of the flow domain; at the north boundary the nondimensional velocity $u = 1$. So we set those velocity nodes located right on the boundaries to the known values

$$u_{1,j} = u_{N+2,j} = 0; \quad v_{i,1} = v_{i,M+2} = 0 \tag{5.123}$$

For the boundary nodes that do not lie on the boundary we set the virtual velocity node values as shown in Figure 5.12.

$$u_{i,1} = -u_{i,2}; \quad u_{i,M+2} = 2 - u_{i,M+1}; \quad v_{1,j} = -v_{2,j}; \quad v_{N+2,j} = -v_{N+1,j} \tag{5.124}$$

For pressure boundary conditions, consider the v-momentum equation

$$\frac{\partial v}{\partial t} + \frac{\partial (u \cdot v)}{\partial x} + \frac{\partial (v \cdot v)}{\partial y} - \frac{1}{Re}\left(\frac{\partial^2 v}{\partial x^2} + \frac{\partial^2 v}{\partial y^2}\right) = -\frac{\partial p}{\partial y} \tag{5.125}$$

If we apply this equation to the south boundary ($y = 0$), all terms on its left side vanish (because $u = v = 0$ along this boundary i.e. along the x-direction, and they do not change with time) except $\partial^2 v/\partial y^2$. If you want, you can discretize

$$\frac{\partial p}{\partial y} = \frac{1}{Re}\frac{\partial^2 v}{\partial y^2} \tag{5.126}$$

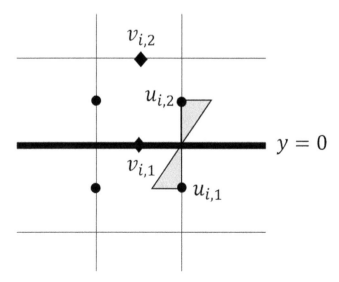

FIGURE 5.12   Application of boundary conditions to nodes that do not lie on the boundary.

and use it as the boundary condition for pressure at this boundary. We, however, simply ignore such diffusion terms since they are small. So we directly set the normal pressure gradient at the solid surfaces to be zero:

$$p_{i,1} = p_{i,2}; \quad p_{i,M+2} = p_{i,M+1}; \quad p_{1,j} = p_{2,j}; \quad p_{N+2,j} = p_{N+1,j} \qquad (5.127)$$

Also we require the normal gradient of pressure corrections at the solid boundaries to vanish

$$p'_{i,1} = p'_{i,2}; \quad p'_{i,M+2} = p'_{i,M+1}; \quad p'_{1,j} = p'_{2,j}; \quad p'_{N+2,j} = p'_{N+1,j} \qquad (5.128)$$

This is because the velocities at all boundaries are known, so no velocity correction is needed at the boundaries. Since the SIMPLE method assumes velocity corrections to be proportional to pressure correction gradients (see Equations (5.103) and (5.104)), the above conditions follow. For example, as Figure 5.13 shows, at $y = 0$ (the south boundary), since $v'_{i,1} = d^y_{i,j}\left(p'_{i,1} - p'_{i,2}\right) \equiv 0$, we should set $p'_{i,1} = p'_{i,2}$.

One difficulty related to using the vanishing p' gradient condition at all boundaries is the system of p' equations becomes singular. This can be understood if we use only one control volume inside the cavity i.e. the cavity itself is the only interior control volume. The finite difference equation of p' of this control volume is then

$$a^P_p p'_{2,2} + a^P_w p'_{1,2} + a^P_e p'_{3,2} + a^P_s p'_{2,1} + a^P_n p'_{2,3} = b^P \qquad (5.129)$$

The boundary condition is $p'_{1,2} = p'_{3,2} = p'_{2,1} = p'_{2,3} = p'_{2,2}$, whence

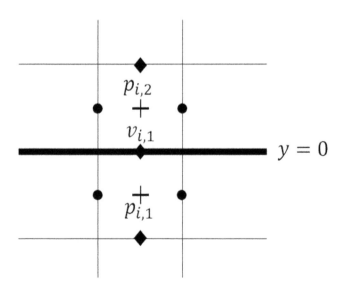

FIGURE 5.13 Application of boundary conditions of pressure.

$$(a_p^p + a_w^p + a_e^p + a_s^p + a_n^p)p'_{2,2} = b^p \tag{5.130}$$

However, Equation (5.113) tells us $a_p^p + a_w^p + a_e^p + a_s^p + a_n^p = 0$, which makes Equation (5.130) singular. The same problem arises for other numbers of control volumes. A simple fix is changing one p' boundary condition to a Dirichlet condition. For example, we may set the p' value at one arbitrary boundary point to zero. Then the whole system is no longer singular. This practice is equivalent to setting that boundary point as the reference point of pressure values of the flow field. Notice that the pressure value is in fact usually unimportant, it is the pressure difference, or gradient value that matters because only pressure gradients show up in the Navier–Stokes equations. Therefore, no matter which reference point we select to set its pressure to zero, or any other constant value, the converged flow fields will be the same, except that the pressure fields differ by a constant.

5.3.4 FLOW FIELD VISUALIZATION

After the velocity field calculation converges, we may want to visualize the flow field. We can plot the streamlines which show the flow direction at each point in the flow domain. The streamlines are contours of the stream function ϕ. It is closely related to the vorticity ω. Their definitions are

$$u = \frac{\partial \phi}{\partial y}, \quad v = -\frac{\partial \phi}{\partial x} \tag{5.131}$$

$$\omega = \frac{\partial v}{\partial x} - \frac{\partial u}{\partial y} \tag{5.132}$$

The most convenient locations to store ϕ and ω are the corners of p control volumes, as shown in Figure 5.14.

The vorticity value at each of such points can be obtained using central differences:

$$\omega_{i,j} = \frac{v_{i+1,j} - v_{i,j}}{\delta x_i} - \frac{u_{i,j+1} - u_{i,j}}{\delta y_j} \tag{5.133}$$

Stream function can be solved from Equation (1.24), repeated here:

$$\frac{\partial^2 \phi}{\partial x^2} + \frac{\partial^2 \phi}{\partial y^2} = -\omega \tag{5.134}$$

The boundary condition for ϕ is that the stream function at all solid walls is the same, which can be conveniently set to zero in the current case. This is because

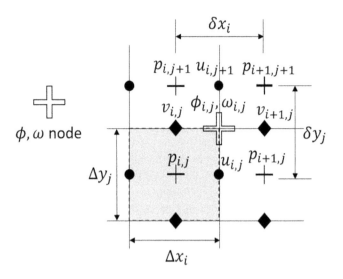

FIGURE 5.14 Stream function and vorticity control volumes.

physically the difference between the stream function values at any two points in a flow field is the volumetric flow flux across an arbitrary line connecting these two points; obviously if we select two points along the walls and use the walls themselves as lines to connect these two points, there is no flow across such lines. These two points, therefore, share the same stream function values.

Another option to visualize the flow field is using MATLAB built-in functions including `streamline`, `quiver`, `surf`, `contour` and many others. You have to pay attention to the format of the inputs and outputs of such functions, as they may produce figures based on input matrices according to a mapping other than our common practice, which is shown in Figure 4.8.

5.3.5   PROCEDURE OF SIMPLE METHOD

The procedure for implementing the SIMPLE method to an unsteady flow problem is as follows:

1. Provide initial values of the velocity field u^1, v^1, guess the pressure field p^1. Set the time moment index n to 1.
2. Assign the flow field to u^o, v^o, and p^o; solve the momentum equations for the intermediate velocity field u^*, v^*.
3. Solve the pressure correction equation for p'.
4. Calculate the corrected flow field u, v, and p.
5. Repeat steps (2) through (4) until convergence is reached, then the flow field is assigned to u^{n+1}, v^{n+1}, and p^{n+1}. Then increase n by one.
6. Repeat steps (2) through (5) until the desired final time moment is reached.
7. Post-process data.

5.3.6   RESULTS AND DISCUSSION

For the lid-driven cavity flow example, we use a 40×40 uniform mesh for the $Re = 100$ case. The iterations are terminated when the sum of the residuals of the p' equation is less than 10^{-6}. The results are shown in Figure 5.15.

The nondimensional stream function (upper left) and vorticity (upper right) contours agree well with the benchmark results (Ghia, Ghia, & Shin, 1982). The normalized velocity values along the vertical and horizontal lines through the cavity center are compared to the benchmark data in the two lower graphs of Figure 5.15.

5.3.7   PROCEDURE OF PROJECTION METHOD

We can also solve the lid-driven cavity flow by using the projection method. To find the velocity field at the new time step (the $(n + 1)^{th}$ step), we need to solve the momentum equation

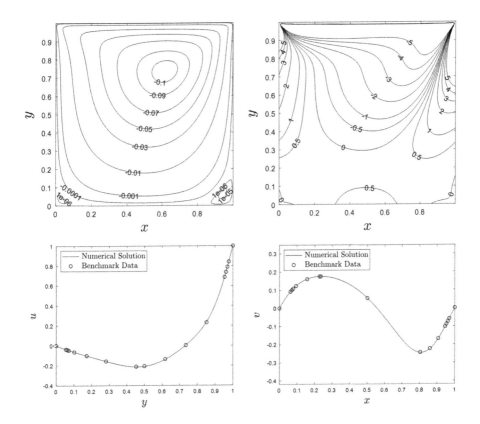

FIGURE 5.15   Numerical results of the lid-driven cavity flow.

$$\frac{\partial u_i}{\partial t} + \frac{\partial (u_j u_i)}{\partial x_j} = \nu \frac{\partial^2 u_i}{\partial x_j \partial x_j} - \frac{1}{\rho} \frac{\partial p}{\partial x_i} \tag{5.135}$$

Notice that this equation uses the Einstein convention. Under this convention, a suffix not repeated in one term must not be repeated in any term and can take any possible value; while a repeated suffix in a term is a dummy suffix and is held to be summed over all possible values. For example, $\partial u_i / \partial x = u_i$ represents equations like $\partial u_1 / \partial x = u_1$, $\partial u_2 / \partial x = u_2$, etc. but $\partial u_i / \partial x_i = 0$ stands for $\partial u_1 / \partial x_1 + \partial u_2 / \partial x_2 + \partial u_3 / \partial x_3 = 0$, which is the continuity equation in the context of fluid mechanics. In Equation (5.135), you can see the convection and diffusion term indeed represent summations of three terms each.

The projection methods solve the velocity field in a two-step procedure. For example, the following version:
Step 1:

$$\frac{u_i^* - u_i^n}{\Delta t} + \frac{\partial (u_j u_i)^n}{\partial x_j} = \nu \frac{\partial^2 u_i^*}{\partial x_j \partial x_j} \tag{5.136}$$

Step 2:

$$\frac{u_i^{n+1} - u_i^*}{\Delta t} = -\frac{1}{\rho} \frac{\partial p^{n+1}}{\partial x_i} \tag{5.137}$$

We first solve the momentum equation for a preliminary velocity field u_i^*, then we take the derivative of each term of Equation (5.137) with respect to x_i i.e. the divergence, and enforce the continuity equation

$$\frac{\partial u_i^{n+1}}{\partial x_i} = 0 \tag{5.138}$$

Then we have

$$-\frac{1}{\Delta t} \frac{\partial u_i^*}{\partial x_i} = -\frac{1}{\rho} \frac{\partial}{\partial x_i} \left(\frac{\partial p^{n+1}}{\partial x_i} \right) \tag{5.139}$$

which solves for p^{n+1} since u_i^* have been solved from Equation (5.136). After p^{n+1} is solved then u_i^{n+1} is easily obtained by using Equation (5.137).

Another version of the projection method is as follows:
Step 1:

$$\frac{u_i^* - u_i^n}{\Delta t} + \frac{\partial (u_j^n u_i^*)}{\partial x_j} = \nu \frac{\partial^2 u_i^*}{\partial x_j \partial x_j} - \frac{1}{\rho} \frac{\partial p^n}{\partial x_i} \tag{5.140}$$

Step 2:

$$\frac{u_i^{n+1} - u_i^*}{\Delta t} = -\frac{1}{\rho}\frac{\partial}{\partial x_i}(p^{n+1} - p^n) = -\frac{1}{\rho}\frac{\partial p'}{\partial x_i} \tag{5.141}$$

Again to enforce continuity we take divergence of Equation (5.141):

$$-\frac{1}{\Delta t}\frac{\partial u_i^*}{\partial x_i} = -\frac{1}{\rho}\frac{\partial}{\partial x_i}\left(\frac{\partial p'}{\partial x_i}\right) \tag{5.142}$$

After we find p' we then update pressure and velocities according to Equation (5.141).

If you decide to do this simulation with a co-located mesh, using the first version of the projection method probably is the simplest. You only need to replace nodal velocities with face velocities in Equation (5.139)(5.142)

5.4 EXAMPLE: NATURAL CONVECTION IN A CAVITY

5.4.1 PROBLEM DESCRIPTION

As Figure 5.16 shows, a square box with well-insulated top and bottom surfaces has different temperatures at the other two surfaces. Because of the temperature difference, the density of air inside the box varies and results in a buoyancy force, which drives the air to flow. Such a phenomenon is known as natural convection. In this case we are interested not only in the velocity field, but also the temperature field and the heat transfer across the isothermal walls.

5.4.2 GOVERNING EQUATIONS AND BOUSSINESQ ASSUMPTION

Because the buoyancy force is due to the air density variation and the consequent change in the gravity force, we have to include gravity in the governing equations. Also the fluid temperature distribution has to be solved since it is the ultimate cause of the flow. The governing equations are

$$\frac{\partial \rho}{\partial t} + \frac{\partial(\rho u)}{\partial x} + \frac{\partial(\rho v)}{\partial y} = 0 \tag{5.143}$$

$$\frac{\partial(\rho u)}{\partial t} + \frac{\partial(\rho u \cdot u)}{\partial x} + \frac{\partial(\rho v \cdot u)}{\partial y} - \left[\frac{\partial}{\partial x}\left(\mu\frac{\partial u}{\partial x}\right) + \frac{\partial}{\partial y}\left(\mu\frac{\partial u}{\partial y}\right)\right] = -\frac{\partial p}{\partial x} \tag{5.144}$$

$$\frac{\partial(\rho v)}{\partial t} + \frac{\partial(\rho u \cdot v)}{\partial x} + \frac{\partial(\rho v \cdot v)}{\partial y} - \left[\frac{\partial}{\partial x}\left(\mu\frac{\partial v}{\partial x}\right) + \frac{\partial}{\partial y}\left(\mu\frac{\partial v}{\partial y}\right)\right] = -\frac{\partial p}{\partial y} - \rho g \tag{5.145}$$

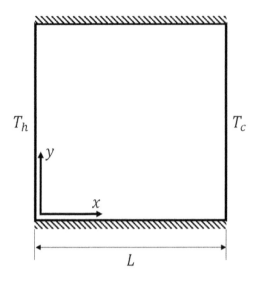

FIGURE 5.16 Natural convection in a cavity example setup.

$$\frac{\partial(\rho e)}{\partial t} + \frac{\partial(\rho u \cdot h)}{\partial x} + \frac{\partial(\rho v \cdot h)}{\partial y} - \left[\frac{\partial}{\partial x}\left(k\frac{\partial T}{\partial x}\right) + \frac{\partial}{\partial y}\left(k\frac{\partial T}{\partial y}\right) \right] = 0 \quad (5.146)$$

which are the continuity equation, the momentum equations and the energy equation. g is the gravitational acceleration (9.81 m/s^2). e and h are the specific internal energy and enthalpy of air. k is the thermal conductivity of air.

An equation of state which relates density to temperature should be provided. For air, this equation is the ideal gas law

$$p = \rho R T \qquad (5.147)$$

where R is the gas constant. For an ideal gas like air we also have

$$e = c_v T; \ h = c_p T \qquad (5.148)$$

where c_v and c_p are specific heats at constant volume and constant pressure, respectively.

Obviously, now the equations are more complicated than what we had in the previous example. To simplify the equations, we might adopt the so-called Boussinesq assumption, which includes the following suppositions:

1. Fluid properties are constant except density; the fluid density variation is also unimportant except in the $-\rho g$ term in the y momentum equation. A constant density ρ_0 is hence used elsewhere except in the $-\rho g$ term. This ρ_0 is the

density at a reference temperature T_0. We will use the cold wall temperature T_c as the reference temperature.

2. We assume that the pressure does not change significantly, so the density variation in the $- \rho g$ term is solely due to temperature variation

$$\rho \approx \rho_c + \frac{\partial \rho}{\partial T} \Delta T = \rho_c - \left(\frac{p}{RT_c^2}\right)(T - T_c) = \rho_c - \left(\frac{\rho_c}{T_c}\right)(T - T_c) \quad (5.149)$$

So the $- \rho g$ term is

$$- \rho g = -\rho_c g \left[1 - \beta(T - T_c)\right] \quad (5.150)$$

where $\beta = 1/T_c$.

The constant term $- \rho_c g$ in $- \rho g$ might be absorbed into an equivalent pressure

$$\hat{p} = p + \rho_c g y \quad (5.151)$$

Notice that $\partial \hat{p}/\partial x = \partial p/\partial x$ and $- \partial \hat{p}/\partial y = -\partial p/\partial x - \rho_c g$.

Based on these assumptions, the governing equations become

$$\frac{\partial u}{\partial x} + \frac{\partial v}{\partial y} = 0 \quad (5.152)$$

$$\rho_c \left[\frac{\partial u}{\partial t} + \frac{\partial (u \cdot u)}{\partial x} + \frac{\partial (v \cdot u)}{\partial y}\right] - \mu\left(\frac{\partial^2 u}{\partial x^2} + \frac{\partial^2 u}{\partial y^2}\right) = -\frac{\partial \hat{p}}{\partial x} \quad (5.153)$$

$$\rho_c \left[\frac{\partial v}{\partial t} + \frac{\partial (u \cdot v)}{\partial x} + \frac{\partial (v \cdot v)}{\partial y}\right] - \mu\left(\frac{\partial^2 v}{\partial x^2} + \frac{\partial^2 v}{\partial y^2}\right) = -\frac{\partial \hat{p}}{\partial y} + \rho_c g \beta (T - T_c) \quad (5.154)$$

$$\frac{1}{K}\frac{\partial T}{\partial t} + \frac{\partial (u \cdot T)}{\partial x} + \frac{\partial (v \cdot T)}{\partial y} - \left(\frac{k}{\rho_c c_p}\right)\left(\frac{\partial^2 T}{\partial x^2} + \frac{\partial^2 T}{\partial y^2}\right) = 0 \quad (5.155)$$

where $K = c_p/c_v$ is the specific heat ratio. We may "absorb" this K into time t of the energy equation. You may think this creates an inconsistency among the time scales used in different equations, but this is fine because we are only interested in the steady-state solution and using different time step sizes in solving different equations might even help convergence.

We define kinematic viscosity $\gamma = \mu/\rho_c$ (I do not use the Greek letter ν so that you do not confuse it with velocity v) and thermal diffusivity $\alpha = k/(\rho_c c_p)$. If we use the cavity size L, α/L, L^2/α, $\rho_c \alpha^2/L^2$ and the temperature difference between the two isothermal walls, $T_h - T_c$, as reference length, velocity, time, pressure, and temperature difference scales, the equations can be normalized as

$$\frac{\partial \tilde{u}}{\partial \tilde{x}} + \frac{\partial \tilde{v}}{\partial \tilde{y}} = 0 \tag{5.156}$$

$$\frac{\partial \tilde{u}}{\partial \tilde{t}} + \frac{\partial (\tilde{u} \cdot \tilde{u})}{\partial \tilde{x}} + \frac{\partial (\tilde{v} \cdot \tilde{u})}{\partial \tilde{y}} - Pr \left(\frac{\partial^2 \tilde{u}}{\partial \tilde{x}^2} + \frac{\partial^2 \tilde{u}}{\partial \tilde{y}^2} \right) = -\frac{\partial \tilde{p}}{\partial \tilde{x}} \tag{5.157}$$

$$\frac{\partial \tilde{v}}{\partial \tilde{t}} + \frac{\partial (\tilde{u} \cdot \tilde{v})}{\partial \tilde{x}} + \frac{\partial (\tilde{v} \cdot \tilde{v})}{\partial \tilde{y}} - Pr \left(\frac{\partial^2 \tilde{v}}{\partial \tilde{x}^2} + \frac{\partial^2 \tilde{v}}{\partial \tilde{y}^2} \right) = -\frac{\partial \tilde{p}}{\partial \tilde{y}} + Ra \cdot Pr \cdot \theta \tag{5.158}$$

$$\frac{\partial \theta}{\partial \tilde{t}} + \frac{\partial (\tilde{u} \cdot \theta)}{\partial \tilde{x}} + \frac{\partial (\tilde{v} \cdot \theta)}{\partial \tilde{y}} - \left(\frac{\partial^2 \theta}{\partial \tilde{x}^2} + \frac{\partial^2 \theta}{\partial \tilde{y}^2} \right) = 0 \tag{5.159}$$

where the dimensionless temperature difference is

$$\theta = \frac{T - T_c}{T_h - T_c} \tag{5.160}$$

In the above equations

$$Pr = \frac{\gamma}{\alpha} \tag{5.161}$$

is a dimensionless physical property called Prandtl number. For air $Pr \approx 0.71$ at room temperature.

Another dimensionless parameter that also arises in the governing equations is the Rayleigh number

$$Ra = \frac{g \beta (T_h - T_c) L^3}{\alpha \gamma} \tag{5.162}$$

We will simulate a case in which $Ra = 1000$. From now on we drop the accents over the variables.

5.4.3   DISCRETIZATION AND BOUNDARY CONDITIONS

We use a staggered mesh. Temperature values are stored at the p nodes. The discretization of the continuity and momentum equations are similar to the practice introduced in Section 5.3, except that the y momentum equation now has an additional source term.

We will derive the finite difference formula of the energy equation. Integrate Equation (5.159) over the $T_{i,j}$ control volume, which of course is also the $\theta_{i,j}$ control volume, as shown in Figure 5.17, we have

$$\int_s^n \int_w^e \left[\frac{\partial \theta}{\partial t} + \frac{\partial}{\partial x}\left(u \cdot \theta - \frac{\partial \theta}{\partial x} \right) + \frac{\partial}{\partial y}\left(v \cdot \theta - \frac{\partial \theta}{\partial y} \right) \right] dxdy = 0 \qquad (5.163)$$

which can be approximated as

$$\frac{\theta_{i,j} - \theta_{i,j}^n}{\Delta t}\Delta x_i \Delta y_j + \left(E_{i,j} - E_{i-1,j} \right)\Delta y_j + \left(F_{i,j} - F_{i,j-1} \right)\Delta x_i = 0 \qquad (5.164)$$

where

$$E_{i,j} = \frac{1}{2}u_{i,j}\left(\theta_{i+1,j} + \theta_{i,j} \right) - \frac{\theta_{i+1,j} - \theta_{i,j}}{\delta x_i} \qquad (5.165)$$

$$E_{i-1,j} = \frac{1}{2}u_{i-1,j}\left(\theta_{i,j} + \theta_{i-1,j} \right) - \frac{\theta_{i,j} - \theta_{i-1,j}}{\delta x_{i-1}} \qquad (5.166)$$

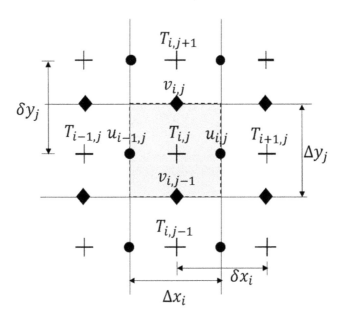

FIGURE 5.17 Temperature control volumes.

$$F_{i,j} = \frac{1}{2} v_{i,j} \left(\theta_{i,j+1} + \theta_{i,j} \right) - \frac{\theta_{i,j+1} - \theta_{i,j}}{\delta y_j} \tag{5.167}$$

$$F_{i,j-1} = \frac{1}{2} v_{i,j-1} \left(\theta_{i,j} + \theta_{i,j-1} \right) - \frac{\theta_{i,j} - \theta_{i,j-1}}{\delta y_{j-1}} \tag{5.168}$$

Therefore, the energy equation becomes

$$a_p^T \theta_{i,j} + a_w^T \theta_{i-1,j} + a_e^T \theta_{i+1,j} + a_s^T \theta_{i,j-1} + a_n^T \theta_{i,j+1} = a_t^T \theta_{i,j}^n \tag{5.169}$$

where

$$a_w^T = -\frac{\Delta y_j}{2} u_{i-1,j} - \frac{\Delta y_j}{\delta x_{i-1}} = a_{wc}^T - a_{wd}^T \tag{5.170}$$

$$a_e^T = \frac{\Delta y_j}{2} u_{i,j} - \frac{\Delta y_j}{\delta x_i} = a_{ec}^T - a_{ed}^T \tag{5.171}$$

$$a_s^T = -\frac{\Delta x_i}{2} v_{i,j-1} - \frac{\Delta x_i}{\delta y_{j-1}} = a_{sc}^T - a_{sd}^T \tag{5.172}$$

$$a_n^T = \frac{\Delta x_i}{2} v_{i,j} - \frac{\Delta x_i}{\delta y_j} = a_{nc}^T - a_{nd}^T \tag{5.173}$$

$$a_t^T = \frac{\Delta x_i \Delta y_j}{\Delta t} \tag{5.174}$$

$$a_p^T = a_{wc}^T + a_{ec}^T + a_{sc}^T + a_{nc}^T + a_{wd}^T + a_{ed}^T + a_{sd}^T + a_{nd}^T + a_t^T \tag{5.175}$$

Boundary conditions are as follows. No-slip velocity and vanishing normal pressure gradient conditions apply at all boundaries. At the insulated surfaces we have zero temperature gradients:

$$\frac{\partial \theta}{\partial y} = 0 \ at \ y = 0 \ and \ 1 \tag{5.176}$$

And at the isothermal surfaces we have Dirichlet conditions:

$$\theta = 1 \ at \ x = 0; \ \theta = 0 \ at \ x = 1 \tag{5.177}$$

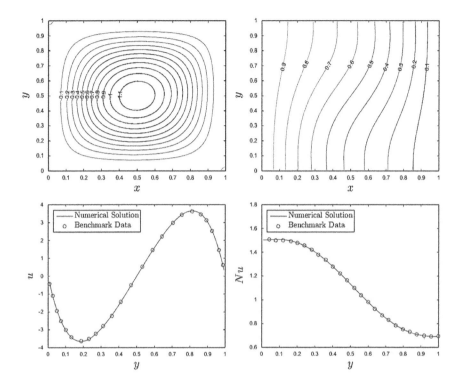

FIGURE 5.18   Numerical results of the natural convection in a cavity example.

These conditions can be implemented with the help of virtual nodes.

5.4.4   RESULTS AND DISCUSSION

By using a 60 × 60 nonuniform mesh, which is refined in regions close to the walls (20 nodes are put within the dimensionless distance 0.3 from each wall in each direction), and a dimensionless time step 0.0001, the calculation continues until the summation of residuals of the continuity equation of all interior nodes drops below 10^{-8}.

The stream lines are shown in the upper left graph of Figure 5.18; the isotherms (the temperature contours) are shown in the upper right graph; the u velocity along cavity bisection $x = L/2$ is compared with the benchmark results (De Vahl Davis, 1983) in the lower left graph and the Nusselt number along the hot wall ($x = 0$) is compared with benchmark results in the lower right graph of Figure 5.18. The Nusselt number is a dimensionless parameter defined as

$$Nu = \frac{hL}{k} \tag{5.178}$$

where h is the convective heat transfer coefficient

$$q = h(T_h - T_c) \tag{5.179}$$

Since the heat flux q (the time rate of heat transfer across per unit surface area) can also be evaluated by using the Fourier's law

$$q = -k\left(\frac{\partial T}{\partial x}\right)_{x=0} \tag{5.180}$$

in which x is the original dimensional x coordinate, one may readily verify that

$$Nu = -\left(\frac{\partial \theta}{\partial x}\right)_{x=0} \tag{5.181}$$

where x is now dimensionless.

Excellent agreement has been achieved between the current simulation and the benchmarkresults.

5.5   EXAMPLE: FLOW OVER A BACKWARD FACING STEP

Flow over a backward facing step is another benchmark case usually used to test CFD codes.

5.5.1   PROBLEM DESCRIPTION

As seen in Figure 5.19, a fully developed laminar flow enters a channel with a backward facing step. Due to the sudden expansion of the channel cross section, flow reversal appears behind the step whose length L_r is sensitive to the flow Reynolds number. We will simulate a case in which the Reynolds number $Re_h = \rho u_m h / \mu = 100$. Here u_m is the mean velocity of the inflow and h is the step height. The height of the channel exit is $H = 2h$.

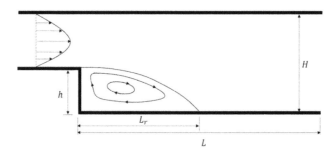

FIGURE 5.19   Flow over a backward facing step example setup.

By using h and u_m as the reference length and velocity, we can normalize the governing equations, which are of the same form as Equations (5.61), (5.62), and (5.63).

To simplify the mesh generation, we choose a computational domain which is the rectangular region on the right of the step and assume that the velocity distribution across the inlet of this region is still the fully developed laminar channel flow distribution. The computational domain has to be long enough so that the outflow boundary is far away from the reversed flow region. Otherwise we will have difficulty in assigning a proper boundary condition at the outlet. We choose a domain length $L = 10h$. Our purpose is to determine the reverse flow bubble length L_r.

5.5.2 BOUNDARY CONDITIONS

We use a fully developed laminar channel flow velocity profile as the inlet velocity condition. As you may derive from the results in Section 2.1.2 in Chapter 2, this velocity profile is

$$u(y) = 6u_m \left[\frac{(2h - y)(y - h)}{h^2} \right]; v = 0 \qquad (5.182)$$

where y is the distance measured from the bottom surface of the computational domain.

At the inlet we may assume the pressure gradient is a constant in the flow direction i.e. we simply extrapolate pressure from the interior nodes:

$$p_{1,j} = 2p_{2,j} - p_{3,j} \qquad (5.183)$$

At the channel exit, since it is far away from the reversed flow region, we may assume the flow velocity no longer changes, which is equivalent to homogeneous Neumann conditions

$$\frac{\partial u}{\partial x} = 0; \frac{\partial v}{\partial x} = 0 \qquad (5.184)$$

Pressure is specified at the exit due to the outflow condition (5.184). The reasoning is as follows. Because $\partial u/\partial x = 0$ at the outflow boundary we must have $\partial v/\partial y = 0$ as continuity demands. Since in the vicinity of the outlet $\partial v/\partial x = 0$ too, we can see $\partial p/\partial y = 0$ must hold at the exit as being obvious from the y-momentum equation:

$$\rho u \frac{\partial v}{\partial x} + \rho v \frac{\partial v}{\partial y} - \left[\frac{\partial}{\partial x} \left(\mu \frac{\partial v}{\partial x} \right) + \frac{\partial}{\partial y} \left(\mu \frac{\partial v}{\partial y} \right) \right] = -\frac{\partial p}{\partial y} \qquad (5.185)$$

That is, pressure is a constant along the outflow boundary. Note that the pressure value indeed does not matter in incompressible flows since we only have pressure gradient in the governing equations. So we can specify an arbitrary constant pressure at the channel outlet, say 0.

Another conclusion we can make is $v = 0$ along the outlet because at the exit $\partial v/\partial y = 0$ i.e. v does not change along the vertical direction; since $v = 0$ at the channel surfaces, it should be zero along the outlet. So at the outlet we can use a Dirichlet condition for v too.

At the solid boundaries no-slip condition holds and pressure gradient normal to boundaries vanishes.

5.5.3   Results and Discussion

We use a nonuniform 60×40 mesh in this simulation. The mesh is refined in the $4 < x < 6$ and $0 < y < 1$ region where the flow reattachment occurs. 20×20 nodes are put in this region. Between $x = 0$ and 4 we use 30 nodes in the x-direction. The SIMPLE method is used for the simulation.

The stream function contours are shown in Figure 5.20. The flow separation bubble length is $L_r/h = 5.1$, which is in good agreement with the value 4.9 from experimental measurements (Armaly, Durst, Peireira, & Schonung, 1983). The numerical result may be further improved if we adopt a longer computational domain and a finer mesh.

Based on numerical experiments, we find that difficulty in convergence may occur if the mesh densities differ too significantly in the x- and y-directions (usually called high aspect ratio control volumes). This is a well-known phenomenon in separated flows. In such flows large velocity gradients exist in both stream-wise and

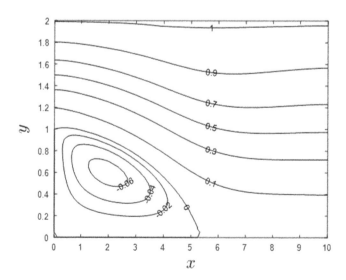

FIGURE 5.20   Numerical results of the flow over a backward facing step example.

normal-to-stream directions. Therefore, we need mesh to be similarly fine in both directions.

5.5.4 SIMPLEC METHOD

We have used the SIMPLE method in many simulations. In fact there are many variations of this method. For example, a method called SIMPLEC (Van Doormaal & Raithby, 1984) was proposed to improve the SIMPLE procedure by providing a more accurate relationship between the velocity and pressure corrections.

Recall the exact relationship between the velocity and pressure corrections are

$$a_p^u u'_{i,j} + \sum a_{nb}^u u'_{nb} = \Delta y_j (p'_{i,j} - p'_{i+1,j})   \tag{5.186}$$

$$a_p^v v'_{i,j} + \sum a_{nb}^v v'_{nb} = \Delta x_i (p'_{i,j} - p'_{i,j+1})   \tag{5.187}$$

In the SIMPLE method, the $\sum a_{nb}^u u'_{nb}$ and $\sum a_{nb}^v v'_{nb}$ terms are neglected, so that the velocity corrections can be directly proportional to the pressure correction differences:

$$u'_{i,j} \approx d_{i,j}^u (p'_{i,j} - p'_{i+1,j})   \tag{5.188}$$

$$v'_{i,j} \approx d_{i,j}^v (p'_{i,j} - p'_{i,j+1})   \tag{5.189}$$

Although this practice brings us significant convenience, it inevitably slows down the convergence speed due to the inaccurate velocity-pressure-correction formula.

The SIMPLEC method came up with a fix by manipulating Equations (5.186) and (5.187):

$$(a_p^u + \sum a_{nb}^u) u'_{i,j} + \sum a_{nb}^u (u'_{nb} - u'_{i,j}) = \Delta y_j (p'_{i,j} - p'_{i+1,j})   \tag{5.190}$$

$$(a_p^v + \sum a_{nb}^v) v'_{i,j} + \sum a_{nb}^v (v'_{nb} - v'_{i,j}) = \Delta x_i (p'_{i,j} - p'_{i,j+1})   \tag{5.191}$$

Then the $\sum a_{nb}^u (u'_{nb} - u'_{i,j})$ and $\sum a_{nb}^v (v'_{nb} - v'_{i,j})$ terms are dropped. This trick helps because the magnitude of $u'_{i,j}$ should be close to u'_{nb}; and $v'_{i,j}$ should be close to v'_{nb}, so these two neglected terms should be smaller than the $\sum a_{nb}^u u'_{nb}$ and $\sum a_{nb}^v v'_{nb}$ terms neglected in the SIMPLE method. As a result, SIMPLEC gives a more accurate velocity-pressure-correction relation and, hence, usually shows better convergence characteristics than SIMPLE does.

To implement the SIMPLEC method, we have to replace Equations (5.105) and (5.106) with

$$d_{i,j}^u = \Delta y_j / \left(a_p^u + \sum a_{nb}^u \right) \tag{5.192}$$

$$d_{i,j}^v = \Delta x_i / \left(a_p^v + \sum a_{nb}^v \right) \tag{5.193}$$

Using this method we do not need under-relaxation for pressure.

5.6 EXAMPLE: FLOW OVER A SQUARE CYLINDER

5.6.1 PROBLEM DESCRIPTION

As shown in Figure 5.21, fluid flows over a square block whose side length is D. The Reynolds number $Re = UD/\nu$ is 40. To calculate the flow field, we choose a computational domain $H = 20D$, $L_u = 5D$, and $L_d = 34D$. The distances between the cylinder from the top and bottom boundaries are the same. Notice that this is an external flow, which means the flow extends to infinity. The computational domain we choose, therefore, is not enclosed by physical walls. The computational domain should be large enough so that the flow no longer varies significantly at its boundaries, to which reasonably simple boundary conditions then can be applied.

By using D and U as the reference length and velocity, we can normalize the governing equations, which are then of the same form as Equations (5.61), (5.62), and (5.63).

5.6.2 MESH AND BOUNDARY CONDITIONS

In this problem we have to enforce no-slip condition at the surface of the square cylinder. For this purpose we set up a mesh as sketched in Figure 5.22, in which dashed lines denote mesh interfaces that coincide with the cylinder surfaces and the

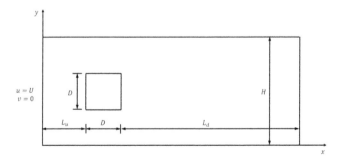

FIGURE 5.21 Flow over a square cylinder example setup.

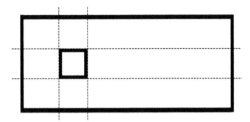

FIGURE 5.22   Sketch of the mesh used for the flow over a square cylinder example.

dark regions represent virtual volumes. Notice that those blank regions should be filled with control volumes. You can see we have a layer of virtual volumes inside the cylinder. The way to set up the boundary/virtual-node values for the cylinder is as follows.

For instance, to enforce the no-slip condition at the west (left) surface of the cylinder we set $u_{i,j} = 0$ since u points are directly located on this surface; and we set $v_{i+1,j} = -v_{i,j}$ for the virtual nodes $(i + 1, j)$. The condition for pressure and pressure correction is a zero gradient condition: $p_{i+1,j} = p_{i,j}$; $p'_{i+1,j} = p'_{i,j}$.

At the flow inlet we directly specify the velocity; at the top and bottom boundaries (notice that they are not solid walls) we use a frictionless boundary condition (FBC): $\frac{\partial u}{\partial y} = 0$; $v = 0$. This condition is also called symmetry boundary condition as it entails a symmetrical flow field at the immediate vicinity of the boundary. At the flow outlet, we use the vanishing x-derivative condition for both u and v velocities.

One may wonder what happens to the "real" nodes inside the square cylinder. Well, since there is no real flow there, we may want to suppress the solving governing equations at such nodes to avoid any trouble that may arise.

When we visualize the flow field using streamlines, we have to calculate the stream function values at vertices of the p control volumes. Since there is no real flow inside the cylinder, we should assign the same stream function value over all the cylinder surface.

5.6.3   RESULTS AND DISCUSSION

We use a 220×170 mesh with refinement in the proximity of the cylinder, including the expected separation bubble region, which is about $3D$ after the cylinder. The streamlines around the cylinder is shown in Figure 5.23.

The calculated separation bubble consists of a pair of counter-rotating vortices, and its dimensionless length L/D is 2.88 which compares well with the published value $L/D = 2.82$ (Dhiman, Chhabra, Sharma, & Eswaran, 2006). For a flow over an immersed body, an important parameter is the drag coefficient

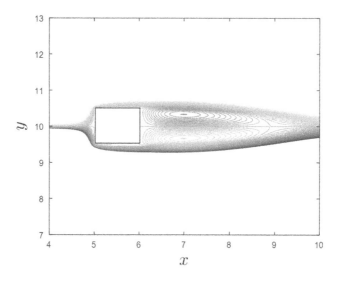

FIGURE 5.23   Streamlines around the square cylinder.

$$C_d = \frac{Drag}{\frac{1}{2}\rho U^2 D} \tag{5.194}$$

The drag experienced by the body is due to the pressure force and shear stress acting on the body surfaces.

To evaluate shear stress we have to invoke the Stokes' stress law, which states that

$$\tau_{ij} = 2\mu\left(S_{ij} - \frac{1}{3}S_{kk}\delta_{ij}\right) \tag{5.195}$$

where τ_{ij} is the deviatoric stress (stress except pressure) pointing to the $\pm j^{th}$ direction and acting on the fluid surface whose outward normal is in the $\pm i^{th}$ direction, see Figure 5.24.

$$S_{ij} = \frac{1}{2}\left(\frac{\partial u_i}{\partial x_j} + \frac{\partial u_j}{\partial x_i}\right) \tag{5.196}$$

is the strain rate tensor and its trace is (Einstein convention applies)

$$S_{kk} = S_{11} + S_{22} + S_{33} = \frac{\partial u}{\partial x} + \frac{\partial v}{\partial y} + \frac{\partial w}{\partial z} \tag{5.197}$$

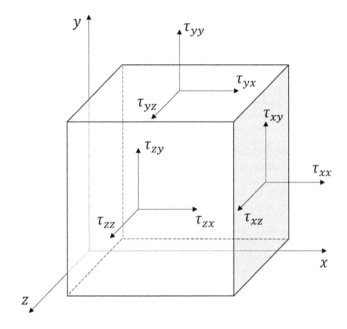

FIGURE 5.24   Stresses at control volume surfaces.

Obviously for incompressible flows S_{kk} vanishes (see Equation (1.9)). In Equation (5.195) δ_{ij} is the Kronecker delta

$$\delta_{ij} = \begin{cases} 1 & i = j \\ 0 & i \neq j \end{cases} \tag{5.198}$$

For an incompressible flow we thence have

$$\tau_{ij} = \mu \left(\frac{\partial u_i}{\partial x_j} + \frac{\partial u_j}{\partial x_i} \right) \tag{5.199}$$

The total stress pointing to the $\pm\, i^{th}$ direction and acting on the fluid surface whose outward normal is in the $\pm\, i^{th}$ direction is therefore

$$\sigma_{ij} = \tau_{ij} - p\delta_{ij} = \mu \left(\frac{\partial u_i}{\partial x_j} + \frac{\partial u_j}{\partial x_i} \right) - p\delta_{ij} \tag{5.200}$$

This is why we have the Navier–Stokes equation (1.12):

$$\frac{\partial(\rho u_j)}{\partial t} + \frac{\partial(\rho u u_j)}{\partial x} + \frac{\partial(\rho v u_j)}{\partial y} + \frac{\partial(\rho w u_j)}{\partial z} = \frac{\partial \sigma_{xj}}{\partial x} + \frac{\partial \sigma_{yj}}{\partial y} + \frac{\partial \sigma_{zj}}{\partial z} + f_j$$

$$= \left\{ \frac{\partial}{\partial x}\left[\mu\left(\frac{\partial u_j}{\partial x} + \frac{\partial u}{\partial x_j}\right)\right] + \frac{\partial}{\partial y}\left[\mu\left(\frac{\partial u_j}{\partial y} + \frac{\partial v}{\partial x_j}\right)\right] \right.$$

$$\left. + \frac{\partial}{\partial z}\left[\mu\left(\frac{\partial u_j}{\partial z} + \frac{\partial w}{\partial x_j}\right)\right] \right\} - \frac{\partial p}{\partial x_j} + \rho g_j + f_j \qquad (5.201)$$

where f_j includes the body forces (per unit volume) except the gravity force. When the fluid viscosity is a constant, this equation reduces to a more familiar form

$$\frac{\partial(\rho u_j)}{\partial t} + \frac{\partial(\rho u u_j)}{\partial x} + \frac{\partial(\rho v u_j)}{\partial y} + \frac{\partial(\rho w u_j)}{\partial z}$$

$$= \mu\left(\frac{\partial^2 u_j}{\partial x^2} + \frac{\partial^2 u_j}{\partial y^2} + \frac{\partial^2 u_j}{\partial z^2}\right) - \frac{\partial p}{\partial x_j} + \rho g_j + f_j \qquad (5.202)$$

The drag force acting on the cylinder, which is in the x-direction is therefore

$$Drag = \sum \sigma_{ix} A\,(\vec{n_A} \cdot \vec{e_i}) \qquad (5.203)$$

where the summation is over all the control volume faces that coincide with the cylinder surface; A is the area of the control volume face which is numerically the same as the length of the face in the current 2-D case; $\vec{n_A}$ is the unit outward normal vector of A and $\vec{e_i}$ is the unit vector in the positive i^{th} direction. You may find that in σ_{xx} only pressure is present since $\tau_{xx} = 0$ at the east and west surfaces of the cylinder; in σ_{yx} along the north and south cylinder surfaces only τ_{yx} is present since $\delta_{yx} = 0$ in Equation (5.200). Therefore, the drag is due to the pressure force acting on the west and east surfaces and shear force acting on the north and south surfaces of the cylinder, which is in agreement with our intuition.

Equation (5.203) is in dimensional form. If we use fluid density ρ, fluid velocity far away from the cylinder, U, and the cylinder side length D to normalize the terms in Equation (5.203), we find that

$$C_d = 2 \sum \sigma_{ix} A\,(\vec{n_A} \cdot \vec{e_i}) \qquad (5.204)$$

in which A and σ_{ij} now are nondimensional e.g.

$$\sigma_{ij} = \tau_{ij} - p\delta_{ij} = \left(\frac{\partial u_i}{\partial x_j} + \frac{\partial u_j}{\partial x_i}\right)/Re - p\delta_{ij} \qquad (5.205)$$

Using this equation we find the drag coefficient is 1.89, which is within about 7% from the published value 1.77 (Dhiman, Chhabra, Sharma, & Eswaran, 2006).

5.6.4   FLOW OVER A SQUARE CYLINDER AT RE = 100

Now let us simulate the flow over a square cylinder again but this time the Reynolds number is 100. This is not a trivial repetition of our last example. In fact now the flow can no longer be steady like the $Re = 40$ case is, but showing a regular oscillation pattern in the flow behind the cylinder, a phenomenon called vortex shedding. Therefore, we have to use the unsteady algorithm of the SIMPLE method summarized in Section 5.3.5 in this simulation. That means we have to somehow know how the flow behaves before we calculate the flow behavior.

Using a 200 × 200 uniform mesh and $H = 20D$, $L_u = 4D$, $L_d = 15D$, and a nondimensional time step size $\Delta t = 0.1$, we find that the flow pattern evolves as shown in Figure 5.25. One pair of vortices gradually prolongs and eventually begins to oscillate and develop into a series of vortices. At the same time the drag coefficient drops as the vortices prolong, then increases as the oscillation is about to begin. Eventually the drag coefficient oscillates at a frequency twice that of the vortex shedding. By inspecting the drag coefficient data (see Figure 5.26), we find the average drag coefficient is 1.59 and the nondimensional vortex shedding frequency is 0.153. These values agree well with the published data of the average drag coefficient being 1.53 and the nondimensional vortex shedding frequency being 0.145 (Sen, Mittal, & Biswas, 2010).

For this case you might discover that the program diverges if we use a Dirichlet condition for the v velocity, say $v = 0$, at the flow outlet, while such a condition works well for the low Reynolds number example. The reason is because this condition conflicts with the real situation as vortex shedding oscillates at the flow exit. We will have to use a more realistic outflow boundary condition like a vanishing velocity gradient or an advection condition given in Equation (3.165). On the contrary v indeed is close to 0 when there is no vortex shedding, for example in the previous $Re = 40$ case.

You may find that it takes a really long time to simulate such an unsteady flow using the SIMPLE method. This, of course, is partly due to our program which is not optimized for maximum execution efficiency. Another reason is we need to carry out numerous iterations in each time step. Such iterations are necessary for a time-accurate solution.

5.7   VERIFICATION AND VALIDATION

There is a joke that has been circulating in the CFD community for some time. It goes like this: what is the difference between a theorist, an experimentalist, and a CFD guy? The answer is: no one believes the results of the theorist, except her/himself; everyone believes the results of the experimentalist, except her/himself; and no one believes the results of the CFD guy, including her/himself.

So how can we tell if those colorful figures and diagrams (CFD) produced by our programs are correct? The answer to this question is verification and validation (Roache, 1998).

Verification of a CFD code/simulation may be described as trying to make sure the code/simulation is "solving the equations right." When we solve differential

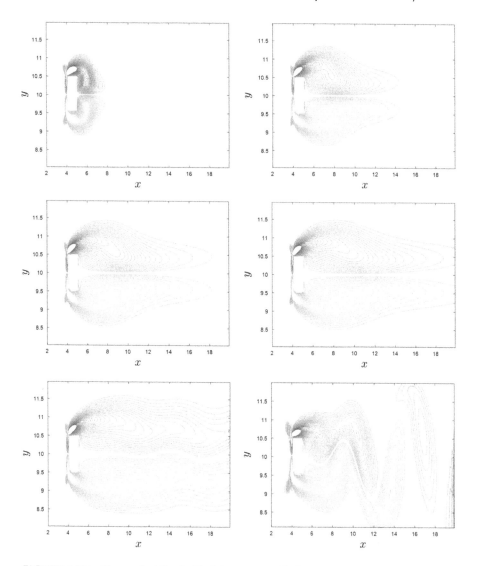

FIGURE 5.25   Vortex shedding behind the square cylinder.

equations numerically, obviously we may use wrong numerical methods; we may make programming mistakes; we may adopt a very coarse mesh; we may choose a very large time step size; we may not run the simulation long enough to really reach convergence etc. All these lead to incorrect solutions. Verification is the process to address such issues. To avoid programming mistakes, for example, we should thoroughly test our codes. A very effective way to do that is using the manufactured solution method (Salari & Knupp, 2000). The idea of this method is to first manufacture a solution, then substitute it into the equation we want to use to test our code; typically a source term arises at this step. Since we can manufacture solutions

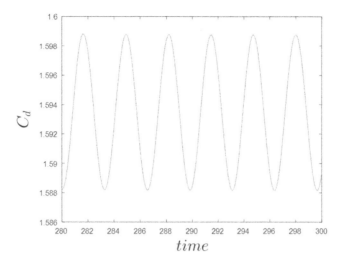

FIGURE 5.26　Drag coefficient.

with any desired features, this method can test codes in a variety of ways, and as a consequence, can detect many types of program bugs e.g. incorrect matrix indices, incorrect parenthesis positions, incorrect loops, etc.

To avoid incorrect solutions due to defective mesh, we typically need to pursue a mesh-independent solution. That is we keep refining the mesh until the results no longer change.

We should also be aware of the possibility that the equations we solve do not reflect the physics of the phenomenon we try to simulate. The effort to make sure that we are "solving the right equations" is called "validation." For example, in the previous section, if we had solved the steady Naiver–Stokes equations for the flow over a cylinder at $Re = 100$ example, we would not have the physically correct result no matter how accurate our code is and how fine a mesh we use, because the real flow is not steady. To make sure we are simulating the real physics, we will have to compare our results with experimental data. It is usually possible to run our codes on cases which are physically close to, but simpler than the case we want to simulate eventually. And very likely there are experimental data available for such simpler cases which we can compare with. Once we gain enough confidence in our codes based on the favorable comparisons with experimental data, we can go ahead to simulate the case in hand.

Despite the joke told at the beginning of this section, we should have strong faith in the fundamental reliability of the CFD methodology, otherwise we will not going to make it, especially at times when our codes refuse to work correctly.

Exercises

1. Reproduce the results shown in Figure 5.5 with the SIMPLE method. Use the same mesh as shown in Figure 5.4.

2. Calculate the filter flow example with the projection method using the same mesh as shown in Figure 5.4.

3. Calculate the filter flow example with the projection method using the same mesh as shown in Figure 5.4. But this time totally drop the pressure term when solving the intermediate velocity.

4. Use the coupled method to solve the filter flow example, again with the same mesh you used for the first three problems.

5. Use the co-located mesh shown in Figure 5.2 and the projection method to solve the filter flow example. You should use the MIM technique.

6. Redo the filter flow example with a co-located mesh and the SIMPLE method. Again apply MIM.

7. Reproduce the lid-driven cavity flow results presented in Section 5.3.

8. Redo the lid-driven cavity flow but this time using the projection method on a staggered mesh.

9. Redo the lid-driven cavity flow with the projection method on a co-located mesh. Use FFT method to solve the PPE; also use SOR to accelerate the PPE solution process.

10. Simulate the lid-driven cavity flow with $Re = 1000$. Using a 60×60 mesh properly refined. Plot stream function and vorticity contours. Also plot the velocities along the cavity bisectors. Compare your CFD solution with the benchmark results (Ghia, Ghia, & Shin, 1982).

11. Reproduce the CFD results of the natural convection in a cavity in Section 5.4.

12. Reproduce the CFD results of the backward facing step case in Section 5.5. Try a longer flow domain and refine mesh at the reattachment region, see if your result improves.

6 Unstructured Mesh

6.1 INTRODUCTION

You probably have already noticed that we only had calculated flows within or without rectangular regions up to this point. The reason of course is the simplicity of creating mesh for such geometries. Unfortunately, many practical flows occur in quite complex geometric setups. One such example is the natural convection in a gap between two concentric circular pipes at two different temperatures (see Figure 6.1).

Obviously, we can no longer use our familiar rectangular mesh to discretize this circular gap, at least not in our familiar way. Some of you may have already come up with a solution to this small problem: why not use the cylindrical coordinate system? Following this idea we can create a mesh like the one shown in Figure 6.2, which in essence maps the circular gap into a rectangle on the r-θ plane, as shown in Figure 6.3 (in which r_i and r_o denote the radii of the two pipes). Then you basically can forget the circular gap since you will only need the rectangular mesh on the r-θ plane to derive your finite difference formulas and in the subsequent solving processes.

The price of being able to use the nice rectangular mesh on the r-θ plane, or the so-called computational plane, is twofold. First, we have to find a way to set up such a mesh, so that the complex flow domain can be converted into a rectangular region on the computational plane. Second, since the computational plane no longer uses Cartesian coordinate system, we have to solve the governing differential equations in forms different from, and usually more complicated than, those under the Cartesian coordinate system. For instance, the continuity equation on the r-θ plane is no longer familiar

$$\frac{\partial u}{\partial x} + \frac{\partial v}{\partial y} = 0 \tag{6.1}$$

but

$$\frac{\partial v_r}{\partial r} + \frac{v_r}{r} + \frac{1}{r}\frac{\partial v_\theta}{\partial \theta} = 0 \tag{6.2}$$

where v_r and v_θ are velocity components along the r- and θ-directions (see Figure 6.2).

Any mesh generated by mapping the physical flow domain into a rectangle or several connected rectangles (or rectangular prisms for 3-D) on a computational plane is called a structured mesh. It is structured in the sense that the connectivity between the control volumes in such a mesh is implied by their indices e.g. control

DOI: 10.1201/9781003138822-6

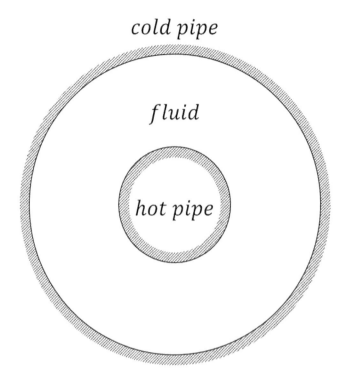

FIGURE 6.1 Geometry of the natural convection in a concentric annulus problem.

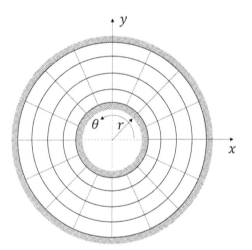

FIGURE 6.2 Mesh under the cylindrical coordinate system.

volume $(i + 1, j)$ is always on the east side of volume (i, j) according to our usual practice in this book. Therefore, it is unnecessary to specially store such connectivity information, which is good. However, to create such a mesh, one always has to pay the two aforementioned prices, which grow rapidly as the flow geometry

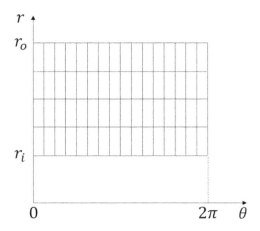

FIGURE 6.3   Rectangular mesh on the *r*-*θ* plane.

becomes more and more complex. It is not uncommon for experienced CFD people to spend hours to produce a structured mesh of reasonable quality, even with the help of a sophisticated meshing software.

One way to alleviate the usually time-consuming mesh generation process is using an unstructured mesh, which is not based on the mapping idea. Instead we employ control volumes of various shape to directly fill the flow domain. For example, we can use triangular volumes to pave the whole circular gap, as shown in Figure 6.4. As the connectivity between these triangles is not related to their indices (you can verify this by numbering these triangles and see if you can tell which triangles are immediate neighbors of a specific triangle, say triangle number 12, without looking at such a figure), such connectivity information has to be stored separately, which is an overhead we have to pay to use the unstructured mesh. But the tradeoff is that we no longer need geometrical mapping. Neither do we need to use special coordinate systems, instead we can simply use the Cartesian coordinate system.

Before we get into more details, it should be understood that control volumes do not necessarily always be rectangles (or rectangular prisms for 3-D). When we integrate a differential equation over a control volume, what we obtain is just the conservation law behind that differential equation. Since a conservation law is a universal truth which is independent of the choice of control volumes, it is clear that control volumes can be of any shape. The simplest shape that can always cover an arbitrary 2-D geometry is triangles and we will use triangular mesh in all flow problems in this chapter.

6.2   TRIANGULAR MESH GENERATION

6.2.1   DELAUNAY TRIANGULATION

To create a triangular mesh like the one in Figure 6.4, we can begin with a set of points within and along the boundaries of the geometry we want to mesh, then connect these points to form triangles. Although great freedom exists in terms of

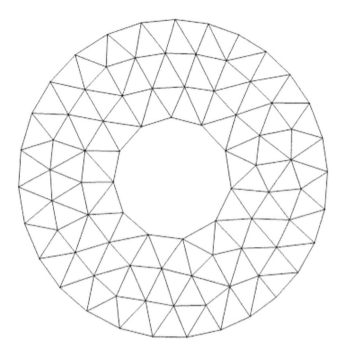

FIGURE 6.4   Triangular mesh.

which two points should be connected, there is only one arrangement that results in nonoverlapping triangles which fill the whole area defined by the points, such that the circumcircle of each triangle contains no other points in its interior. This unique triangulation is called Delaunay triangulation. It ensures the triangles to be as close as possible to equilateral triangles, which is a desirable feature. The left triangulation in Figure 6.5, for example, is a Delaunay triangulation but the right one is not as there is one point within each circumcircle.

MATLAB has a built-in function "delaunay" which implements the Delaunay triangulation. We only need to provide a bunch of points to initiate the process. How to generate a batch of points that fall in the geometry of interest and at the same time represent the desired mesh density distribution is a problem. Another issue is that the triangular mesh thus formed might be of poor quality, as you can see an example in Figure 6.6. Hence we also have to find a way to improve the mesh quality.

There are many algorithms available to address these two concerns and generate high-quality triangular mesh. Here we will briefly discuss a very concise yet surprisingly effective mesh generation algorithm developed by Persson and Strang (Persson & Strang, 2004).

6.2.2   MESH GENERATION ALGORITHM OF PERSSON AND STRANG

The algorithm begins with creating an array of equally xml:spaced points in a rectangular region that is able to cover the whole geometry which needs to be

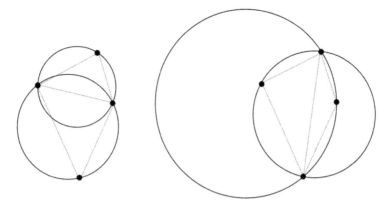

FIGURE 6.5 Delaunay triangulation.

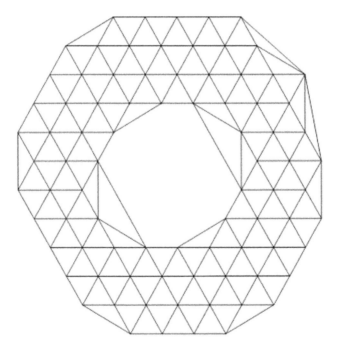

FIGURE 6.6 Low-quality Delaunay triangular mesh.

meshed. Then all points outside the geometry are deleted. This is done with the help of a signed distance function which has to be provided by the user. This function returns a negative value if a point falls inside the given geometry, zero if it is on the geometry boundary and a positive number if it is outside the geometry. For example, the signed distance function for a unit circle centered at the origin is

$$f(x, y) = \sqrt{x^2 + y^2} - 1 \qquad (6.3)$$

The signed distance function is a type of level-set function (Osher & Sethian, 1988), which is an important tool in multiphase flow simulations, as will be discussed in Section 7.4 of Chapter 7.

The points inside the flow domain are then further removed based on the desired mesh density distribution. This is realized by creating a random number between 0 and 1 for each point, and a point is discarded if the corresponding random number is greater than the desired mesh density at that location, which is also a positive number between 0 and 1. The points that remain are then triangulated by using MATLAB built-in Delaunay triangulation function. The quality of such a triangulation, for instance the one shown in Figure 6.6, can be improved by moving the points around and re-triangulation. The idea is based on an analogy between a triangular mesh and a mechanical truss structure. The edges of triangles correspond to the bars of a truss and the points correspond to the joints. A nice mesh corresponds to a truss at an equilibrium state. In order that the edges are of desired length and the points can spread out across the whole geometry, a repulsive force is assumed to exist along each bar:

$$F = \begin{cases} l_0 - l, & if \ \ l < l_0 \\ 0, & if \ \ l \geq l_0 \end{cases} \qquad (6.4)$$

in which l_0 is the desired length of the bar and l is its current length. Each joint is then moved around by solving the differential equation

$$\frac{d\vec{r}}{dt} = \sum \vec{F} \qquad (6.5)$$

where $\vec{r}$ is the position vector of the joint and the right side of the equation is the total force acting on this joint. Here t is a time-like variable and the time derivative is added to facilitate the updating of joint locations. If points are moved quite significantly, they will be re-triangulated. This procedure continues until the total force imposed on every joint vanishes i.e. when equilibrium is reached. The result is a triangular mesh of high quality. This program can be found in Appendix A.7 Triangular Mesh Generation Program of Persson and Strang. The outputs of this program are two matrices. One matrix contains the coordinates of each vertex; the other stores the indices of the three vertices of each triangle. Appendix A.8 includes some very useful accessory functions that can be used to generate complex geometries. More details of the program can be found in Persson & Strang, (2004).

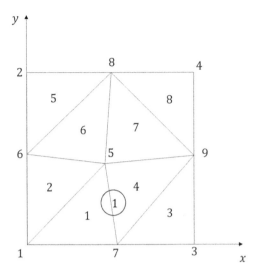

FIGURE 6.7 Simple triangular mesh for a square geometry.

6.2.3 CONNECTIVITY AND GEOMETRY INFORMATION

It is necessary to collect the connectivity and geometry information of the mesh. For example, we will end up with a mesh shown in Figure 6.7 if we use the uniform mesh density function and a mesh size $h = 0.5$ in the program provided in Appendix A.7 to mesh a square with unity side length.

Your code then should be able to collect such information, including: the coordinates of the centroid of each triangle; the area of each triangle; the indices of the three edges and corresponding three neighboring control volumes of each triangle; the indices of the two triangles that share each edge; the length of each edge; the unit normal vector of each edge; the distance between the two centroids straddling each edge and the unit vector along the line connecting these two centroids; the position of the intersection of an edge and the line connecting the two centroids straddling the edge, and the total number of edges as well as the total number of boundary edges. Only basic analytic geometry/vector analysis is needed for calculating these parameters, although great care has to be practiced to avoid mistakes. For instance, if we focus on the edge number one (index circled in Figure 6.7), its length should be 0.4994; the distance between the centroids of triangle number one and triangle number 4 is 0.3727; the unit vector along the line connecting these two centroids is [0.8943, 0.4474].

You will want to write and test this part of code first.

6.3 SOLVING GENERAL CONVECTION–DIFFUSION EQUATION WITH UNSTRUCTURED MESH

6.3.1 THE GENERAL CONVECTION–DIFFUSION EQUATION

The general convection–diffusion equation is

$$\frac{\partial(\rho\phi)}{\partial t} + \sum_{j=1}^{3} \frac{\partial(\rho u_j \phi)}{\partial x_j} = \sum_{j=1}^{3} \frac{\partial}{\partial x_j}(\Gamma\frac{\partial\phi}{\partial x_j}) + S \qquad (6.6)$$

Recall that this form of equation was derived by applying the conservation law of a property $\Phi = m\phi$ to an infinitesimal cubic control volume (refer to Section 1.2.1 in Chapter 1). Therefore, it is not unexpected that this form of equation is not very convenient if we are going to integrate it over a triangular control volume. It will be advantageous to use another form of the equation:

$$\frac{\partial(\rho\phi)}{\partial t} + \nabla\cdot(\rho\vec{u}\phi) = \nabla\cdot(\Gamma\nabla\phi) + S \qquad (6.7)$$

where ∇ is the vector differential operator

$$\nabla = \sum_{j=1}^{3} \frac{\partial}{\partial x_j}\vec{e_j} \qquad (6.8)$$

Here $\vec{e_j}$ denotes the unit vector along the x_j-coordinate. The gradient of ϕ, $\nabla\phi$ is therefore

$$\nabla\phi = \sum_{j=1}^{3} \frac{\partial\phi}{\partial x_j}\vec{e_j} \qquad (6.9)$$

The divergence of a vector $\vec{v}$, $\nabla\cdot\vec{v}$ is the dot product of ∇ and $\vec{v}$:

$$\nabla\cdot\vec{v} = \sum_{j=1}^{3} \frac{\partial}{\partial x_j}\vec{e_j}\cdot \sum_{j=1}^{3} v_j\vec{e_j} = \sum_{j=1}^{3} \frac{\partial v_j}{\partial x_j} \qquad (6.10)$$

At this point this ∇ seems no more than a fancy notation to replace the otherwise tedious derivative and summation symbols, but the real nice thing about ∇ is that numerous useful identities can be made universal with its use, which means they are independent of coordinate systems and the shape of control volumes. For example

$$\int_V \nabla\phi dV = \int_A \phi\vec{n}dA \qquad (6.11)$$

which transforms a volume integral into a surface integral. A is the surface that encloses the volume V and $\vec{n}$ is the unit vector perpendicular to surface A and pointing outward of A. Another similar identity is

$$\int_V \nabla \cdot \vec{v} \, dV = \int_A \vec{v} \cdot \vec{n} \, dA \tag{6.12}$$

Both Equations (6.11) and (6.12) will be termed Gauss theorems in this book, although actually only Equation (6.12) is associated with Gauss (Deen, 1998).

6.3.2 Discretization of the General Convection–Diffusion Equation

To discretize the general convection–diffusion equation on a triangular mesh, we can apply the finite volume method to the triangular control volumes or cells. As Figure 6.8 shows, we may use centroids of the triangles as representative node points. This means the value of any variable at the centroid of a triangle is assumed to be the volumetric average of the variable over that triangle:

$$\phi_P = \frac{\int_V \phi \, dV}{V_P} \tag{6.13}$$

where V_P is the volume of the triangular cell. For a 2-D case, V_P is numerically equal to the area of the triangle. Another option would be utilizing the vertices of triangles as node points, one then needs to create control volumes around such nodes, which is more challenging in programming than using the current method.

Now let us integrate the general convection–diffusion equation, Equation (6.7), with the control volume P (see Figure 6.8):

$$\int_V \frac{\partial(\rho\phi)}{\partial t} dV + \int_V \nabla \cdot (\rho\vec{u}\phi) dV = \int_V \nabla \cdot (\Gamma\nabla\phi) dV + \int_V S dV \tag{6.14}$$

For simplicity we assume ρ and Γ are both constants.

The integral of the transient term can be approximated by

$$\int_V \frac{\partial(\rho\phi)}{\partial t} dV \approx \rho \frac{\phi_P^{n+1} - \phi_P^n}{\Delta t} V_P \tag{6.15}$$

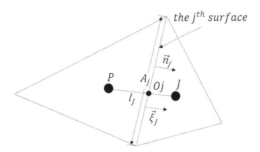

FIGURE 6.8   Two neighboring triangular control volumes.

In virtue of Gauss theorem, the integral of convection term becomes

$$\int_V \nabla \cdot (\rho \vec{u} \phi) dV = \int_A \rho \phi \vec{u} \cdot \vec{n} \, dA \approx \sum_j \rho \phi_j \, \vec{u_j} \cdot \vec{n_j} A_j = \sum_j F_j \phi_j \qquad (6.16)$$

where A_j is the area of the jth surface and the summation is over all three surfaces (edges in 2-D) of the triangular control volume. For a 2-D case, A_j is numerically equal to the length of the jth edge. $\vec{n_j}$ is the unit vector perpendicular with the jth surface and it should always point from the current control volume, volume P, to the neighboring volume across the jth surface, volume J. Notice that

$$F_j = \rho \vec{u_j} \cdot \vec{n_j} A_j \qquad (6.17)$$

is the mass flow rate or mass flux across the jth surface. F_j is designated to be positive if the fluid flows out of the current control volume through the jth surface.

You can see we have two types of variables in the formulas: those evaluated at the centroids like ϕ_P and those assessed at the volume surfaces like $\vec{u_j}$. It is possible to employ the idea similar to what we used in Section 5.1.1 in Chapter 5 and store velocities at the cell surfaces. We will, however, not take this route but choosing to store all variables at the centroids i.e. using co-located meshes. For this reason, we have to evaluate the variables at the cell surfaces.

The surface velocity is estimated with a simple average of the centroid velocity values:

$$\vec{u_j} \approx \frac{\vec{u_P} + \vec{u_J}}{2} \qquad (6.18)$$

After the surface velocities are calculated, the mass flux F_j across each surface can be easily computed according to Equation (6.17).

The last piece of information needed to evaluate the convection term is the surface ϕ value, ϕ_j, which should be obtained by using one of the numerical schemes introduced in Chapter 3. For example,

FUS:

$$\phi_j = \begin{cases} \phi_P & , \quad if \quad F_j > 0 \\ \phi_J & , \quad if \quad F_j < 0 \end{cases} \qquad (6.19)$$

CDS:

$$\phi_j = \frac{\phi_P + \phi_J}{2} \qquad (6.20)$$

SUS:

$$\phi_j = \begin{cases} \phi_P + (\nabla\phi)_P \cdot (\vec{r}_{Oj} - \vec{r}_P) , & \text{if } F_j > 0 \\ \phi_J + (\nabla\phi)_J \cdot (\vec{r}_{Oj} - \vec{r}_j) , & \text{if } F_j < 0 \end{cases} \tag{6.21}$$

in which $\vec{r}$ is the position vector.

These three schemes may be unified in one single expression as

$$F_j\phi_j = \frac{1}{2}\left(F_j + \gamma|F_j|\right)\phi_P + \frac{1}{2}\left(F_j - \gamma|F_j|\right)\phi_J + \frac{\omega}{2}\left(F_j + \gamma|F_j|\right)(\nabla\phi)_P$$
$$\cdot\left(\vec{r}_{Oj} - \vec{r}_P\right) + \frac{\omega}{2}\left(F_j - \gamma|F_j|\right)(\nabla\phi)_J \cdot \left(\vec{r}_{Oj} - \vec{r}_j\right) \tag{6.22}$$

It is CDS if $\gamma = \omega = 0$; FUS if $\gamma = 1$, $\omega = 0$; and SUS if $\gamma = \omega = 1$. One may set up γ and ω e.g. according to Equation (3.136) or Equation (3.137) to form hybrid schemes. If the convection term is treated implicitly, the ϕ terms in Equation (6.22) should be evaluated at the $n + 1$ moment while those terms involving $\nabla\phi$ are typically treated explicitly to avoid too many unknown ϕ values in each finite difference equation. $\nabla\phi$ at the control volume centers can be evaluated with the help of the Gauss theorem and the assumption that the value of any variable at the cell center is the volumetric average of that variable over the control volume:

$$(\nabla\phi)_P = \frac{\int_V \nabla\phi dV}{V_P} = \frac{\int_A \phi\vec{n} dA}{V_P} \approx \frac{\Sigma_j \phi_j A_j \vec{n}_j}{V_P} \tag{6.23}$$

The above practice of evaluating surface variable values usually does not give us second-order accuracy even with CDS or SUS (although the accuracy is typically still higher than first-order). The reason is because we calculate ϕ_j from the ϕ values at nodes P and J. And the position where ϕ_j is calculated is the intersection of the line PJ and the jth surface i.e. point Oj, which is normally not the center of the jth surface. This decreases the accuracy. One way to fix this problem and giving true second-order accuracy (Ferziger & Peric, 2002; Yu, Ozoe, & Tao, 2005) is as follows. Drawing a line perpendicular to the jth surface and passing its middle point Mj, then finding the two points P' and J' along this line so that PP' and JJ' are parallel to the jth surface, as shown in Figure 6.9.

ϕ_j is then evaluated at the surface center Mj with using ϕ values at "nodes" P' and J', instead of P and J. ϕ values at P' and J' are

$$\phi_{P'} = \phi_P + (\nabla\phi)_P \cdot (\vec{r}_{P'} - \vec{r}_P); \quad \phi_{J'} = \phi_J + (\nabla\phi)_J \cdot (\vec{r}_{J'} - \vec{r}_j) \tag{6.24}$$

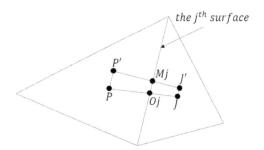

FIGURE 6.9   Improved surface variable evaluation method.

For simplicity, we will still adhere to the less accurate method throughout the rest of this chapter.

Now let us turn our attention to the integrated diffusion term in Equation (6.14). Applying Gauss theorem we have

$$\int_V \nabla \cdot (\Gamma \nabla \phi) dV = \int_A \Gamma \nabla \phi \cdot \vec{n} \, dA \approx \sum_j \Gamma A_j (\nabla \phi)_j \cdot \vec{n_j} \qquad (6.25)$$

If the diffusion term is to be treated implicitly as what we usually did in order to avoid too stringent restriction on time step size, $(\nabla \phi)_j \cdot \vec{n_j}$, the derivative of ϕ along the $\vec{n_j}$ direction, then should be computed in a deferred-correction manner (Ferziger & Peric, 2002):

$$(\nabla \phi)_j \cdot \vec{n_j} = (\nabla \phi)_j \cdot \vec{\xi_j} + (\nabla \phi)_j \cdot \left(\vec{n_j} - \vec{\xi_j} \right)$$
$$\approx \frac{\phi_J - \phi_P}{l_J} + (\nabla \phi)_j \cdot \left(\vec{n_j} - \vec{\xi_j} \right) \qquad (6.26)$$

in which $(\nabla \phi)_j \cdot \vec{\xi_j}$ is the derivative of ϕ along the direction pointing from point P to point J (see Figure 6.8). $\vec{\xi_j}$ is the unit vector along this direction and l_J is the distance between these two points. The ϕ terms in Equation (6.26) are evaluated implicitly at the $(n + 1)^{th}$ moment and the very last term, the so-called cross-diffusion term, is handled explicitly, with $(\nabla \phi)_j$ being calculated as

$$(\nabla \phi)_j \approx \frac{(\nabla \phi)_P + (\nabla \phi)_J}{2} \qquad (6.27)$$

which are based on values at the previous i.e. the nth moment.

If you decide to treat the diffusion term explicitly, you can still follow the above procedure but by evaluating every term in Equation (6.26) explicitly at the nth

moment. What you should avoid is using Equation (6.27) alone to estimate the $(\nabla\phi)_j$ in Equation (6.25) because such a practice will lead to odd–even decoupling in the same way as the central difference's promoting the checkerboard pressure distribution on a co-located mesh (see Section 5.1 Odd-Even Decoupling). For example, if we calculate the integrated diffusion term over the third control volume on a simple 1-D uniform mesh as shown in Figure 6.10 with Equation (6.27), assuming the interface area is unity, we have

$$
\begin{aligned}
\int_V \nabla\cdot(\Gamma\nabla\phi)dV = \int_A \Gamma\nabla\phi\cdot\vec{n}\,dA &\approx \Sigma_j \Gamma A_j (\nabla\phi)_j\cdot\vec{n}_j \\
&= \Gamma\left[(\nabla\phi)_e - (\nabla\phi)_w\right] \\
&\approx \Gamma\left[\frac{(\nabla\phi)_3+(\nabla\phi)_4}{2} - \frac{(\nabla\phi)_2+(\nabla\phi)_3}{2}\right] \\
&= \Gamma\frac{(\nabla\phi)_4-(\nabla\phi)_2}{2}
\end{aligned}
\tag{6.28}
$$

which equals to

$$
\Gamma\frac{\frac{\phi_5-\phi_3}{2\Delta x} - \frac{\phi_3-\phi_1}{2\Delta x}}{2} = \frac{\Gamma}{4\Delta x}(\phi_5 - 2\phi_3 + \phi_1)
\tag{6.29}
$$

if the central difference is provoked to assess the gradients at node points. You may verify that the equation we used to evaluate such gradients, Equation (6.23), does reduce to a central difference on a uniform Cartesian mesh.

Therefore, if we directly use Equation (6.27) to evaluate $(\nabla\phi)_j$, the diffusion term of an odd-indexed control volume will be only related to the other odd-indexed control volumes and has nothing to do with any even-indexed control volumes. If we use such a method to calculate a pure diffusion problem we will end up with a sawtooth/checkerboard type of solution, since we are essentially working on two separate sets of control volumes, namely the odd-indexed ones and the even-indexed ones. Such an odd–even decoupling is eliminated by Equation (6.26) as it introduces ϕ values at two neighboring volumes (one odd-indexed and the other even-indexed) into the $(\nabla\phi)_j$ formula. In fact if we apply Equation (6.26) to the above 1-D case, we find the integrated diffusion term is

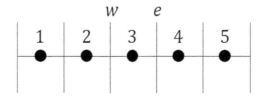

FIGURE 6.10 Simple 1-D uniform mesh illustrating possible odd–even decoupling.

$$\frac{\Gamma}{\Delta x}(\phi_4 - 2\phi_3 + \phi_2) \tag{6.30}$$

which does not have a decoupling problem.

The integration of source term is easily evaluated with

$$\int_V S dV = S_P V_P \tag{6.31}$$

in the spirit of Equation (6.13), where S_P is the S value at point P.

Now putting all terms together, the discretized convection–diffusion equation is

$$a_{P,P}\phi_P^{n+1} + \sum_J a_{P,J}\phi_J^{n+1} = b_P \tag{6.32}$$

where

$$\begin{cases} a_{P,P} = \frac{\rho V_P}{\Delta t} + C_{P,P} + \Gamma D_{P,P} = \frac{\rho V_P}{\Delta t} + \frac{1}{2}\sum_j \left(F_j + \gamma|F_j|\right) + \Gamma \sum_{j,J}\frac{A_j}{l_j} \\[2mm] a_{P,J} = C_{P,J} + \Gamma D_{P,J} = \frac{1}{2}\left(F_j - \gamma|F_j|\right) - \Gamma\frac{A_j}{l_j} \\[2mm] b_P = \frac{\rho\phi_P^n V_P}{\Delta t} + S_P V_P + b_{SUS} + \Gamma b_{CD} \end{cases} \tag{6.33}$$

And

$$b_{SUS} = -\frac{\omega}{2}\left\{\sum_j\left[\left(F_j + \gamma|F_j|\right)\left(\vec{r}_{Oj} - \vec{r}_P\right)\right]\right\}\cdot(\nabla\phi)_P$$
$$-\frac{\omega}{2}\sum_{j,J}\left[\left(F_j - \gamma|F_j|\right)\left(\vec{r}_{Oj} - \vec{r}_J\right)\cdot(\nabla\phi)_J\right] \tag{6.34}$$

is the term arising from applying SUS, while

$$b_{CD} = \sum_j\left[A_j(\nabla\phi)_j\cdot\left(\vec{n}_j - \vec{\xi}_j\right)\right] \tag{6.35}$$

is due to the cross-diffusion term in Equation (6.26).

The discretized Equation (6.32) is needed for all N_P nodes inside the flow boundary. The system of such equations can be written in a matrix format:

$$[A_P]\{\Phi\} = ([C] + \Gamma[D])\{\Phi\} = \{B_P\} \tag{6.36}$$

where the coefficient matrix $[C]$ takes account of the effects of convection and $\Gamma[D]$ is owing to diffusion. Matrix $[D]$ only needs to be constructed once. Matrix $[C]$, on the other hand, has to be updated continuously with the updated velocity field. These matrices can be created by following Equation (6.33). The vector $\{B_P\}$ is calculated with using Equations (6.33), (6.34), and (6.35).

Once you begin to work on these matrices, you should immediately realize the necessity of collecting the connectivity and geometry information of the triangular mesh, as mentioned in Section 6.2.3.

6.3.3 BOUNDARY CONDITIONS

The three most commonly used boundary conditions are
Dirichlet condition:

$$\phi_b = f \tag{6.37}$$

where ϕ_b is the ϕ value at the boundary and f can be a constant or a known function;
Neumann condition:

$$(\nabla\phi)_b \cdot \vec{n}_b = f \tag{6.38}$$

Mixed (Robin) condition:

$$a\phi_b + b(\nabla\phi)_b \cdot \vec{n}_b = f \tag{6.39}$$

where a and b are known constants. Obviously, both Dirichlet and Neumann conditions can be recovered from the mixed condition by choosing proper a and b values.

To apply such boundary conditions, we can use virtual nodes as before. A virtual node is simply the centroid of the neighboring boundary control volume mirrored by the boundary, as shown in Figure 6.11. Evidently the unit vector $\vec{\xi}_V$ is coincident with the unit normal $\vec{n}_b$.

A second-order accurate finite difference formula for the boundary condition is then

$$a\frac{\phi_V + \phi_P}{2} + b\frac{\phi_V - \phi_P}{l_V}\left(\frac{l_V}{2}\right)^c = f \tag{6.40}$$

If $a = 1$, $b = 0$, this formula reproduces the Dirichlet condition; if $b = 1$, $a = c = 0$, it reduces to the Neumann condition; if $a = b = c = 1$, it specifies the virtual node value ϕ_V directly, which is sometimes desired e.g. when we set up the inflow condition with an upwind scheme.

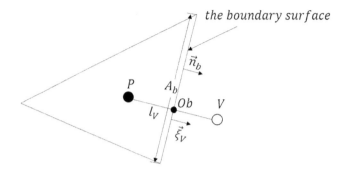

FIGURE 6.11 Virtual node for a boundary cell.

Equation (6.40) can be simplified to

$$a_{V,V}\phi_V + a_{V,P}\phi_P = b_V \tag{6.41}$$

where

$$\begin{cases} a_{V,V} = \dfrac{a}{2} + \dfrac{b}{l_V}\left(\dfrac{l_V}{2}\right)^c \\[2mm] a_{V,P} = \dfrac{a}{2} - \dfrac{b}{l_V}\left(\dfrac{l_V}{2}\right)^c \\[2mm] \quad b_V = f \end{cases} \tag{6.42}$$

The number of virtual nodes is the same as the number of boundary edges. Suppose there are N_V such nodes, then the N_V boundary conditions in the form (6.41) again can be recast into the matrix format:

$$[A_V]\{\Phi\} = \{B_V\} \tag{6.43}$$

Notice that matrix $[A_V]$ and vector $\{B_V\}$ typically only need to be created once.

Now we may combine finite difference Equations (6.36) and (6.43), and they look like

$$[A]\{\Phi\} = \begin{bmatrix} A_P \\ A_V \end{bmatrix}\begin{Bmatrix} \Phi_P \\ \Phi_V \end{Bmatrix} = \begin{Bmatrix} B_P \\ B_V \end{Bmatrix} = \{B\} \tag{6.44}$$

in which $[A_P] = [C] + \Gamma[D]$ is a $N_P \times (N_P + N_V)$ matrix, and $[A_V]$ is a $N_V \times (N_P + N_V)$ matrix. $\{\Phi_P\}$ and $\{\Phi_V\}$ are column unknown vectors that contain the unknown ϕ values at the N_P interior nodes and N_V virtual nodes, respectively.

Under-relaxation can be incorporated into the formulas in a way similar to that introduced in Section 5.3.2 in Chapter 5.. We only need to replace the diagonal elements of $[A_P]$, say $a_{P,P}$, with

$$\tilde{a}_{P,P} = \frac{a_{P,P}}{\alpha} \tag{6.45}$$

and replacing elements of $\{B_P\}$, say b_P, with

$$\tilde{b}_P = b_P + \frac{1-\alpha}{\alpha} a_{P,P} \phi_P^n \tag{6.46}$$

in which α is the under-relaxation factor.

Equation (6.44) can be solved by methods introduced in Section 4.3 in Chapter 4, and notice that the coefficient matrix $[A]$ is no longer banded due to the unstructured mesh.

6.3.4 EXAMPLE: HEAT TRANSFER OVER A CORNER

Let us test our code with a case related to heat transfer.

As Figure 6.12 shows, air at temperature $T_{in} = 10$ °C flows over a square corner of size $L = 0.1$ m. Uniform heat flux $q = 10$ W/m^2 is present over the whole corner surface. Suppose the velocity field in the corner is described by

$$u = x, \quad v = -y \tag{6.47}$$

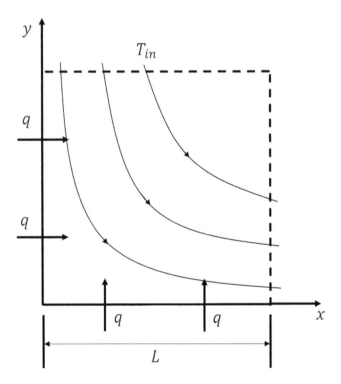

FIGURE 6.12 Heat transfer over a corner example setup.

We want to determine the steady-state temperature distribution in the corner.

The governing equation for the temperature distribution is

$$\nabla \cdot (\vec{u} T) - \alpha \nabla \cdot (\nabla T) = 0 \tag{6.48}$$

Comparing it with the general convection–diffusion equation, you can see the diffusion coefficient $\Gamma = \alpha$ and the source term $S = 0$. For air, its thermal diffusivity $\alpha = 2.2 \times 10^{-5} \, m^2/s$ at room temperature. We can set $\rho = 1$ and Δt a large number (10^{20}, say) in our unsteady general convection–diffusion equation code for this case.

We have a Dirichlet condition at the north boundary and Neumann conditions at the west and south boundaries:

$$(\nabla T) \cdot \vec{n} = q/k \tag{6.49}$$

in which the thermal conductivity of air $k = 0.026 \, W/(m \cdot K)$. Notice that $\vec{n}$ is the unit normal pointing to the outside of the flow domain so we do not have the negative sign before the temperature gradient. At the east boundary we assume a vanishing temperature gradient:

$$(\nabla T) \cdot \vec{n} = 0 \tag{6.50}$$

We use a mesh comprising about 5000 triangular cells, refined close to the boundaries, as shown in Figure 6.13. The result agrees well with what we may obtain with a structured mesh, as shown in Figure 6.14 in which the temperature contours are displayed. We observe, however, iterations are needed to carry out this simulation with the unstructured mesh due to the deferred-correction terms like the cross-diffusion term (see Equation (6.35)). If your code cannot reproduce the correct result, you are recommended to run a pure diffusion problem e.g. the laminar duct flow example we discussed in Chapter 4 first. If your code succeeds in simulating the duct flow, the problem is then likely due to the portion dealing with convection.

6.4 SOLVING NAVIER–STOKES EQUATIONS WITH UNSTRUCTURED MESH

6.4.1 The SIMPLE Procedure

When we solve Navier–Stokes equations for a flow field, we first recognize that the momentum equation is of the general convection–diffusion equation form:

$$\frac{\partial (\rho \vec{u})}{\partial t} + \nabla \cdot (\rho \vec{u} \vec{u}) = \nabla \cdot (\mu \nabla \vec{u}) - \nabla p + \vec{f} \tag{6.51}$$

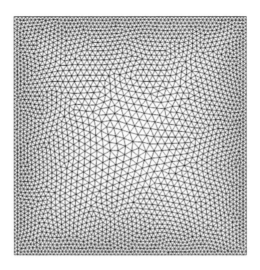

FIGURE 6.13 Triangular mesh for the heat transfer over a corner example.

Therefore, we can solve the velocity field with the same code we developed to solve the heat transfer over a corner example. Here we assume the physical properties like ρ, μ, etc. are constants. $\vec{f}$ is the body force per unit volume of the fluid. Notice that in the convection term $\nabla\cdot(\rho\vec{u}\,\vec{u})$ the first $\vec{u}$ is the latest solved flow field while the second $\vec{u}$ is the unknown to be solved. The pressure gradient in the source term can be evaluated with Equation (6.23):

$$(\nabla p)_P = \frac{\int_V \nabla p\, dV}{V_P} = \frac{\int_A p\,\vec{n}\, dA}{V_P} \approx \frac{\Sigma_j\, p_j A_j \vec{n}_j}{V_P} \tag{6.52}$$

in which the surface pressure p_j is obtained with a simple average of node values:

$$p_j \approx \frac{p_P + p_J}{2} \tag{6.53}$$

Such a practice, however, may cause a checkerboard pressure/velocity distribution for the current co-located mesh. We have to fix this problem with the momentum interpolation method (see Section 5.2.4 in Chapter 5). To apply this method we only need to make a little change in the mass flux term in the continuity equation. The continuity equation is

$$\nabla\cdot(\rho\vec{u}) = 0 \tag{6.54}$$

If we integrate it over a control volume, it becomes

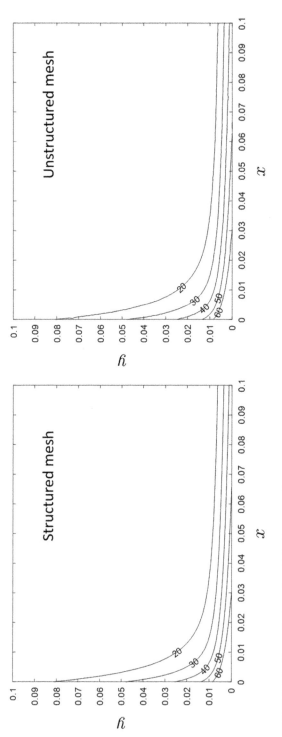

FIGURE 6.14 Numerical results of the heat transfer over a corner example.

$$\int_V \nabla \cdot (\rho \vec{u}) dV = \int_A \rho \vec{u} \cdot \vec{n} \, dA \approx \sum_j \rho \vec{u_j} \cdot \vec{n_j} A_j = \sum_j F_j = 0 \tag{6.55}$$

The momentum interpolation method is then implemented by adding a correction term to the mass flux in Equation (6.55):

$$\tilde{F_j} = \rho A_j \left\{ \vec{u_j} \cdot \vec{n_j} + \frac{1}{2} \left(\frac{V_P}{a_{P,P}^{u,v}} + \frac{V_J}{a_{J,J}^{u,v}} \right) \left[\frac{p_P - p_J}{l_J} + (\nabla p)_j \cdot \vec{\xi_j} \right] \right\} \tag{6.56}$$

in which $a_{P,P}^{u,v}$ and $a_{J,J}^{u,v}$ are the coefficients before u_P (or v_P) and u_J (or v_J) in the finite difference momentum equations. The continuity equation thus becomes

$$\sum_j \tilde{F_j} = 0 \tag{6.57}$$

The finite difference momentum and continuity equations can be solved by various algorithms discussed in Chapter 5, for example the SIMPLE method, which proceeds as follows.

We first solve the velocity field $\vec{u}^*$ from the momentum equations based on the flow field $\vec{u}^n$ and p^n calculated from the most recent iteration. That is, the coefficient matrices and constant vectors are evaluated with $\vec{u}^n$ and p^n. The velocity field $\vec{u}^*$ thus solved needs correction so that it may satisfy the continuity equation:

$$\sum_j \tilde{F_j} = \sum_j (\tilde{F_j^*} + F_j') = 0 \tag{6.58}$$

where $\tilde{F_j^*}$ is the mass flux calculated with using $\vec{u}^*$ and $p^* = p^n$. The adjustment in mass flux is

$$F_j' = \rho A_j \vec{u}_j' \cdot \vec{n_j} = \rho A_j \left[\frac{1}{2} \left(\frac{V_P}{a_{P,P}^{u,v}} + \frac{V_J}{a_{J,J}^{u,v}} \right) \left(\frac{p_P' - p_J'}{l_J} \right) \right] \tag{6.59}$$

in virtue of Equation (6.56) and the SIMPLE method practice. The continuity equation then becomes an equation for pressure corrections:

$$a_{P,P}^p p_P' + \sum_J a_{P,J}^p p_J' = b_P^p \tag{6.60}$$

and

$$
\begin{cases}
a_{P,P}^{p} = -\Sigma_J a_{P,J}^{p} = \frac{1}{2}\rho \Sigma_{j,J} \left[\left(\frac{V_P}{a_{P,P}^{u,v}} + \frac{V_J}{a_{J,J}^{u,v}} \right) \frac{A_j}{l_j} \right] \\
\qquad a_{P,J}^{p} = -\frac{1}{2}\rho \left(\frac{V_P}{a_{P,P}^{u,v}} + \frac{V_J}{a_{J,J}^{u,v}} \right) \frac{A_j}{l_j} \\
\qquad\qquad b_P^{p} = -\Sigma_j \tilde{F}_j^{*}
\end{cases}
\tag{6.61}
$$

For the boundary control volumes, we have to provide proper boundary conditions. For a boundary with a specified velocity e.g. a solid surface, we require

$$
p_V' - p_P' = 0 \tag{6.62}
$$

because no velocity correction is needed at such boundaries (see Equation (6.59)). This is equivalent to $(\nabla p') \cdot \vec{n} = 0$ at the boundary which entails

$$
(\nabla p')_V + (\nabla p')_P = 0 \tag{6.63}
$$

The related conditions for pressure are

$$
p_V - p_P = 0; \ (\nabla p)_V + (\nabla p)_P = 0 \tag{6.64}
$$

which correspond to $(\nabla p) \cdot \vec{n} = 0$ at the boundary. The rationale behind this condition has been discussed in Section 5.3.3 in Chapter 5.

At the outlet of a fully developed flow, the velocity derivatives along the flow direction vanish, which implies

$$
p_V' + p_P' = 0 \tag{6.65}
$$

so that there is no pressure correction and the pressure is constant along such a boundary. The rationale behind such practices has been discussed in Section 5.5.2. The conditions for gradients of pressure and pressure corrections are

$$
(\nabla p')_V - (\nabla p')_P = 0; \ (\nabla p)_V - (\nabla p)_P = 0 \tag{6.66}
$$

This can be understood by considering two nodes close to the boundary. To apply the designated p' or p values at the boundary, the corresponding ghost node values should be such that Equation (6.66) is satisfied, as depicted in Figure 6.15, in which the vertical location of each circle denotes the magnitude of the p' or p value at that node.

Since the virtual node values are set passively with the changing neighboring boundary node values, velocity correction at the boundary, if needed, should be the

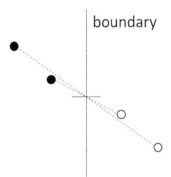

FIGURE 6.15 Pressure boundary conditions.

responsibility of the boundary nodes, not the virtual nodes. For this reason, we should set the $V/a^{u,v}$ value of a virtual node in Equations (6.56), (6.59), and (6.61) to be equal to that of its neighboring boundary node.

Equations in the form of Equation (6.60), with boundary conditions in the form of Equations (6.62) and/or (6.65), can be written in a matrix format, Equation (6.44). If the Neumann condition is applied to p' at all bound-aries, one should make change to the coefficient matrix as discussed in Section 5.3.3.

Once p' values are solved, we may calculate the gradient $\nabla p'$. We then can go ahead to correct the flow field with

$$\begin{cases} \overrightarrow{u}_P^{\,n+1} = \overrightarrow{u}_P^{\,*} - \alpha_{u,v}\left(\frac{V_P}{a_{P,P}^{u,v}}\right)(\nabla p')_P \\ p_P^{n+1} = p_P^* + \alpha_p p_P' \end{cases} \tag{6.67}$$

at every node, including the virtual nodes. The α values are the under-relaxation factors.

The flow field should be updated repeatedly until the desired accuracy is reached.

6.4.2 EXAMPLE: LID-DRIVEN CAVITY FLOW

Let us test our unstructured mesh Navier–Stokes solver with the lid-driven cavity flow, which has been studied in Section 5.3 in Chapter 5. We consider a square cavity of side length H filled with a viscous fluid that is driven to flow by the moving lid as shown in Figure 6.16. The Reynolds number is $Re = \rho U H/\mu = 100$.

We use a mesh with about 2900 triangles, refined close to the boundaries. One small issue that needs attention, which had been pointed out in the previous sub-section, is the singularity due to the Neumann conditions of p' enforced at all boundaries. We can fix this problem by setting the p' value at a specific node to

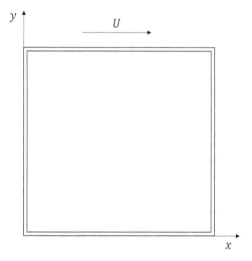

FIGURE 6.16   Lid-driven cavity flow example setup.

zero. The streamlines shown in Figure 6.17 and velocity profiles displayed in Figure 6.18 are very close to what we obtained with structured mesh (cf. Section 5.3.6).

6.4.3   EXAMPLE: NATURAL CONVECTION IN CONCENTRIC CYLINDRICAL ANNULUS

Let us study the very first problem we proposed in this chapter: the natural convection in a horizontal concentric cylindrical annulus, as depicted in Figure 6.19.

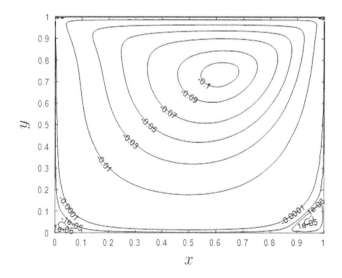

FIGURE 6.17   Streamlines of the flow in the cavity.

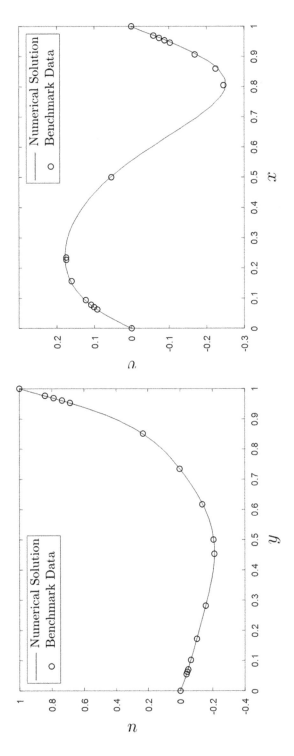

FIGURE 6.18 Velocity profiles along bisectors of the cavity.

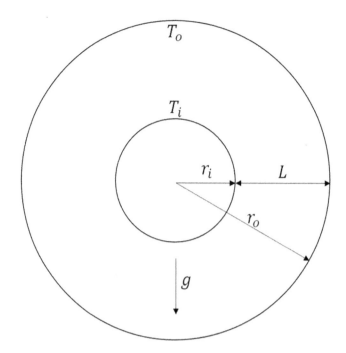

FIGURE 6.19 Natural convection in a concentric annulus example setup.

The radii of the inner and outer cylinders are r_i and r_o, respectively. The distance between the two cylinders is $L = r_o - r_i$. The temperatures of the inner and outer cylinders are T_i and T_o ($T_i > T_o$), respectively. The annulus is filled with fluid whose density varies with temperature and is driven to flow due to the buoyancy force. As discussed in Section 5.4.2, a natural convection is characterized by two dimensionless parameters: the Rayleigh number Ra and Prandtl number Pr, which are defined as

$$Ra = \frac{g\beta(T_i - T_o)L^3}{\alpha\nu}, \quad Pr = \frac{\nu}{\alpha} \tag{6.68}$$

Here g is the gravitational acceleration, β is the thermal expansion coefficient of the fluid, α is the thermal diffusivity of the fluid, and ν is the kinematic viscosity of the fluid.

The dimensionless governing equations are

$$\frac{\partial \tilde{u}}{\partial \tilde{x}} + \frac{\partial \tilde{v}}{\partial \tilde{y}} = 0 \tag{6.69}$$

$$\frac{\partial(\tilde{u}\cdot\tilde{u})}{\partial \tilde{x}} + \frac{\partial(\tilde{v}\cdot\tilde{u})}{\partial \tilde{y}} - Pr\left(\frac{\partial^2 \tilde{u}}{\partial \tilde{x}^2} + \frac{\partial^2 \tilde{u}}{\partial \tilde{y}^2}\right) = -\frac{\partial \tilde{p}}{\partial \tilde{x}} \tag{6.70}$$

$$\frac{\partial(\tilde{u}\cdot\tilde{v})}{\partial\tilde{x}} + \frac{\partial(\tilde{v}\cdot\tilde{v})}{\partial\tilde{y}} - Pr\left(\frac{\partial^2\tilde{v}}{\partial\tilde{x}^2} + \frac{\partial^2\tilde{v}}{\partial\tilde{y}^2}\right) = -\frac{\partial\tilde{p}}{\partial\tilde{y}} + Ra\cdot Pr\cdot\theta \tag{6.71}$$

$$\frac{\partial(\tilde{u}\cdot\theta)}{\partial\tilde{x}} + \frac{\partial(\tilde{v}\cdot\theta)}{\partial\tilde{y}} - \left(\frac{\partial^2\theta}{\partial\tilde{x}^2} + \frac{\partial^2\theta}{\partial\tilde{y}^2}\right) = 0 \tag{6.72}$$

where the dimensionless variables are defined as

$$\tilde{x} = \frac{x}{L}, \quad \tilde{y} = \frac{y}{L}, \quad \tilde{u} = \frac{uL}{\alpha}, \quad \tilde{v} = \frac{vL}{\alpha}, \quad \tilde{p} = \frac{pL^2}{\rho_o\alpha^2}, \quad \theta = \frac{T - T_o}{T_i - T_o} \tag{6.73}$$

The boundary conditions are

$$\begin{cases} \tilde{u} = \tilde{v} = 0; \ \theta = 1 \ , \quad \tilde{r} = \tilde{r}_i = \frac{r_i}{L} \\ \tilde{u} = \tilde{v} = 0; \ \theta = 0 \ , \quad \tilde{r} = \tilde{r}_o = \frac{r_o}{L} \end{cases} \tag{6.74}$$

From now on we drop the tildes from the variables.

We will simulate a case with $\tilde{r}_i = 0.625$, $Ra = 4.7 \times 10^4$, and $Pr = 0.71$.

The mesh we use contains about 6500 triangles and is refined close to the boundaries as well as close to the top portion of the annulus to resolve the higher temperature gradients present there, as shown in Figure 6.20. When we carry out this simulation, we should begin with a low Ra and after the code converges we set the Rayleigh number to a higher value and then continue the calculation based on the just converged results. We repeat this procedure until the desired Rayleigh number is reached. This practice effectively prevents divergence due to the strong coupling between velocity and temperature at high Rayleigh numbers.

The calculated streamlines (left) and temperature contours (right) are shown in Figure 6.21.

The temperature distribution in the lower half of the annulus tends to stabilize the fluid as the hot fluid is above the cold fluid. The temperature distribution in the upper half, on the other hand, promotes fluid flow as the hot fluid is under the cold fluid. A plume, therefore, rises up at the top of the inner cylinder and goes downward after it touches the outer cylinder. As a result, two counter-rotating vortices form inside the annulus. The temperature distribution along three lines in the annulus are shown in Figure 6.22, which compares well with the experimental data (Kuehn & Goldstein, 1976).

6.5 OTHER MEANS TO HANDLE COMPLEX BOUNDARIES

Trianglular mesh is not the only means to realize an unstructured mesh, and unstructured mesh is certainly not the only means to simulate flows with complex

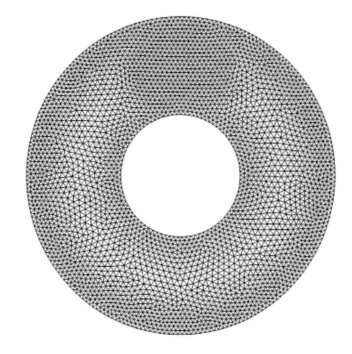

FIGURE 6.20   Triangular mesh for the natural convection in a concentric annulus example.

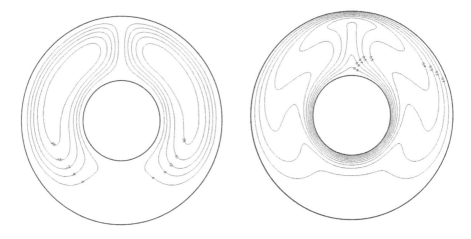

FIGURE 6.21   Streamlines and temperature contours in the concentric annulus.

boundaries. For example, the very first successful finite difference solution of a fluid flow (to the best of my knowledge) was the work of A. Thom who solved the flow past a circular cylinder (Thom, 1933). A cut-cell mesh similar to the one shown in Figure 6.23 was employed in his work. Such a mesh, as its name implies, is formed

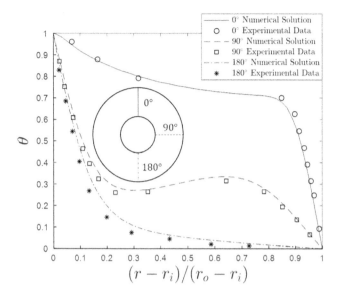

FIGURE 6.22 Temperature distribution along three lines in the annulus.

by cutting the otherwise rectangular mesh cells with the boundary. Finite difference method can be implemented on such a cut-cell mesh relatively straightforwardly.

Another type of mesh, the so-called overlapping mesh, can also be used to solve the same problem. One such mesh is shown in Figure 6.24. An overlapping mesh is one in which different types of mesh originate from different boundaries that overlap in part of the flow domain. With this method we can use whatever mesh deemed convenient for a specific boundary. For example, we can use the polar cylindrical mesh for the cylinder boundary and a rectangular mesh for the other

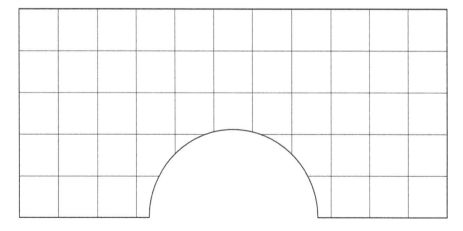

FIGURE 6.23 Cut-cell mesh.

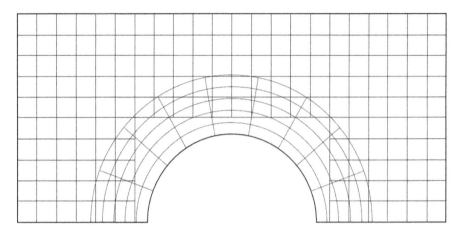

FIGURE 6.24 Overlapping mesh.

boundaries. Data at the boundary of the rectangular mesh in the overlap region can be interpolated from the interior nodes of the cylindrical mesh and vice versa.

Calculus tells us that any shape can be filled by infinite infinitesimal rectangles. One can follow this idea and create a mesh like the one shown in Figure 6.25, in which rectangular cells are split into smaller and smaller cells close to the curved cylinder boundary. The complex boundary is eventually approximated by a series of "step" grids.

It is also possible to just use the rectangular mesh (see Figure 6.26) to simulate the flow past a cylinder. We can treat the cylinder as part of the fluid and apply a proper body force to each cell inside the cylinder to prevent this part of fluid from flowing. This method is called the immersed boundary method (IBM) (Peskin, 2002).

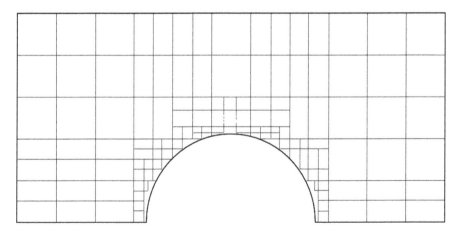

FIGURE 6.25 Cartesian mesh.

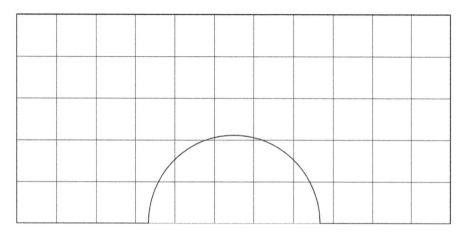

FIGURE 6.26   Immersed boundary method.

Another idea of using a rectangular mesh to simulate a flow with complex boundaries is similar to the one we will be discussing in Chapter 7. We can regard the cylinder as a fluid with an infinite viscosity and set the velocity of this super sticky fluid zero. The cylinder surface then becomes an interface between this fictitious fluid and the real fluid around the cylinder. The techniques we will introduce in the next chapter are able to effectively bring the real fluid velocity down to zero at the interface, which is equivalent to the no-slip condition we want to apply at the cylinder boundary.

It is even possible to solve flow fields without using a mesh. Think about this fact: all fluids are made of microscopic particles and these particles move according to certain physical laws like the second law of Newton. So we can simulate a fluid flow by tracking the movement of its constituent particles. We do not need any mesh for such calculations.

Exercises

1. First let us do a simulation of the heat transfer over a corner example (Section 6.3.4) with a structured mesh. Notice that since the velocity field is given, you only need to solve the transport equation of temperature

$$\frac{\partial (u \cdot T)}{\partial x} + \frac{\partial (v \cdot T)}{\partial y} - \alpha \left(\frac{\partial^2 T}{\partial x^2} + \frac{\partial^2 T}{\partial y^2} \right) = 0 \qquad (6.75)$$

Try to reproduce the results shown in Figure 6.14.

2. Now write a MATLAB function that can retrieve geometrical information from a 2-D triangular mesh. The function inputs are the p and t matrices created by the program in Appendix A.7. Your function should output all the information mentioned in Section 6.2.3.

3. Write a MATLAB function that can calculate $\nabla\phi$ at both a node point and at an edge if given the nodal ϕ values and necessary geometry information.
4. Redo problem 1 with an unstructured mesh.
5. Redo the lid-driven cavity flow simulation using an unstructured mesh. $Re = 100$.

7 Multiphase Flow

7.1 INTRODUCTION

A multiphase flow is one in which two or more distinct phases coexist. Multiphase flows are extremely common in our daily life as the two most abundant fluids we encounter everyday, namely air and water, are usually present as different phases. Examples include rain drops falling through atmosphere; air bubbles rising in water; water flows with a free surface like the water flowing from a drinking fountain or from a water tap, the air–water flow close to the surface of a lake, and the air–water flow around a boat sailing in the lake. Multiphase flows can result from phase change as well. For instance, water vapor bubbles form in boiling water on a stovetop; they also can form along the surface of a fast spinning underwater propeller (a phenomenon known as "cavitation").

Multiphase flows are ubiquitous in industrial applications as well. For example, bubble columns in which liquid circulation is driven by gas bubbles are widely used in chemical plants. In air-conditioning systems and refrigerators the refrigerant experiences repeated phase changes and, as a result, the flow in such systems is usually a multiphase flow. Although most examples mentioned so far are gas–liquid flows, multiphase flows involving a solid phase are also common both in nature and in industry, for example the melting of ice and solidification of water which we witness during changing seasons. Such processes are also essential in the metallurgical industry. Another example is a device frequently used in the chemical industry called fluidized bed in which solid particles are suspended by gas or liquid flows.

Compared with single-phase flows, multiphase flows have many unique features, such as surface tension, deformation, breaking and merging of phase interfaces, etc.

Understanding multiphase flows is important, yet challenging, due to a few reasons. The material properties like density, viscosity, and thermal conductivity, etc. are not uniform in such flows. Instead, these properties undergo a sudden jump across the phase interface. The position and movement of the interface are nevertheless unknown beforehand and have to be determined as part of the solution procedure, which makes theoretical analysis of multiphase flows difficult. Experimental studies of multiphase flows are not easy either because their behavior is affected by so many factors such as the density ratio, the viscosity ratio, the Reynolds number, the Morton number, the Eötvös number (we will give more details of these numbers later), and many others if phase change is involved. It is quite unrealistic to cover a wide range of these parameters and test their combinations with experiments (e.g. how can you vary the density ratio of two fluids while keeping the viscosity ratio a constant?). Numerical simulations of multiphase flows have, therefore, become very attractive since the origin of computational fluid

DOI: 10.1201/9781003138822-7

dynamics (CFD). As a matter of fact, one of the earliest successful Navier–Stokes solving approach, the marker and cell (MAC) method (Harlow & Welch, 1965) was intended to deal with multiphase flows.

Over the past few decades, many new methods have been developed to address multiphase flow problems. Among these methods are the volume of fluid (VOF) method (Hirt & Nichols, 1981), the level-set method (Sussman, Smereka, & Osher, 1994), the front-tracking method (Unverdi & Tryggvason, 1992), the phase-field method (Jacqmin, 1999), the lattice Boltzmann method (LBM) (Shan & Chen, 1993), and the smoothed particle hydrodynamics (SPH) method (Gingold & Monaghan, 1977). Multiphase flow is an area which is still under very active research. It is therefore not surprising that each of these methods have many variations and different practices almost at every turn. More novel approaches for multiphase flow simulation will certainly appear in the years to come. In this chapter we focus on the fundamentals of the VOF method and level-set method for two-phase flow examples. We will also restrict ourselves to 2-D cases.

7.2 VOF METHOD

7.2.1 INTERFACE REPRESENTATION

Let us consider a fluid flow comprising two immiscible phases, say a liquid phase and a gas phase, which are separated by a sharp interface. Both phases are incompressible and at this point we assume there is neither a phase change nor a chemical reaction in the flow. Obviously, it is crucial to be able to accurately identify and track the interface in such flows. This can be done explicitly by tracking massless point particles located along the interface and advecting with the local fluid speed. It is also possible to use a marker function to label different phases and represent the interface implicitly.

The simplest marker function is a step function:

$$C(x, y) = \begin{cases} 1 & \text{if phase 1 is present at point } (x, y) \\ 0 & \text{if phase 2 is present at point } (x, y) \end{cases} \tag{7.1}$$

Which phase is phase 1 or phase 2 is arbitrary: you can designate either one of the two phases as phase 1. The interface is then identified implicitly by the step change of the marker function (see Figure 7.1).

If we use a mesh to discretize the C function, some mesh cells may contain both phases (viz. they contain part of the phase interface). The C value at the center of such a cell, say the (i, j) cell might be interpreted as the volume average of the marker function in this cell:

$$C_{i,j} = \frac{\int_V C dV}{V} = \frac{1 \cdot V_1 + 0 \cdot V_2}{V} = \frac{V_1}{V} \tag{7.2}$$

where V_1 and V_2 are the volume of phase 1 and phase 2 in this cell, respectively, and

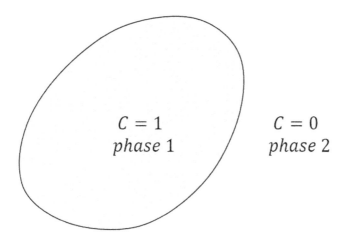

$C = 1$
$phase\ 1$

$C = 0$
$phase\ 2$

FIGURE 7.1 Step marker function.

$V = V_1 + V_2$ is the total volume of this cell. Obviously $0 < C < 1$ in such interface-containing cells. C therefore attains a physical meaning, which is the volume fraction of phase 1. For this reason this method is called the VOF method.

Now let us answer a practical programming question: if the interface is known, how should we have a computer calculate the accurate C values on a specific mesh? For example, suppose we are simulating the motion of an air bubble rising in water. The bubble is initially a circle with known radius and center position, so we know whether a given point is within or without this bubble. How can we find the volume fraction of the gas phase in each mesh cell as shown in Figure 7.2? You can try thinking about this question for a few minutes before you proceed to the paragraph below.

Well, a simple method is to split every cell into, say 100 equal subcells and count the number of subcells whose center is inside the circle. Dividing this number by the total number of subcells gives us a pretty good estimate of the volume fraction in this cell.

7.2.2 INTERFACE RECONSTRUCTION

Now let us consider another related question: if we are given a discretized C field, say the one shown in Figure 7.3, how can we reconstruct the interface i.e. find the exact location of the interface?

A popular method is using a straight line segment to approximate the interface in each cell with a $0 < C < 1$. In this so-called piecewise linear interface calculation (PLIC) method (Youngs, 1982) the orientation of the line segment is determined first and then the exact location of the segment is decided based on the cell C value. Let us see how to find the orientation of the interface first.

If we use a small volume to enclose a segment of the interface as shown in Figure 7.4, the volume integral of the gradient of C over this small volume is

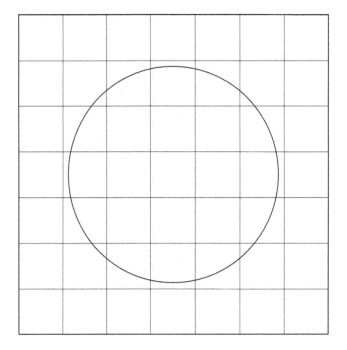

FIGURE 7.2 Air bubble in water.

0.52	0.16	0	0
1	0.97	0.26	0
1	1	0.78	0
1	1	0.86	0

FIGURE 7.3 Discretized C field.

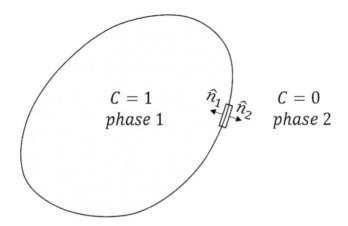

FIGURE 7.4   Differential volume enclosing an interface segment.

$$\int_V \nabla C dV = \int_A C \hat{n} dA \qquad (7.3)$$

which is nothing but the Gauss theorem (Equation (6.11)) we introduced in Section 6.3.1. Here we use $\hat{n}$ to denote the unit outward normal, which is the vector perpendicular to the interface segment, pointing away from it, and with unit length. When the volume is small enough and the interface segment contained in the volume is close to a straight line, we have

$$\int_V \nabla C dV = \int_A C \hat{n} dA \approx (1 \cdot \hat{n}_1 + 0 \cdot \hat{n}_2) \delta A = \hat{n}_1 \delta A \qquad (7.4)$$

where $\hat{n}_1$ and $\hat{n}_2$ are the unit normal vectors of the interface segment pointing to phase 1 and phase 2, respectively. δA is the area of the interface segment (or length of the segment in 2-D). If we apply this result to control volume (i, j) that contains an interface segment, as shown in Figure 7.3, we have

$$\int_V \nabla C dV = (\nabla C)_{i,j} \delta V_{i,j} \approx (\hat{n}_1)_{i,j} \delta A_{i,j} \qquad (7.5)$$

where $\delta V_{i,j}$ is the volume of the mesh cell. Therefore,

$$(\hat{n}_1)_{i,j} \approx (\nabla C)_{i,j} \frac{\delta V_{i,j}}{\delta A_{i,j}} \qquad (7.6)$$

Notice that $\hat{n}_1$ gives us the information about the orientation of the interface segment and this information in fact is solely contained in the C gradient. We can thus conclude that

$$\hat{n}_1 \approx \frac{(\nabla C)_{i,j}}{|(\nabla C)_{i,j}|} \tag{7.7}$$

In case you forget what a gradient is, in 2-D

$$\nabla C = \frac{\partial C}{\partial x}\vec{e}_x + \frac{\partial C}{\partial y}\vec{e}_y \tag{7.8}$$

where $\vec{e}_x$ and $\vec{e}_y$ are the unit vectors along the x- and y-directions. So a gradient is a vector. From now on we will simply use $\hat{n}$ in place of $\hat{n}_1$.

To calculate such a gradient at the center of a certain cell, say the (i, j) cell, we can use central differences to approximate the two derivatives in Equation (7.8). More accurate formulae involving extended neighboring cell values are

$$\left(\frac{\partial C}{\partial x}\right)_{i,j} \approx \frac{(C_{i+1,j+1} + 2C_{i+1,j} + C_{i+1,j-1}) - (C_{i-1,j+1} + 2C_{i-1,j} + C_{i-1,j-1})}{8\Delta x} \tag{7.9}$$

and

$$\left(\frac{\partial C}{\partial y}\right)_{i,j} \approx \frac{(C_{i-1,j+1} + 2C_{i,j+1} + C_{i+1,j+1}) - (C_{i-1,j-1} + 2C_{i,j-1} + C_{i+1,j-1})}{8\Delta y} \tag{7.10}$$

where Δx and Δy are the mesh sizes as usual. These apparently intimidating formulas are indeed nothing more than the simple average of central differences at four corners of the mesh cell.

As an example, for the cell shown in Figure 7.3 whose $C = 0.26$, the derivatives of C calculated from Equations (7.9) and (7.10) are

$$\frac{\partial C}{\partial x} = -\frac{0.3875}{\Delta x}; \quad \frac{\partial C}{\partial y} = -\frac{0.3}{\Delta y} \tag{7.11}$$

Assuming $\Delta x = \Delta y$, the unit normal vector is then

$$\hat{n} = \frac{\frac{\partial C}{\partial x}\vec{e}_x + \frac{\partial C}{\partial y}\vec{e}_y}{\sqrt{\left(\frac{\partial C}{\partial x}\right)^2 + \left(\frac{\partial C}{\partial y}\right)^2}} = \frac{-0.3875\vec{e}_x - 0.3\vec{e}_y}{\sqrt{(-0.3875)^2 + (-0.3)^2}} = -0.79\vec{e}_x - 0.61\vec{e}_y \tag{7.12}$$

This result is reasonable since the normal should point into phase 1 region which is on the southwest side of this cell.

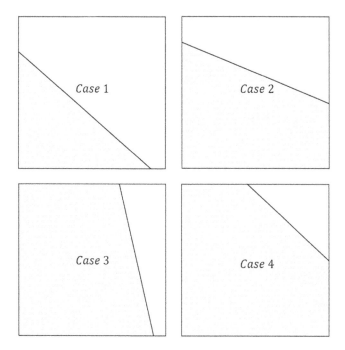

FIGURE 7.5   Four interface configuration cases.

Now let us move on to the second step of the interface reconstruction process, which is to determine the exact location of the interface in a cell. If both $\hat{n}$ components of a certain cell are negative, i.e. $\hat{n}_x < 0$ and $\hat{n}_y < 0$, the interface approximated by a line segment can only intersect the cell surfaces in four possible ways (or cases), as illustrated in Figure 7.5.

Note that the volume fraction of phase 1 (denoted by the shaded region) is equal to the C value of this cell and this value has to be within certain limits for each case.

For example, if the angle β made by the interface with the south surface in case 1 (see Figure 7.5) is such that

$$\tan \beta < \frac{\Delta y}{\Delta x} \tag{7.13}$$

which is the situation depicted in the top left panel of Figure 7.6, the shaded area cannot exceed a limiting value shown in the top right panel, otherwise it is no longer a case 1 but case 2 configuration (see Figure 7.5). The condition that an interface belongs to case 1 when Equation (7.13) holds is therefore

$$C \Delta x \Delta y \le \frac{1}{2} \Delta x (\Delta x \tan \beta) \tag{7.14}$$

Similarly, if

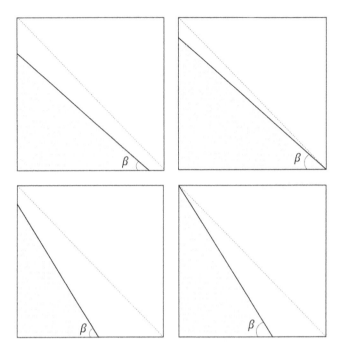

FIGURE 7.6 Different situations of case 1 interface configuration.

$$\tan \beta \geq \frac{\Delta y}{\Delta x} \tag{7.15}$$

as portrayed in the lower left panel of Figure 7.6, the condition for the disposition to be case 1 is

$$C \Delta x \Delta y \leq \frac{1}{2} \Delta y \left(\frac{\Delta y}{\tan \beta} \right) \tag{7.16}$$

See the bottom right panel of Figure 7.6.

We can find similar conditions for the other three cases following the same idea. It is convenient to define

$$\alpha = \frac{\Delta x}{\Delta y} \tan \beta \tag{7.17}$$

Notice that

$$\tan \beta = \left| \frac{\hat{n}_x}{\hat{n}_y} \right| \tag{7.18}$$

so α values can be easily computed as we have already known $\hat{n}$. The two conditions (7.14) and (7.16) then can be rewritten as

$$\begin{cases} 0 < C \leq \frac{\alpha}{2} & \text{if } \alpha < 1 \\ 0 < C \leq \frac{1}{2\alpha} & \text{if } \alpha \geq 1 \end{cases} \tag{7.19}$$

As an example, for the cell in Figure 7.3 where $C = 0.26$, we have

$$\alpha = \frac{\Delta x}{\Delta y} \tan \beta = \tan \beta = \left| \frac{-0.79}{-0.61} \right| = 1.29 \geq 1 \tag{7.20}$$

(refer to Equation (7.12) and assume $\Delta x = \Delta y$) and the interface profile belongs to case 1 because

$$0 < C = 0.26 \leq \frac{1}{2\alpha} = 0.39 \tag{7.21}$$

The conditions one may use to determine which case a certain interface layout belongs to are summarized in Table 7.1.

Once the interface configuration case is determined, the interface reconstruction can be completed by calculating the intercepts the interface makes with the cell edges. Let us still use case 1 for example. If we denote the fraction of the west and south surfaces of the cell that lie within phase 1 r_w and r_s, respectively (see Figure 7.7), obviously

$$\tan \beta = \frac{r_w \Delta y}{r_s \Delta x} \rightarrow \alpha = \frac{\Delta x}{\Delta y} \tan \beta = \frac{r_w}{r_s} \tag{7.22}$$

and

TABLE 7.1
Conditions of the Four Interface Configuration Cases

Case	Conditions	
	$\alpha < 1$	$\alpha \geq 1$
1	$0 < C \leq \frac{\alpha}{2}$	$0 < C \leq \frac{1}{2\alpha}$
2	$\frac{\alpha}{2} < C \leq 1 - \frac{\alpha}{2}$	————
3	————	$\frac{1}{2\alpha} < C \leq 1 - \frac{1}{2\alpha}$
4	$1 - \frac{\alpha}{2} < C < 1$	$1 - \frac{1}{2\alpha} < C < 1$

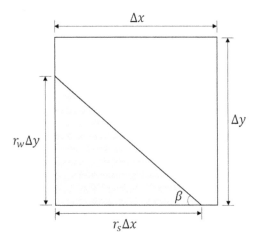

FIGURE 7.7   Side fractions.

$$C\Delta x\Delta y = \frac{1}{2}(r_s\Delta x)(r_w\Delta y) \tag{7.23}$$

From these two equations we can solve for r_w and r_s, which are

$$r_w = \sqrt{2C\alpha}, \quad r_s = \sqrt{\frac{2C}{\alpha}} \tag{7.24}$$

Let us use again the cell with $C = 0.26$ in Figure 7.3 as an example. Since $\alpha = 1.29$ and $C = 0.26$, easily you can find $r_w = 0.82$ and $r_s = 0.63$.

Side fractions for other cases can be found analogously. The results are summarized in Table 7.2.

TABLE 7.2

Side Fractions in the Four Interface Configuration Cases

Case	r_n	r_e	r_w	r_s
1	0	0	$\sqrt{2C\alpha}$	$\sqrt{\frac{2C}{\alpha}}$
2	0	$C - \frac{\alpha}{2}$	$C + \frac{\alpha}{2}$	1
3	$C - \frac{1}{2\alpha}$	0	1	$C + \frac{1}{2\alpha}$
4	$1 - \sqrt{\frac{2(1-C)}{\alpha}}$	$1 - \sqrt{2(1-C)\alpha}$	1	1

7.2.3 INTERFACE ADVECTION

In the VOF method, the interface is identified by the step change of the marker function C. Therefore, we only need to know how the C field advects with the fluid flow in order to track the interface.

Notice that C is a "label" of each fluid particle. If it is a phase 1 (e.g. gas) particle, its C value is 1; if it is a phase 2 (e.g. liquid) particle, its C value is 0. Obviously the phase of a given fluid particle never changes since we assume the two phases are immiscible and there is no phase change. As a result, C value of a given particle also does not change as it moves around with the flow. This fact implies that the material derivative of C vanishes i.e.

$$\frac{DC}{Dt} = \frac{\partial C}{\partial t} + \vec{u} \cdot \nabla C = \frac{\partial C}{\partial t} + u\frac{\partial C}{\partial x} + v\frac{\partial C}{\partial y} = 0 \qquad (7.25)$$

since the material derivative of C is the time rate of change of C of a moving particle (see Equation (1.7) in Section 1.2.1 in Chapter 1). This is the governing equation of the interface advection.

Equation (7.25) can be rewritten as

$$\frac{\partial C}{\partial t} + \frac{\partial (uC)}{\partial x} + \frac{\partial (vC)}{\partial y} - C\left(\frac{\partial u}{\partial x} + \frac{\partial v}{\partial y} \right) = 0 \qquad (7.26)$$

or in a more compact form

$$\frac{\partial C}{\partial t} + \nabla \cdot (\vec{u} C) = C\nabla \cdot \vec{u} \qquad (7.27)$$

Notice that although $\nabla \cdot \vec{u} = 0$ for an incompressible flow, we still keep this term on the right side of Equation (7.27) to take care of the possible small nonzero $\nabla \cdot \vec{u}$ from numerical calculations.

If we integrate Equation (7.27) over the $(i, j)^{th}$ control volume, we have

$$\int_V \frac{\partial C}{\partial t} dV + \int_V \nabla \cdot (\vec{u} C) dV = \int_V C\nabla \cdot \vec{u} \, dV \qquad (7.28)$$

Applying the Gauss divergence theorem to the second term, it becomes

$$\int_V \frac{\partial C}{\partial t} dV + \int_A \vec{n} \cdot \vec{u} \, CdA = \int_V C\nabla \cdot \vec{u} \, dV \qquad (7.29)$$

It can be approximated as

$$\frac{C_{i,j}^{m+1} - C_{i,j}^m}{\Delta t}\Delta x \Delta y + \sum_{k=1}^4 \vec{n}_k \cdot \vec{u}_k C_k A_k = C_{i,j}(\nabla \cdot \vec{u})_{i,j}\Delta x \Delta y \qquad (7.30)$$

Where m and $m + 1$ are two consecutive time moments. $\vec{n}_k$ and $\vec{u}_k$ are the unit outward normal vector and the velocity vector at the kth surface of the cell. C_k and A_k are the C value and area (length in 2-D) of the kth surface. The C flux across the kth surface is defined as

$$F_k = \vec{n}_k \cdot \vec{u}_k C_k \qquad (7.31)$$

Notice that the flux is positive if it is an outflow. One may then write Equation (7.30) as

$$\frac{C_{i,j}^{m+1} - C_{i,j}^m}{\Delta t} = -\frac{F_e + F_w}{\Delta x} - \frac{F_n + F_s}{\Delta y} + C_{i,j}(\nabla \cdot \vec{u})_{i,j} \qquad (7.32)$$

This equation is a pure convection (advection) equation, and what it advects is a field that contains abrupt step changes (discontinuities). We have learned from Section 3.3 that special numerical schemes like the TVD or ENO schemes are needed to preserve the boundedness of the C field. Such high-order accurate bounded schemes however still introduce numerical diffusion into the solution, which gradually smears the C field. That is, after solving this equation for certain number of time steps, we may observe many layers of cells with $0 < C < 1$ i.e. the interface grows thick. This contradicts our assumption that the interface should be always sharp. Moreover, we begin to lose track of the interface because the interface can be hidden anywhere in those layers. One way to circumvent this problem is to solve the pure convection equation geometrically.

Since what Equation (7.25) tells us is that both phases simply advect with the local fluid speed and bear their C values with them, we may directly calculate the amount of each phase getting into or out of a control volume.

For example, if we have a case 1 interface configuration in a cell and suppose the fluid velocity at its east surface is u_e and $u_e > 0$, clearly in Δt period the fluid right to the dashed line shown in Figure 7.8 will leave this cell and enter its east neighboring cell. This cell thus loses some volume (or area in 2-D) of phase 1 to its east neighboring cell. The amount of this volume exchange is

$$V_e = \begin{cases} 0 & \text{if } u_e\Delta t < (1 - r_s)\Delta x \\ 0.5\tan\beta[u_e\Delta t - (1 - r_s)\Delta x]^2 & \text{if } (1 - r_s)\Delta x \le u_e\Delta t < \Delta x \end{cases} \qquad (7.33)$$

Here we have to require that $u_e\Delta t < \Delta x$, otherwise we will run out of fluid in this cell after Δt. Similar considerations lead to the following restriction on the time step size Δt, which is summarized by using the Courant or CFL number (see Equation (3.75)):

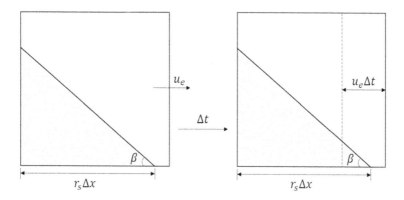

FIGURE 7.8   Volume transfer through a surface.

$$CFL = \max\left[\frac{\max(|u|)\Delta t}{\Delta x}, \frac{\max(|v|)\Delta t}{\Delta y}\right] < 1 \tag{7.34}$$

The C flux leaving this cell, say the (i, j) cell through its east surface, is therefore

$$F_e(i, j) = u_e C_e = u_e\left(\frac{V_e}{u_e \Delta t \Delta y}\right) = \frac{V_e}{\Delta t \Delta y} \tag{7.35}$$

Of course its east neighboring cell, the $(i + 1, j)$ cell, gains the same amount of flux via its west surface:

$$F_w(i + 1, j) = -F_e(i, j) \tag{7.36}$$

We add a negative sign to this flux because it is an inflow from the perspective of the $(i + 1, j)$ cell. So here is another point that deserves attention: we always evaluate such fluxes based on information provided by the upstream, or "donor" cell. The downstream, or "acceptor" cell, simply receives whatever amount of C the donor cell gives to it.

Now that a rectangular cell has four surfaces, we need to do the same type of computation four times. And each time the calculation should be based on the upstream donor cell. Notice that a cell cannot be solely a donor since it cannot create fluid to give to its neighbors. Similarly a cell also cannot be always an acceptor. The cell shown in Figure 7.9, for example, is a donor in terms of the fluxes across its north and east surfaces while an acceptor when we evaluate the fluxes through its west and south surfaces. The point is that donors and acceptors always appear in pairs. We will not miss counting any fluxes if we always update the fluxes of both donor and acceptor after we calculate every outgoing flux. For example, after we assess the flux across the north and east surfaces of the cell shown in Figure 7.9, we should not only keep a record of these values for this cell, we also save them (with adding a negative sign) as the fluxes across the south/west surfaces of its north/east neighboring cells.

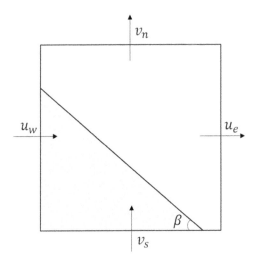

FIGURE 7.9 Bookkeeping of fluxes.

The formulas you can use to evaluate the volume transfer of phase 1 across surfaces of an interface-containing donor cell are summarized in Table 7.3 and Table 7.4.

After we obtain the fluxes across all four surfaces of a cell, we can evaluate the overall change in its C value according to Equation (7.32):

$$\frac{C_{i,j}^{m+1} - C_{i,j}^{m}}{\Delta t} = -\frac{F_e + F_w}{\Delta x} - \frac{F_n + F_s}{\Delta y} + C_{i,j}\left(\frac{u_e - u_w}{\Delta x} + \frac{v_n - v_s}{\Delta y}\right) \quad (7.37)$$

There are a few issues here.

First, we may double-count some fluxes. For example, if fluid flows out of a cell through its east and south surfaces, as depicted in Figure 7.10, we will count the fluid right to the vertical dashed line as F_e and that below the horizontal dashed line as F_s. But that means the fluid in the corner area marked by letter A is counted twice.

Second problem. In Equation (7.37) the $C_{i,j}$ in the last term does not have a superscript with it. So should we use $C_{i,j}^{m}$ or $C_{i,j}^{m+1}$ for this term?

Third problem. How should we reach second-order accuracy in time?

In fact these three problems can be solved simultaneously. As we have learned from previous chapters, second-order accuracy in time can be achieved if we use a fractional time step method and alternate explicit and implicit schemes in those fractional steps (see e.g. Section 4.2.3). To avoid double-counting fluxes, we can advect the C field horizontally in one fractional step and vertically in another fractional step, which in the mean time guarantees a second-order accuracy in time if we alternate the advection direction in each time step (Strang, 1968). These considerations result in the following procedure of advancing the C field.

TABLE 7.3

Formulas for Evaluating Volume Transfer through East and West Surfaces

Case	V_w (if $u_w < 0$)	V_e (if $u_e > 0$)
1	if $|u_w|\Delta t < r_s\Delta x$	if $u_e\Delta t < (1 - r_s)\Delta x$
	$|u_w|\Delta t\,(r_w\Delta y - 0.5|u_w|\Delta t \tan\beta)$	0
	else	else
	$C\Delta x\Delta y$	$0.5\tan\beta\,[u_e\Delta t - (1 - r_s)\Delta x]^2$
2	$|u_w|\Delta t\,(r_w\Delta y - 0.5|u_w|\Delta t \tan\beta)$	$u_e\Delta t\,(r_e\Delta y + 0.5u_e\Delta t \tan\beta)$
3	if $|u_w|\Delta t < r_n\Delta x$	if $u_e\Delta t < (1 - r_s)\Delta x$
	$|u_w|\Delta t\Delta y$	0
	elseif $|u_w|\Delta t < r_s\Delta x$	elseif $u_e\Delta t < (1 - r_n)\Delta x$
	$|u_w|\Delta t\Delta y - 0.5\tan\beta\,(|u_w|\Delta t - r_n\Delta x)^2$	$0.5\tan\beta\,[u_e\Delta t - (1 - r_s)\Delta x]^2$
	else	else
	$C\Delta x\Delta y$	$u_e\Delta t\Delta y - (1 - C)\Delta x\Delta y$
4	if $|u_w|\Delta t < r_n\Delta x$	if $u_e\Delta t < (1 - r_n)\Delta x$
	$|u_w|\Delta t\Delta y$	$u_e\Delta t\,(r_e\Delta y + 0.5u_e\Delta t \tan\beta)$
	else	else
	$|u_w|\Delta t\Delta y - 0.5\tan\beta\,(|u_w|\Delta t - r_n\Delta x)^2$	$u_e\Delta t\Delta y - (1 - C)\Delta x\Delta y$

TABLE 7.4

Formulas for Evaluating Volume Transfer through North and South Surfaces

Case	V_s (if $v_s < 0$)	V_n (if $v_n > 0$)
1	if $|v_s|\Delta t < r_w\Delta y$	if $v_n\Delta t < (1 - r_w)\Delta y$
	$|v_s|\Delta t\,(r_s\Delta x - 0.5|v_s|\Delta t \tan\beta)$	0
	else	else
	$C\Delta x\Delta y$	$0.5\tan\beta\,[v_n\Delta t - (1 - r_w)\Delta y]^2$
2	if $|v_s|\Delta t < r_e\Delta y$	if $v_n\Delta t < (1 - r_w)\Delta y$
	$|v_s|\Delta t\Delta x$	0
	elseif $|v_s|\Delta t < r_w\Delta y$	elseif $v_n\Delta t < (1 - r_e)\Delta y$
	$|v_s|\Delta t\Delta x - 0.5\tan\beta\,(|v_s|\Delta t - r_e\Delta y)^2$	$0.5\tan\beta\,[v_n\Delta t - (1 - r_w)\Delta y]^2$
	else	else
	$C\Delta x\Delta y$	$v_n\Delta t\Delta x - (1 - C)\Delta x\Delta y$
3	$|v_s|\Delta t\,(r_s\Delta x - 0.5|v_s|\Delta t \tan\beta)$	$v_n\Delta t\,(r_n\Delta x + 0.5v_n\Delta t \tan\beta)$
4	if $|v_s|\Delta t < r_e\Delta y$	if $v_n\Delta t < (1 - r_e)\Delta y$
	$|v_s|\Delta t\Delta x$	$v_n\Delta t\,(r_n\Delta x + 0.5v_n\Delta t \tan\beta)$
	else	else
	$|v_s|\Delta t\Delta x - 0.5\tan\beta\,(|v_s|\Delta t - r_e\Delta y)^2$	$v_n\Delta t\Delta x - (1 - C)\Delta x\Delta y$

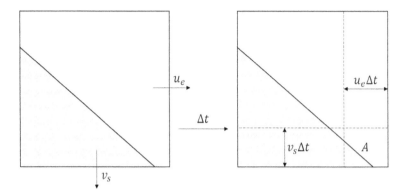

FIGURE 7.10 Possibility of double-counting fluxes.

First, evaluate F_e and F_w according to the C field at the mth moment and advect C horizontally.

$$\frac{C_{i,j}^* - C_{i,j}^m}{\Delta t} = -\frac{F_e^m + F_w^m}{\Delta x} + C_{i,j}^*\left(\frac{u_e - u_w}{\Delta x}\right) \tag{7.38}$$

and then reconstruct the interface as well as evaluating F_n and F_s based on the intermediate C^* field and then advance C to the new time moment:

$$\frac{C_{i,j}^{m+1} - C_{i,j}^*}{\Delta t} = -\frac{F_n^* + F_s^*}{\Delta y} + C_{i,j}^*\left(\frac{v_n - v_s}{\Delta y}\right) \tag{7.39}$$

Notice that we should alternate the advection direction at each time step. C is treated implicitly in the first fractional step and explicitly in the second (Puckett, Almgren, Bell, Marcus, & Rider, 1997).

The last subtlety involved in the interface reconstruction and advection procedure is that all formulas given in Table 7.1, Table 7.2, Table 7.3, and Table 7.4 are for the situation when both components of the interface normal vector $\hat{n}$ are negative (case 1) and special care is needed if the interface is oriented differently.

For example, if the interface segment in a cell looks like the one shown in the lower left panel of Figure 7.11, we can first flip it vertically then horizontally to turn it into the "standard" configuration. Then all the formulas can still be used but we need to use the negative u_w value of the real-world cell (the lower left one) in the place of u_e in such formulas as it is clear from this figure. Similar changes have to be made for the other surface velocities. Once the result is obtained, we have to flip backwards to the original real cell and assign the result to it properly. For example, suppose u_e of the real-world cell is positive, and we want to find F_e for this cell. We will calculate F_w of the "standard" cell by substituting u_w in the F_w formula with the $-u_e$ value of the real-world cell. After we find this flux, we will interpret it as the flux leaving the real cell via its east surface i.e. F_e.

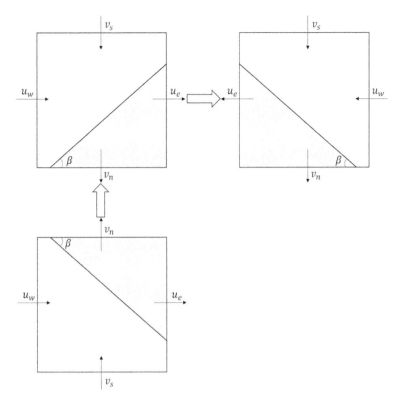

FIGURE 7.11   Process of transforming an interface configuration to the standard layout.

Let us solve one example.

7.2.4   EXAMPLE: INTERFACE TRANSPORTATION BY UNIFORM VELOCITY

The C field corresponding to a gas bubble immersed in a liquid is shown in Figure 7.12. The radius of the bubble is 0.3 and it is initially centered at (0.5, 0.5). The fluid velocity is uniform: $u = v = 1$. The mesh size is $\Delta x = \Delta y = 0.25$ and the units of all these parameters are consistent. The gas phase will be regarded as phase 1.

According to Equation (7.34), if we choose $CFL = 0.5$, the time step size is

$$\Delta t = CFL \cdot \min\left[\frac{\Delta x}{\max(u)}, \frac{\Delta y}{\max(v)} \right] = 0.5\frac{0.25}{1} = 0.125 \qquad (7.40)$$

We will focus on the $i = 2$, $j = 3$ cell (the cell whose west and north neighbors both have a $C = 0.092$) and see what happens in one time step.

First we reconstruct the interface for the cell we choose. According to Equations (7.7), (7.9), and (7.10) we find $\hat{n}_x = 0.7071$ and $\hat{n}_y = -0.7071$. Since $\hat{n}_x > 0$, we will

0	0.092	0.092	0
0.092	0.948	0.948	0.092
0.092	0.948	0.948	0.092
0	0.092	0.092	0

FIGURE 7.12 C field corresponding to a gas bubble in a liquid.

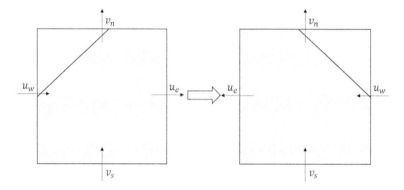

FIGURE 7.13   Transformation of the interface configuration to the standard layout.

flip the cell horizontally to turn it into the standard configuration as shown in Figure 7.13.

Then using equations in Table 7.1 and Table 7.2, we find that this standard interface arrangement belongs to case 4 and $r_n = r_e = 0.6778$. You do not really have to do anything special at this step because only the absolute values of $\hat{n}_x$ and $\hat{n}_y$ are needed in those formulas so far.

Then we calculate the horizontal fluxes across the surfaces of the chosen cell. It is a donor in terms of F_e because $u_e > 0$. This flux corresponds to F_w of its flipped image cell (see Figure 7.13). Using $-u_e = -1$ in place of u_w in equations of Table 7.3, you can find the gas volume moved out of this image cell across its west surface is $V_w = =|u_w|\Delta t \Delta y = 0.0313$. Therefore, $F_w = \frac{V_w}{\Delta t \Delta y} = 1$ for the flipped cell, which is F_e of the original cell. At this point, you should update the F_w value of its east neighboring cell, which should be -1. Similarly, F_w of the current (2,3) cell should be updated when we calculate F_e of its west neighboring (1,3) cell. We find that $F_w(2, 3) = -F_e(1, 3) = -0.1846$.

Using Equation (7.38) we can find the intermediate C^* value of the current cell which is

0	0.0087	0.1672	0.0087
0	0.5404	1	0.5404
0	0.5404	1	0.5404
0	0.0087	0.1672	0.0087

FIGURE 7.14 $C*$ field.

$$C_{2,3}^* = \frac{C_{2,3}^0 - \Delta t \frac{F_e^0(2,3) + F_w^0(2,3)}{\Delta x}}{\left[1 - \Delta t\left(\frac{u_e - u_w}{\Delta x}\right)\right]} = \frac{0.948 - 0.125 * \frac{1 - 0.1846}{0.25}}{1 - 0} = 0.5404 \quad (7.41)$$

Here the superscript 0 indicates the values based on the initial C distribution.

Similar calculations give the following $C*$ field, according to which we can construct the interface again (Figure 7.14).

Then we compute the vertical fluxes based on the reconstructed interface. F_n^* of the (2,3) cell is 0.3907 and its F_s^* is updated when we calculate F_n^* of its south neighboring cell, the (2,2) cell. The result is $F_s^*(2,3) = -F_n^*(2,2) = -0.6901$. And now we can find the C value of the selected cell at $t = \Delta t = 0.125$ moment:

$$C_{2,3}^1 = C_{2,3}^* - \Delta t \frac{F_n^*(2,3) + F_s^*(2,3)}{\Delta y} + C_{2,3}^* \Delta t\left(\frac{v_n - v_s}{\Delta y}\right)$$

$$= 0.5404 - 0.125 * \frac{0.3907 - 0.6901}{0.25} + 0 = 0.6901 \quad (7.42)$$

The whole C field at this moment is shown in Figure 7.15. In the next time step, one should first advect the C field vertically then horizontally.

The initial and final states of the interface are illustrated in the left panel of Figure 7.16. The interface is clearly being transported to the northeast as expected. The very rough circle shape is due to the very coarse mesh as well as the PLIC interface reconstruction procedure, which is only first-order accurate in terms of geometry reproduction precision. More accurate (but more complex) interface reconstruction methods can be found in literature e.g. Pilliod & Puckett, (2004). The result of the same example obtained with using PLIC on a 40 × 40 mesh is shown in the right panel of Figure 7.16. One advantage of the VOF method is that the volume of each phase is very well conserved, even with a very coarse mesh. For example with the 4 × 4 mesh, the area of the gas phase is always $(\sum C)\Delta x \Delta y = 0.2832$ m^2 as you can easily verify.

0	0.204	0.667	0.204
0	0.69	1	0.69
0	0.204	0.667	0.204
0	0	0	0

FIGURE 7.15   *C* field after one time step.

7.2.5   FLOW FIELD CALCULATION

Since the interface advects with the fluid flow, we need to solve for the flow field to complete the multiphase flow calculation. The velocity and pressure of a multiphase flow are governed by the Navier–Stokes equations as usual. If both phases are incompressible, we still have the familiar form of continuity equation

$$\nabla \cdot \vec{u} = \frac{\partial u}{\partial x} + \frac{\partial v}{\partial y} + \frac{\partial w}{\partial z} = 0 \qquad (7.43)$$

You might be a little concerned how the varying density across the phase interface disappears from the continuity equation. Consider a fluid particle, its mass is constant as long as its identity is preserved. Its volume is also constant although it may be made of more than one phase because each phase is incompressible. Therefore, the density of this particle never changes as it moves around, in other words the material derivative of its density is zero:

$$\frac{D\rho}{Dt} = \frac{\partial \rho}{\partial t} + \vec{u} \cdot \nabla \rho = \frac{\partial \rho}{\partial t} + u\frac{\partial \rho}{\partial x} + v\frac{\partial \rho}{\partial y} + w\frac{\partial \rho}{\partial z} = 0 \qquad (7.44)$$

In its most general form the continuity equation is

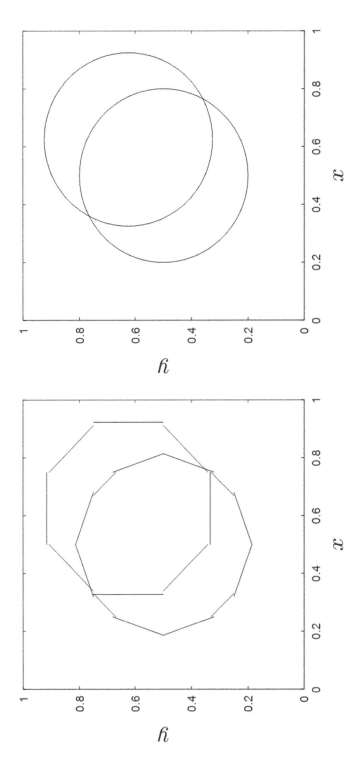

FIGURE 7.16 Initial and final states of the interface.

$$\frac{\partial \rho}{\partial t} + \frac{\partial (\rho u)}{\partial x} + \frac{\partial (\rho v)}{\partial y} + \frac{\partial (\rho w)}{\partial z} = 0 \qquad (7.45)$$

That is

$$\frac{\partial \rho}{\partial t} + u\frac{\partial \rho}{\partial x} + v\frac{\partial \rho}{\partial y} + w\frac{\partial \rho}{\partial z} + \rho\left(\frac{\partial u}{\partial x} + \frac{\partial v}{\partial y} + \frac{\partial w}{\partial z}\right) = 0 \qquad (7.46)$$

In order that both Equations (7.44) and (7.46) are satisfied, we must conclude that Equation (7.43) is still valid for an incompressible multiphase flow.

Due to the nonuniformity of viscosity, we have to use the general form of momentum equation, Equation (5.201) for multiphase flow simulations:

$$\frac{\partial (\rho u_i)}{\partial t} + \frac{\partial (\rho u u_i)}{\partial x} + \frac{\partial (\rho v u_i)}{\partial y} + \frac{\partial (\rho w u_i)}{\partial z}$$

$$= \left\{\frac{\partial}{\partial x}\left[\mu\left(\frac{\partial u_i}{\partial x} + \frac{\partial u}{\partial x_i}\right)\right] + \frac{\partial}{\partial y}\left[\mu\left(\frac{\partial u_i}{\partial y} + \frac{\partial v}{\partial x_i}\right)\right]\right.$$

$$\left. + \frac{\partial}{\partial z}\left[\mu\left(\frac{\partial u_i}{\partial z} + \frac{\partial w}{\partial x_i}\right)\right]\right\} - \frac{\partial p}{\partial x_i} + \rho g_i + f_i \qquad (7.47)$$

The left side of this equation can be written as

$$\frac{\partial (\rho u_i)}{\partial t} + \frac{\partial (\rho u u_i)}{\partial x} + \frac{\partial (\rho v u_i)}{\partial y} + \frac{\partial (\rho w u_i)}{\partial z}$$

$$= u_i\left[\frac{\partial \rho}{\partial t} + \frac{\partial (\rho u)}{\partial x} + \frac{\partial (\rho v)}{\partial y} + \frac{\partial (\rho w)}{\partial z}\right]$$

$$+ \rho\left(\frac{\partial u_i}{\partial t} + u\frac{\partial u_i}{\partial x} + v\frac{\partial u_i}{\partial y} + w\frac{\partial u_i}{\partial z}\right) \qquad (7.48)$$

$$= \rho\left(\frac{\partial u_i}{\partial t} + u\frac{\partial u_i}{\partial x} + v\frac{\partial u_i}{\partial y} + w\frac{\partial u_i}{\partial z}\right)$$

$$= \rho\left[\frac{\partial u_i}{\partial t} + \frac{\partial (u u_i)}{\partial x} + \frac{\partial (v u_i)}{\partial y} + \frac{\partial (w u_i)}{\partial z}\right]$$

with the help of Equation (7.45) and Equation (7.43). Equation (7.47) hence can be written as

$$\rho\left[\frac{\partial u_i}{\partial t} + \frac{\partial (u u_i)}{\partial x} + \frac{\partial (v u_i)}{\partial y} + \frac{\partial (w u_i)}{\partial z}\right]$$

$$= \left\{\frac{\partial}{\partial x}\left[\mu\left(\frac{\partial u_i}{\partial x} + \frac{\partial u}{\partial x_i}\right)\right] + \frac{\partial}{\partial y}\left[\mu\left(\frac{\partial u_i}{\partial y} + \frac{\partial v}{\partial x_i}\right)\right]\right.$$

$$\left. + \frac{\partial}{\partial z}\left[\mu\left(\frac{\partial u_i}{\partial z} + \frac{\partial w}{\partial x_i}\right)\right]\right\} - \frac{\partial p}{\partial x_i} + \rho g_i + f_i \qquad (7.49)$$

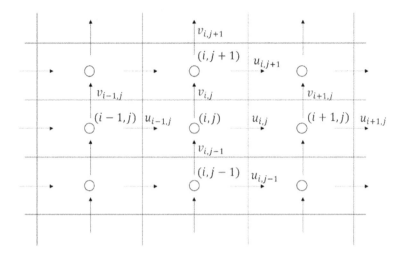

FIGURE 7.17   Staggered mesh used for multiphase flow calculations.

Equations (7.43) and (7.49) can be solved with methods introduced in Chapter 5. For example we may use the projection method (see Section 5.2.3) to solve these equations on a staggered mesh (see Section 5.1.2).

The staggered mesh is shown in Figure 7.17. All scalars including pressure, density, viscosity, marker function, etc. are stored at center of the pressure control volumes. Velocities are located at the pressure cell surfaces as usual.

Due to the presence of two phases, all these flow field variables may experience a sudden jump across the phase interface. To capture such sharp changes, one may solve the Navier–Stokes equations in each phase separately which requires boundary conditions to be applied at the interface for each phase. An alternative is to treat all phases as one single fluid with varying physical properties. For the cells that contain both phases, an overall density will be assigned to this fictitious fluid:

$$\rho = \frac{m}{V} = \frac{m_1 + m_2}{V} = \frac{\rho_1 V_1 + \rho_2 V_2}{V} = \rho_1 C + \rho_2 (1 - C) \qquad (7.50)$$

where the subscripts 1 and 2 denote phase 1 and phase 2, respectively.

The treatment of viscosity is analogous to what discussed in Section 2.4.2. A harmonic mean of the viscosities of two phases usually gives more accurate results than an algebraic mean. The reasoning is as follows.

Focus on an interface-containing control volume. Suppose the phase interface is horizontal as shown in Figure 7.18. The height of the phase 1 is then $C\Delta y$ where C is the volume fraction of phase 1. The shear stress acting on this control volume should be

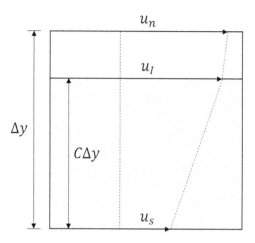

FIGURE 7.18 Sketch of the phase interface.

$$\tau_{yx} = \mu \frac{\partial u}{\partial y} \approx \mu \frac{u_n - u_s}{\Delta y} \tag{7.51}$$

where μ is the effective fluid viscosity in this control volume. This shear stress also can be evaluated with using the velocity profile of phase 2 in the cell alone:

$$\tau_{yx} = \mu_2 \left(\frac{\partial u}{\partial y} \right)_2 \approx \mu_2 \frac{u_n - u_I}{(1 - C)\Delta y} \tag{7.52}$$

Similarly we may calculate the shear stress from the perspective of phase 1

$$\tau_{yx} = \mu_1 \left(\frac{\partial u}{\partial y} \right)_1 \approx \mu_1 \frac{u_I - u_s}{C\Delta y} \tag{7.53}$$

Equating these expressions we have

$$\frac{u_n - u_I}{\frac{1-C}{\mu_2}\Delta y} = \frac{u_I - u_s}{\frac{C}{\mu_1}\Delta y} = \frac{u_n - u_s}{\left(\frac{1-C}{\mu_2} + \frac{C}{\mu_1} \right)\Delta y} = \frac{u_n - u_s}{\frac{1}{\mu}\Delta y} \tag{7.54}$$

which entails

$$\frac{1}{\mu} = \frac{1-C}{\mu_2} + \frac{C}{\mu_1} \rightarrow \mu = \frac{1}{\frac{1-C}{\mu_2} + \frac{C}{\mu_1}} \tag{7.55}$$

Since the fluid density and viscosity have now been determined, we are ready to solve the flow field with the projection method. Suppose we have already obtained the flow field at the mth moment. Let us see the steps to compute the flow field at the $(m + 1)^{th}$ moment.

We first discretize the momentum equations over the u and v control volumes without including the pressure gradient term. This step should be relatively straightforward by following the same procedure described in Chapter 5. The only notable new features are the prolonged diffusion terms and the space-dependent density and viscosity. Simple average can be used at spots where physical properties are not immediately known. For instance, at the east face of the (i, j) pressure control volume, the density is estimated as

$$\rho_e = \frac{\rho_{i,j} + \rho_{i+1,j}}{2} \tag{7.56}$$

The viscosity at the north face of the (i, j) control volume for u velocity, which is the northeastern corner of the (i, j) pressure control volume, can be evaluated as

$$\mu_{ne} = \frac{1}{2}\left(\frac{\mu_{i,j} + \mu_{i,j+1}}{2} + \frac{\mu_{i+1,j} + \mu_{i+1,j+1}}{2}\right) \tag{7.57}$$

The additional diffusion terms are easy to discretize.

To avoid possible unphysical results due to the property jump across phase interface, high-order bounded schemes introduced in Chapter 3 are recommended to approximate the convection terms.

From such discretized momentum equations an intermediate velocity field u^* and v^* can be solved.

Then we add the pressure derivatives back to the momentum equations

$$\frac{u^{m+1} - u^*}{\Delta t} = -\frac{1}{\rho^m}\frac{\partial p^{m+1}}{\partial x}, \quad \frac{v^{m+1} - v^*}{\Delta t} = -\frac{1}{\rho^m}\frac{\partial p^{m+1}}{\partial y} \tag{7.58}$$

where the superscripts m and $m + 1$ are the mth and $(m + 1)^{th}$ moments at which the variables are evaluated.

By requiring that the continuity equation is satisfied at the $(m + 1)^{th}$ moment, we end up with a Poisson equation for pressure

$$\frac{\partial}{\partial x}\left(\frac{1}{\rho^m}\frac{\partial p^{m+1}}{\partial x}\right) + \frac{\partial}{\partial y}\left(\frac{1}{\rho^m}\frac{\partial p^{m+1}}{\partial y}\right) = \frac{1}{\Delta t}\left(\frac{\partial u^*}{\partial x} + \frac{\partial v^*}{\partial y}\right) \tag{7.59}$$

from which the pressure field at the $(m + 1)^{th}$ moment can be solved. The velocity at the new time moment can be computed by using Equation (7.58).

Once we have the updated flow field, the phase interface can be advected as discussed in the last section and the C field as well as the physical properties then all can be advanced to the new time moment.

The following is the flowchart of the solution process (Figure 7.19).

Notice that since the diffusion terms are treated explicitly, the time step size has to satisfy the condition below to ensure stability (see Section 3.1.4):

$$\Delta t \leq \frac{\rho}{2\mu \left(\frac{1}{\Delta x^2} + \frac{1}{\Delta y^2} \right)} \tag{7.60}$$

This condition has to be combined with Equation (7.34), repeated below, to determine the proper time step size we can use.

$$CFL = \max \left[\frac{\max(|u|)\Delta t}{\Delta x}, \frac{\max(|v|)\Delta t}{\Delta y} \right] < 1 \tag{7.61}$$

If a TVD scheme is used for the convection terms, you should use Harten's lemma (see Section 3.3.8) to derive another CFL condition that preserves the TVD property.

Second-order accuracy in time can be realized by, for example, using the two-step Runge–Kutta scheme introduced in Section 3.1.7.

7.2.6 EXAMPLE: DAM-BREAK PROBLEM

The dam-break problem is a classical test case of multiphase codes. Liquid water is initially confined in a compartment formed by the vertical wall of a tank and a partition plate which simulates a dam. The setup is shown in Figure 7.20. The partition is then suddenly removed and water is allowed to flow out to mimic a dam-break scenario. We can calculate the flow of water and air in the tank and predict the position versus time of the leading edge of the water.

This phenomenon is mainly due to gravity, which causes water to accelerate. Therefore, we should have

$$\rho \left[\frac{\partial u}{\partial t} + \frac{\partial u^2}{\partial x} + \frac{\partial (vu)}{\partial y} \right] = \mathcal{O}(\rho g) \tag{7.62}$$

If we use the initial width L of the water column as the length scale, easily you can find the proper velocity scale should be $\sqrt{gL}$ and the proper time scale is $\sqrt{L/g}$. We therefore define the nondimensional time as

$$\tilde{t} = t \sqrt{\frac{2g}{L}} \tag{7.63}$$

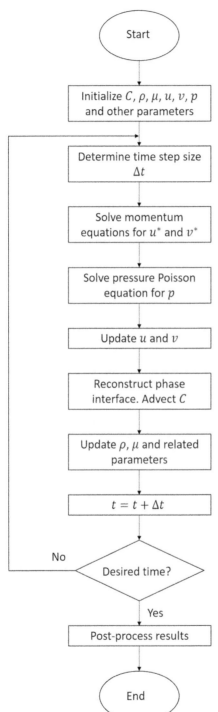

FIGURE 7.19 Flowchart of the VOF method solution process.

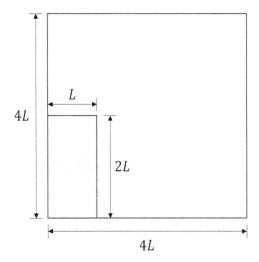

FIGURE 7.20 Dam-break problem example setup.

Based on the same reasoning, it is clear that this phenomenon should not be very sensitive to properties like density and viscosity, since none of them show up in the dimensionless velocity and time scales. Therefore, we can choose these parameters with some flexibility. In the present simulation we use the density and viscosity values of water and air at $25°C$. A slip condition is applied at the tank walls e.g. at the lower wall we set

$$\frac{\partial u}{\partial y} = 0, \quad v = 0 \qquad (7.64)$$

instead of our familiar no-slip condition. This practice is due to the relatively coarse mesh we use, which would not be able to resolve the large velocity gradient at walls if no-slip conditions were implemented.

What boundary condition should we use for C at the tank walls? Well, at the tank wall we have another phase (the solid wall), so the real physics of the three-phase interaction is very complex. Fortunately, such interactions are not important in this example since the flow behavior is mainly determined by gravity. For this reason, we may simply set $C = 1$ or $\nabla C \cdot \vec{n} = 0$ ($\vec{n}$ is the normal vector of the wall) at the boundaries.

The results using a 50×50 mesh are shown in Figure 7.21.

The history of waterfront position along the tank floor is shown in Figure 7.22, in which the experimental data (Hu & Sueyoshi, 2010) is compared with the current VOF solution.

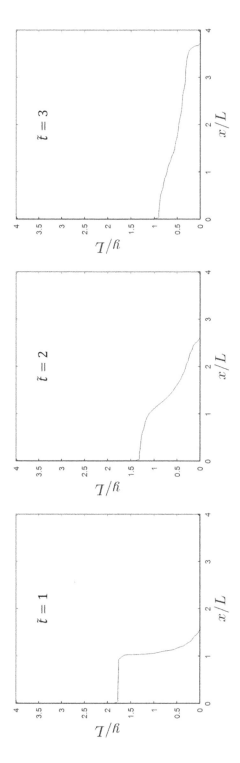

FIGURE 7.21 Evolution of the air–water interface.

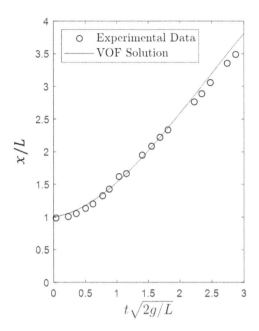

FIGURE 7.22   History of the waterfront.

7.2.7   SURFACE TENSION

Liquid molecules at an interface find themselves in a different environment from the other molecules in the bulk. In fact molecules at a liquid–gas interface have higher energy than those away from it. Since every system has a tendency to minimize its energy (more precisely put: every system minimizes its Helmholtz energy at constant volume and temperature, as thermodynamics requires), an interface always try to shrink to the least possible area. For example, a water drop usually assumes a spherical shape because among all shapes the sphere has the smallest surface area for a given volume. An interface thus behaves like an elastic skin. Every portion of an interface feels pulling or tensile forces perpendicular to its perimeter but tangent to the interface, as shown in Figure 7.23.

This force is the so-called surface tension force. The surface tension force per unit length of the perimeter of an interface segment is a physical property of the two fluids involved. For example, for the air–water interface, this property is about $\sigma \approx 0.072 \ N/m$ at 25 °C. The overall effect of the surface tension force along the interface perimeter is a net force pointing to the concave side of the interface segment, which equals (Brackbill, Kothe, & Zemach, 1992)

$$\vec{\delta F} = \sigma \kappa \hat{n} \delta A \tag{7.65}$$

where κ is the local interface curvature, $\hat{n}$ is the unit normal vector of the interface segment and δA is the area of the segment. The curvature is (Brackbill, Kothe, & Zemach, 1992)

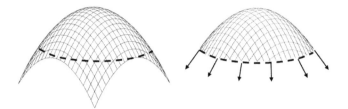

FIGURE 7.23 Surface tension.

$$\kappa = -\nabla \cdot \hat{n} = -\left(\frac{\partial \hat{n}_x}{\partial x} + \frac{\partial \hat{n}_y}{\partial y}\right) \tag{7.66}$$

Notice that no matter how we designate the direction of $\hat{n}$, the surface tension direction found from Equations (7.65) and (7.66) is always correct.

The surface tension force has to be added to the momentum equation if it has nontrivial influences on the transport phenomenon. We then have a small problem: the force terms in a momentum equation indeed are forces acting on per unit volume of fluid. The surface tension force, however, is proportional to the area of the interface, not to the volume of fluid. To circumvent this problem we may adopt the continuum surface force (CSF) model (Brackbill, Kothe, & Zemach, 1992) in which the surface tension is reformulated as a volume force

$$\delta\vec{F}_v = \sigma\kappa\nabla C\delta V \tag{7.67}$$

where C is the familiar volume fraction and δV is the (small) fluid volume we consider. It is very easy to show that the total surface tension force in a differential volume evaluated by the CSF model is close to the correct value:

$$\int_{\delta V} d\vec{F}_v = \int_{\delta V} \sigma\kappa\nabla C dV \approx \sigma\kappa \int_{\delta V} \nabla C dV = \sigma\kappa \int_{\delta A} C\hat{n} dA \approx \sigma\kappa\hat{n}\delta A \tag{7.68}$$

in virtue of Equation (7.4). Once we add the surface tension force per unit volume

$$\vec{f} = \sigma\kappa\nabla C \tag{7.69}$$

to the right side of the momentum equation

$$\begin{aligned}
\rho\left[\frac{\partial u_i}{\partial t} + \frac{\partial(uu_i)}{\partial x} + \frac{\partial(vu_i)}{\partial y} + \frac{\partial(wu_i)}{\partial z}\right] \\
= \left\{\frac{\partial}{\partial x}\left[\mu\left(\frac{\partial u_i}{\partial x} + \frac{\partial u}{\partial x_i}\right)\right] + \frac{\partial}{\partial y}\left[\mu\left(\frac{\partial u_i}{\partial y} + \frac{\partial v}{\partial x_i}\right)\right]\right. \\
+ \left.\frac{\partial}{\partial z}\left[\mu\left(\frac{\partial u_i}{\partial z} + \frac{\partial w}{\partial x_i}\right)\right]\right\} - \frac{\partial p}{\partial x_i} + \rho g_i + f_i
\end{aligned} \tag{7.70}$$

we can simulate multiphase flows with surface tension effects. One should have no difficulty to calculate $\vec{n} = \nabla C$ in the surface tension formula. Then the interface curvature can be found by using

$$\kappa = -\nabla \cdot \hat{n} = \frac{1}{|\vec{n}|} \left[\frac{1}{|\vec{n}|} \left(\vec{n}_x \frac{\partial |\vec{n}|}{\partial x} + \vec{n}_y \frac{\partial |\vec{n}|}{\partial y} \right) - \left(\frac{\partial \vec{n}_x}{\partial x} + \frac{\partial \vec{n}_y}{\partial y} \right) \right] \qquad (7.71)$$

The way to evaluate the derivatives in this formula is similar to what was used to compute ∇C, Equations (7.9) and (7.10).

Due to the inclusion of the surface tension into the momentum equation, another time step size restriction should apply (Brackbill, Kothe, & Zemach, 1992):

$$\Delta t \leq \frac{1}{2} \sqrt{\frac{(\rho_1 + \rho_2)\Delta x^3}{\pi \sigma}} \qquad (7.72)$$

7.2.8   EXAMPLE: EXCESS PRESSURE IN A WATER DROP

An example that surface tension plays a pivotal role is the excess pressure in a liquid drop. As mentioned, the surface tension drives a liquid drop to contract, which results in a higher-than-environment pressure inside the drop. For a 2-D drop (a cylindrical drop), the excess pressure can be calculated by the Laplace formula

$$\Delta p = p_l - p_g = \frac{\sigma}{R} \qquad (7.73)$$

We will directly solve Navier–Stokes equations to find the pressure within a water drop suspending in air. For the current simulation, the computational domain is 0.08 m in both x- and y-directions. A circular water drop of radius $R = 0.02$ m is centered at the domain, which is otherwise occupied by air. The surface tension coefficient is $\sigma = 0.072$ N/m. Gravity is absent in this example. A 30 × 30 mesh is employed. The calculated pressure distribution inside the drop at $t = 1$ s is shown in Figure 7.24.

The result is obviously not good as the excess pressure should be uniform in the drop but the predicted distribution is spiky at certain points; the excess pressure should be $\sigma/R = 3.6$ Pa yet the calculated pressure at the drop center is only about 2.67 Pa. The reason for such a unsatisfactory solution is mainly due to the poor evaluation of interface curvature. The C field has a step change at the interface, which is not friendly to finite difference evaluation of its derivatives. The situation is even worse when evaluating the second derivatives of C. This is exactly what happens when we estimate the interface curvature. Mesh refinement does not fix this problem because a finer mesh indeed makes the jump in C even sharper. The erroneous curvature estimate gives rise to incorrect and spiky surface tension force, which not only causes the unsatisfactory pressure distribution but also results in

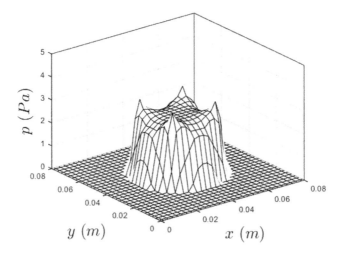

FIGURE 7.24 Pressure distribution in the water drop.

significant spurious currents close to the interface, eventually leading to unphysical interface deformation and breakup.

This issue can be mitigated by utilizing a smoothed C field to perform relevant calculations. We will smear or diffuse the C field in several layers of control volumes around the interface so that we have a smoothed $\tilde{C}$ field instead of a sharp step C function across the interface (see Figure 7.25).

A simple and widely used way to realize the smoothing is weighted averaging neighboring C values. For example, we may average C over the adjacent 3×3 mesh to obtain $\tilde{C}$:

$$\tilde{C}_{i,j} = \frac{1}{4}C_{i,j} + \frac{1}{8}(C_{i+1,j} + C_{i-1,j} + C_{i,j+1} + C_{i,j-1})$$
$$+ \frac{1}{16}(C_{i+1,j+1} + C_{i+1,j-1} + C_{i-1,j+1} + C_{i-1,j-1})$$

$$(7.74)$$

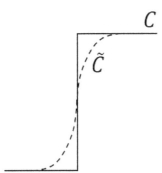

FIGURE 7.25 Step and smoothed C profiles across the interface.

Another method to diffuse the C field is solving a pseudo unsteady diffusion equation (Guillaument, Vincent, & Caltagirone, 2015):

$$\frac{\partial \tilde{C}}{\partial \tau} - \Gamma \nabla^2 \tilde{C} = \frac{\partial \tilde{C}}{\partial \tau} - \Gamma \left(\frac{\partial^2 \tilde{C}}{\partial x^2} + \frac{\partial^2 \tilde{C}}{\partial y^2} \right) = 0 \qquad (7.75)$$

which is subject to an "initial condition"

$$\tilde{C} = C \text{ at } \tau = 0 \qquad (7.76)$$

Notice that here τ is not the physical time t but an artificial time-like variable. As "time" τ increases, the nonuniformity in the C field is gradually smoothed out due to the diffusion term. Well, there is only one nonuniformity in the C field: the jump across interface. Therefore, the longer we advance Equation (7.75) in time, the wider smoothed $\tilde{C}$ field we will obtain on both sides of the interface. In fact we can control the thickness of this diffused $\tilde{C}$ zone by adjusting τ and the (artificial) diffusion coefficient Γ. A simple order-of-magnitude analysis of Equation (7.75) shows the diffused-zone thickness

$$L \sim \sqrt{\Gamma \tau} \qquad (7.77)$$

For example, if we want to have a smooth $\tilde{C}$ field over one grid cell distance on each side of the interface, we may choose $\Gamma = \Delta x \Delta y$ and advance the solution of Equation (7.75) until $\tau = 1$. The original C field and the smoothed $\tilde{C}$ field are shown in Figure 7.26. It is not surprising that the interface curvature κ evaluated from such a smooth $\tilde{C}$ field is much better than the one evaluated from the step C function. Notice that we still use ∇C instead of $\nabla \tilde{C}$ in Equation (7.69).

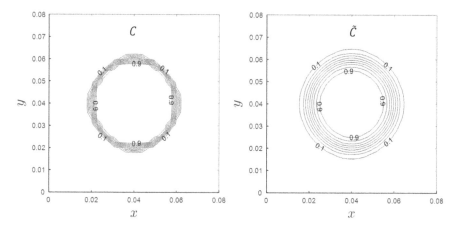

FIGURE 7.26 Original and smoothed C fields of the water drop.

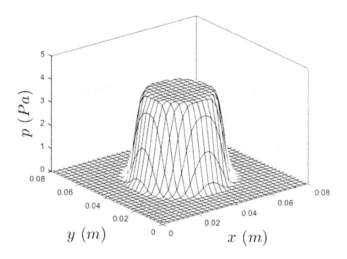

FIGURE 7.27   Pressure distribution in the water drop with the smoothed C field.

Now you probably begin to appreciate this point: differential equations are not just something we can (or have to) solve with numerical methods; differential equations *are* numerical methods.

The excess pressure distribution in the water drop by using this method is shown in Figure 7.27. Now the pressure distribution is uniform and within 5% of the exact solution.

We can see that the step C function adopted in the VOF method causes many problems. One might wonder why we do not choose a smooth marker function like the one shown on the right panel of Figure 7.26 in the first place. This is the basic idea behind the level-set method.

7.3   LEVEL-SET METHOD

7.3.1   INTERFACE REPRESENTATION

In the level-set method, the interface is described by a set of points where a continuous marker function takes a particular value. For example, a circular interface centered at the origin with radius R can be represented by the set of points (x, y) at which the marker function

$$\phi(x, y) = \sqrt{x^2 + y^2} - R \qquad (7.78)$$

equals zero. The 3-D surface and 2-D contours of this function are shown in Figure 7.28.

The level-set function is not unique. For example, we also can use the zero level of $f(x, y) = x^2 + y^2 - R^2$ to represent the same circular interface. However, function (7.78) has a distinct advantage: it is a signed distance function. Its value is

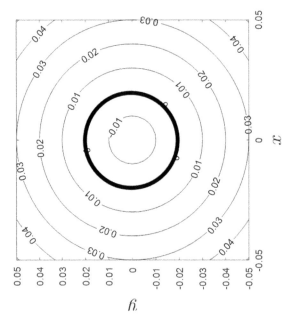

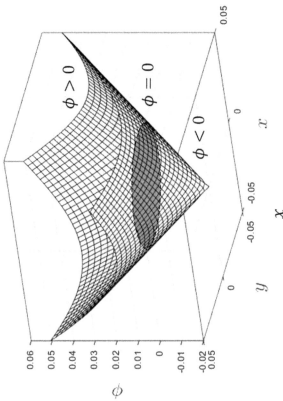

FIGURE 7.28 Level-set function.

the least distance between a point (x, y) and the interface represented by the function. And its sign tells us on which side of the interface the point is located. This feature allows us to compute many quantities very easily. For instance, if the interface is maintained at a constant temperature T_I, and I want to estimate the temperature gradient magnitude at a point (x, y) close to the interface, I may simply use

$$|\nabla T(x, y)| \approx \left| \frac{T(x, y) - T_I}{\phi(x, y)} \right| \tag{7.79}$$

For the same reason you can infer an important feature of the signed distance function

$$|\nabla \phi| = 1 \tag{7.80}$$

From now on we always use the zero level of a signed distance function to represent the interface.

7.3.2 INTERFACE ADVECTION

Since the interface is now associated to one level of a continuous level-set function ϕ, and because the interface always advects with the local flow velocity $\vec{u}$, it is natural to solve an unsteady convection equation of ϕ to track the interface:

$$\frac{\partial \phi}{\partial t} + \vec{u} \cdot \nabla \phi = 0 \tag{7.81}$$

Unlike the volume fraction function C in the VOF method, the level-set function ϕ is continuous, which is amiable to finite difference treatment. Of course we still have to choose a stable scheme with high-order accuracy to solve such a pure convection equation. Some popular options include second-order ENO scheme, the TVD schemes, and the fifth-order WENO scheme. A high-order scheme in time is also needed.

Let us use the second-order ENO scheme (see Section 3.3.7) and a second order Runge–Kutta scheme (see Section 3.1.7) to solve the following example.

An interface is initially a circle of radius 0.2π and centered at $(0.5\pi, 0.75)$ in a domain which spans from 0 to π in both x- and y-directions. The velocity field is $u = -\sin x \cos y$, $v = \cos x \sin y$, which satisfies the continuity equation of an incompressible flow. We use a 40×40 mesh, $\Delta t = 0.0025$ to calculate the evolution of the interface until $t = 2.5$.

The initial and final status of interface is shown in the left panel of Figure 7.29.

Although this result seems good, a few problems do exist. First, the volume enclosed by the interface should be a constant since the fluids are incompressible, the calculated volume however drifts away from its initial value gradually. For this example, the volume within the interface decreases for about 0.85% during the

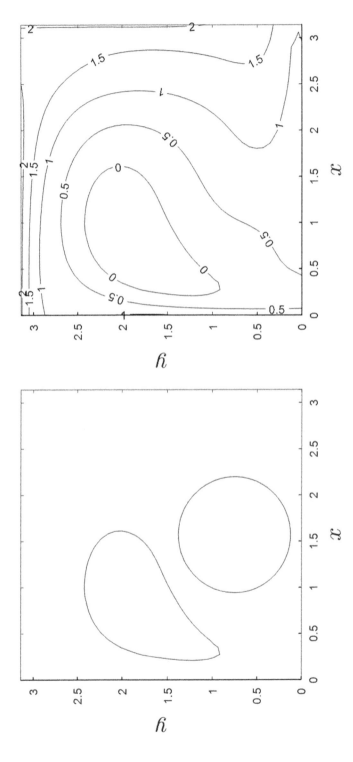

FIGURE 7.29 Interface advection by a flow field.

simulated time period. The second problem is that the ϕ values stop to be signed distances from the interface after advection. This is obvious from the calculated ϕ contours at the end of the advection process, as shown in the right graph of Figure 7.29. The points along the same contour, say the $\phi = 0.5$ contour, now have different distances from the interface.

The reason for the varying-volume problem is due to the unavoidable numerical diffusion involved in finite difference formulas which smooths out the ϕ distribution.

The reason for the second problem is because, unless velocity is uniform, ϕ contours are always distorted by the flow field over time. As a result ϕ no longer reflects the distance from the interface. And consequently $|\nabla\phi|$ is no longer unity everywhere.

7.3.3 REINITIALIZATION OF LEVEL-SET FUNCTION

To restore or reinitialize ϕ back to a signed distance function, one can explicitly calculate the distance of each point in the flow field from the closest interface and use it as the ϕ value at that point (Sun & Tao, 2010). Another way to reinstate the level-set function is solving a pseudo unsteady convection equation (Sussman, Smereka, & Osher, 1994)

$$\frac{\partial\phi}{\partial\tau} = S(\phi_0)(1 - |\nabla\phi|) \tag{7.82}$$

subject to the "initial condition"

$$\phi = \phi_0 \quad at \quad \tau = 0 \tag{7.83}$$

where ϕ_0 is the ϕfield to be reinitialized. Here τ is not the physical time but a time-like variable. $S(\phi_0)$ is the sign function:

$$S(\phi_0) = \begin{cases} 1 & if \ \phi_0 > 0 \\ 0 & if \ \phi_0 = 0 \\ -1 & if \ \phi_0 < 0 \end{cases} \tag{7.84}$$

As you can see, as long as $|\nabla\phi| \neq 1$, the right side of Equation (7.82) is not zero and ϕ will alter with τ until $|\nabla\phi|$ becomes unity again at the "steady state." Although it does not look like one, this equation is indeed a convection equation. The reason is because

$$|\nabla\phi| = \frac{|\nabla\phi|^2}{|\nabla\phi|} = \frac{\nabla\phi\cdot\nabla\phi}{|\nabla\phi|} = \frac{\nabla\phi}{|\nabla\phi|}\cdot\nabla\phi = \hat{n}\cdot\nabla\phi \tag{7.85}$$

in which $\hat{n}$ is the unit normal to the ϕ contours. Equation (7.82) can be written as

$$\frac{\partial \phi}{\partial \tau} + [S(\phi_0)\hat{n}] \cdot \nabla \phi = S(\phi_0) \tag{7.86}$$

which is an unsteady convection equation with a "velocity" $S(\phi_0)\hat{n}$. This fictitious velocity is of unit magnitude normal to the ϕ contours. The role of $S(\phi_0)$ is to advect contours in the correct direction no matter where they are: they will be carried away from the interface if they are too "crowded" than they should i.e. $|\nabla \phi| > 1$; they will be drawn closer to the interface if they are too "coarse" than they should i.e. $|\nabla \phi| < 1$.

Like any unsteady convection equations, Equation (7.86) can be solved with a high-order bounded scheme. In order not to disturb the interface and aggravating the mass conservation problem, we may "freeze" the interface by only reinitializing ϕ in those cells that do not contain interface. One can apply this procedure in every time step after ϕ is advected. Less frequent reinitialization e.g. every 10 time steps, is also common.

The ϕ contours for the present example with applying reinitialization at the end of every time step is shown in the left image of Figure 7.30.

A simple way to conserve the mass (or volume) is solving another differential equation at the end of the reinitialization process:

$$\frac{\partial \phi}{\partial \tau} = V(\tau) - V_0 \tag{7.87}$$

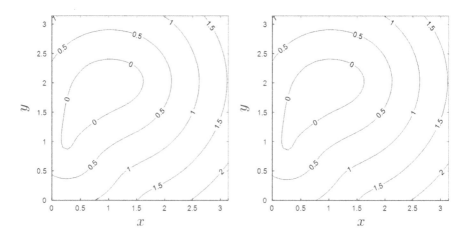

FIGURE 7.30 ϕ contours.

in which τ is again a pseudo time, $V(\tau)$ is the volume of a certain phase, say phase 1 at τ moment, and V_0 is the target volume (usually the initial volume if there is no phase change) of that phase. Obviously the "steady state" is reached when constant volume is preserved. The ϕ contours for the current example using this technique are shown in the right graphic of Figure 7.30. It can be verified that the volume is now conserved.

As I have mentioned at the very beginning of this chapter, a multiphase flow solution method usually has many variations. This is the case for the reinitialization procedures as well. Numerous methods have been and are still being developed to improve mass conservation during reinitialization. Some notable works include the sub-cell fix method (Russo & Smereka, 2000) and the constrained reinitialization method (Hartmann, Meinke, & Schröder, 2008).

7.3.4 FLOW FIELD CALCULATION

The last piece of puzzle in the level-set method is the calculation of flow field. Like in the VOF method, we may treat all phases as one single fluid with varying density and viscosity. To avoid numerical difficulties related to the sharp change of physical properties across interface, we will relax such abrupt transitions with the help of a smoothed step function

$$H(\phi) = \begin{cases} 0 & \text{if } \phi < -\epsilon \\ \frac{1}{2}\left[\left(1 + \frac{\phi}{\epsilon}\right) + \frac{1}{\pi}\sin\left(\pi\frac{\phi}{\epsilon}\right)\right] & \text{if } |\phi| \le \epsilon \\ 1 & \text{if } \phi > \epsilon \end{cases} \tag{7.88}$$

where ϵ is the half thickness of the diffused transition region. Typically, we set $\epsilon = 1.5\Delta x$ where Δx is the mesh size. H function is shown in the left graph of Figure 7.31. Obviously H behaves just like our familiar C function in the VOF method.

The fluid density is then given as

$$\rho = \rho_1 H + \rho_2(1 - H) \tag{7.89}$$

where ρ_1 and ρ_2 are the densities of phases located on the two sides of the interface corresponding to positive and negative ϕ values, respectively. The fluid viscosity can be estimated as

$$\mu = \frac{1}{\dfrac{H}{\mu_1} + \dfrac{1-H}{\mu_2}} \tag{7.90}$$

with the same reasoning that led to Equation (7.55). Moreover, we may estimate the total volume of a certain phase, say the one corresponding to negative ϕ values with

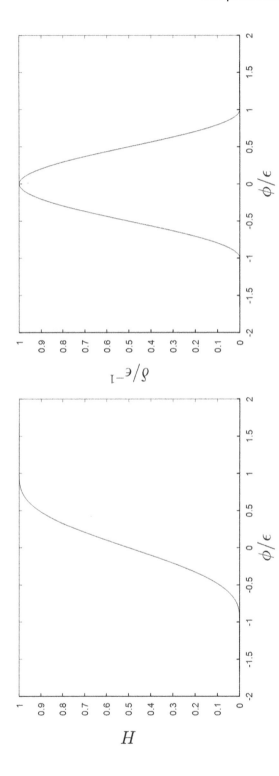

FIGURE 7.31 H and δ functions.

$$V = \sum_{i,j} (1 - H_{i,j})\Delta V_{i,j} \tag{7.91}$$

where $\Delta V_{i,j}$ is the volume of the (i, j) cell.

The surface tension force can be evaluated again with CSF model (Brackbill, Kothe, & Zemach, 1992 which states that surface tension force per unit volume of fluid is

$$\vec{f} = \sigma\kappa\nabla C \tag{7.92}$$

where C is a step function that jumps from 0 to 1 across the interface. Since we use H to approximate the step C function, it is natural to replace ∇C with ∇H:

$$\vec{f} \approx \sigma\kappa\nabla H = \sigma\kappa\frac{dH}{d\phi}\nabla\phi = \sigma\kappa\delta\nabla\phi \tag{7.93}$$

in which δ is

$$\delta(\phi) = \frac{dH(\phi)}{d\phi} = \begin{cases} \frac{1}{2\epsilon}\left[1 + \cos\left(\frac{\pi\phi}{\epsilon}\right)\right] & \text{if } |\phi| < \epsilon \\ 0 & \text{otherwise} \end{cases} \tag{7.94}$$

This function is shown in the right panel of Figure 7.31. It is a smoothed Dirac delta function.

The interface curvature κ can be evaluated fairly easily since

$$\hat{n} = \frac{\nabla\phi}{|\nabla\phi|} \tag{7.95}$$

therefore

$$\kappa = -\nabla\cdot\hat{n} = -\nabla\cdot\left(\frac{\nabla\phi}{|\nabla\phi|}\right) = -\frac{\phi_{xx}\phi_y^2 + \phi_{yy}\phi_x^2 - 2\phi_{xy}\phi_x\phi_y}{(\phi_x^2 + \phi_y^2)^{\frac{3}{2}}} \tag{7.96}$$

in which ϕ_x, etc. are derivatives of ϕ with respect to variable(s) in the subscripts. You can use central differences to approximate such derivatives.

Due to the smooth nature of the ϕ function, the interface normal as well as curvature calculated from the level-set method is usually more accurate than using the VOF method.

The flowchart of solving multiphase flows with level-set method and projection method is shown in Figure 7.32, which is almost the same as that of VOF method.

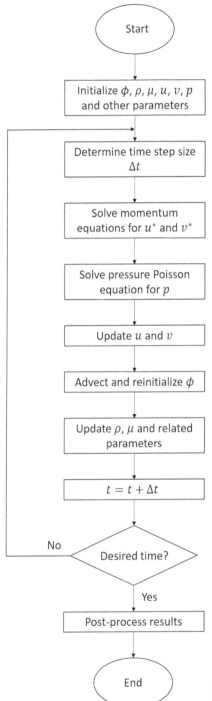

FIGURE 7.32 Flowchart of the level-set method solution process.

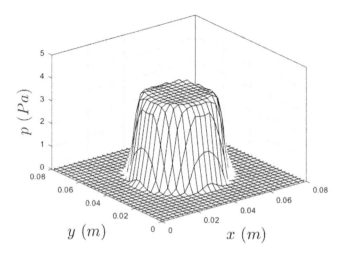

FIGURE 7.33   Pressure distribution in the water drop with the level-set method.

The excess pressure distribution in a circular water drop example (see Section 7.2.8) is recalculated by using the level-set method. We still use a 30 × 30 uniform mesh and the result is shown in Figure 7.33. The distribution is quite uniform and the excess pressure solved is within 1% of the analytical value.

The dam-break problem is also revisited using the level-set method. At the boundary, we may set the virtual node ϕ values, say ϕ_1 as follows

$$\phi_1 = \phi_2 + sign(\phi_2)|\phi_2 - \phi_3| \tag{7.97}$$

In this way the virtual node ϕ values always have correct signs.

The interface shapes and locations at three different moments are shown in Figure 7.34.

The water front along the tank floor versus time is shown in Figure 7.35.

7.3.5   Example: Gas Bubble Rising Problem

In this example a single gas bubble rises in a quiescent viscous liquid. Such a process is crucial in numerous industrial applications e.g. the rise of steam bubbles in water boilers. The bubble deforms at the same time as it moves due to the combined effects of viscous fluid flow, surface tension, and buoyancy force. The variation of bubble shape, in turn, affects the fluid flow and surface tension distribution. It is therefore an involved question—does a terminal steady state of bubble shape and speed exist? Intensive analysis (Grace, 1973;Bhaga & Weber, 1981) of experimental data indicated that the physics of the ascending gas bubble is determined by four dimensionless parameters, which are: the density ratio ρ_l/ρ_g, the

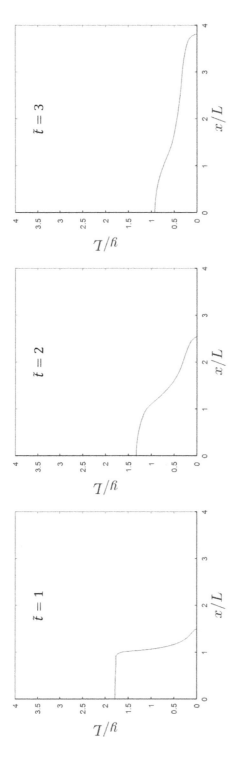

FIGURE 7.34 Evolution of the air–water interface with the level-set method.

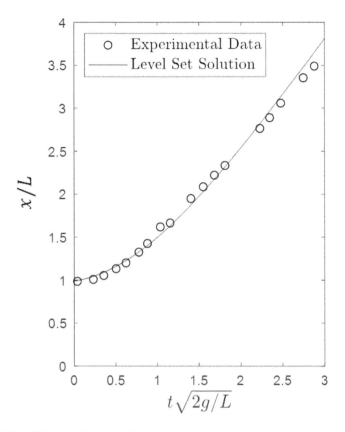

FIGURE 7.35   History of the waterfront with the level-set method.

viscosity ratio μ_l/μ_g, the Eötvös number, and the Morton number. The Eötvös number is defined as

$$Eo = \frac{\rho_l g d_e^2}{\sigma} \tag{7.98}$$

which is the ratio of the buoyancy force to the surface tension force. d_e is the equivalent diameter of the bubble. The Morton number is

$$Mo = \frac{g \mu_l^4}{\rho_l \sigma^3} \tag{7.99}$$

In fact, if we define

$$\tilde{x} = \frac{x}{d_e}, \; \tilde{y} = \frac{y}{d_e}, \; \tilde{u} = \frac{u}{\sqrt{gd_e}}, \; \tilde{v} = \frac{v}{\sqrt{gd_e}}, \; \tilde{t} = \frac{t}{\sqrt{d_e/g}}, \; \tilde{\rho} = \frac{\rho}{\rho_l}, \; \tilde{\mu} = \frac{\mu}{\mu_l}, \; \tilde{p}$$
$$= \frac{p}{\rho_l g d_e}, \; \tilde{\kappa} = \frac{\kappa}{d_e^{-1}}, \; \tilde{g}_x = \frac{g_x}{g}, \; \tilde{g}_y = \frac{g_y}{g} \tag{7.100}$$

The 2-D Navier–Stokes equations can be rendered in a dimensionless form involving these numbers:

$$\tilde{\rho}\left[\frac{\partial \tilde{u}_i}{\partial \tilde{t}} + \frac{\partial(\tilde{u}\tilde{u}_i)}{\partial \tilde{x}} + \frac{\partial(\tilde{v}\tilde{u}_i)}{\partial \tilde{y}}\right]$$
$$= \left(\frac{Mo}{Eo^3}\right)^{\frac{1}{4}}\left\{\frac{\partial}{\partial \tilde{x}}\left[\tilde{\mu}\left(\frac{\partial \tilde{u}_i}{\partial \tilde{x}} + \frac{\partial \tilde{u}}{\partial \tilde{x}_i}\right)\right] + \frac{\partial}{\partial \tilde{y}}\left[\tilde{\mu}\left(\frac{\partial \tilde{u}_i}{\partial \tilde{y}} + \frac{\partial \tilde{v}}{\partial \tilde{x}_i}\right)\right]\right\} \tag{7.101}$$
$$- \frac{\partial \tilde{p}}{\partial \tilde{x}_i} + \tilde{\rho}\tilde{g}_i + \frac{1}{Eo}\tilde{\kappa}\frac{\partial H}{\partial \tilde{x}_i}$$

in which $\tilde{\rho}$ is either 1 or ρ_g/ρ_l, and $\tilde{\mu}$ is either 1 or μ_g/μ_l depends on which phase it is.

In the current example, we simulate a case with $\rho_l/\rho_g = \mu_l/\mu_g = 1000$, $Eo = 10$, and $Mo = 0.1$. The gas bubble is initially a circle with diameter d_e and the computational domain size is $5d_e \times 15d_e$. A 50×150 uniform mesh is used and the free-slip condition is applied to all boundaries. A level-set-projection method code is used for this simulation.

The bubble shapes at eight different moments ($\tilde{t} = 0$, 1.8314, 3.6628, 5.4942, 7.3256, 9.157, 10.9884, and 12.8198) are shown in Figure 7.36. The bubble evolves to an oblate ellipsoidal shape eventually. The bubble terminal rising speed u_∞ is made dimensionless by defining a Reynolds number

$$Re = \frac{\rho_l u_\infty d_e}{\mu_l} \tag{7.102}$$

The Reynolds number obtained from the current simulation is about 4.8. The ultimate bubble shape agrees with experimental observation (Bhaga & Weber, 1981) yet the predicted Reynolds number might suffer quite significant error (~15%)

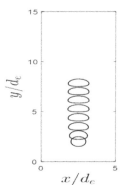

FIGURE 7.36 Bubble shapes and locations at eight different time moments.

compared with the experimental value. This might be partially attributed to the lack of 3-D flow dynamics in the current 2-D simulation.

7.4 MULTIPHASE FLOW WITH PHASE CHANGE

7.4.1 INTRODUCTION

As phase change occurs, the mass of each phase is no longer conserved. For example, if a liquid evaporates and joins the vapor (gas) phase, mass transfers from the liquid to the gas phase, as shown in Figure 7.37. According to the energy conservation principle and Fourier's law, we have

$$\dot{m} = \frac{k_l \frac{\partial T_l}{\partial n}|_I - k_g \frac{\partial T_g}{\partial n}|_I}{h_{lg}} \tag{7.103}$$

where $\dot{m}$ is the time rate of mass transfer across per unit interface area due to phase change. k is the thermal conductivity and $\partial T/\partial n = \hat{n}\cdot\nabla T$ is the temperature gradient (the subscript "I" denotes that the temperature gradients are evaluated at the interface). h_{lg} is the latent heat of vaporization.

According to fluid mechanics, the time rate of mass flow across per unit area is

$$\dot{m} = \rho_l(\vec{u_I} - \vec{u}_{l,I})\cdot\hat{n} = \rho_g(\vec{u_I} - \vec{u}_{g,I})\cdot\hat{n} \tag{7.104}$$

where $\vec{u_I}$ is the interface velocity, while $\vec{u}_{l,I}$ and $\vec{u}_{g,I}$ are the velocities of the liquid and gas phases at the interface, respectively. Therefore, there is a difference between the interface velocity and the velocity of each phase; there is also a difference between the velocities of two phases. The situation is sketched in Figure 7.38. There is a pressure jump across the interface too. Although temperature is still continuous and does not jump across the interface, it usually experiences a very quick change close to the interface.

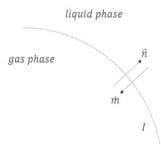

FIGURE 7.37 Sketch of mass transfer across the phase interface during phase change.

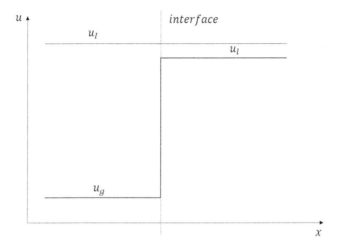

FIGURE 7.38 Velocity profiles across the phase interface.

To simulate a flow with phase change accurately, one must be able to predict the mass transfer $\dot{m}$ and interfacial velocity $\overrightarrow{u_i}$ precisely. The former relies on the exact evaluation of temperature gradient of each phase in the immediate vicinity of the interface and the latter depends on the correct calculation of fluid velocity values close to the interface since

$$\overrightarrow{u_i} \cdot \hat{n} = \overrightarrow{u_{l,I}} \cdot \hat{n} + \frac{\dot{m}}{\rho_l} = \overrightarrow{u_{g,I}} \cdot \hat{n} + \frac{\dot{m}}{\rho_g} \tag{7.105}$$

These requirements, in short, mean that the drastic transitions of temperature and velocity at the interface should be preserved as they are. But a sharp change of a variable, no matter the variable is a physical property like density or viscosity, or it is a flow property like velocity and temperature, usually causes some numerical difficulties. For this reason we have applied "diffusion" to such variable jumps in the past. This practice, however, compromises the variable profiles close to the interface and results in inaccurate $\dot{m}$ and $\overrightarrow{u_i}$ evaluations.

To resolve this dilemma, one may carry out the calculation of different phases separately and apply boundary conditions at the interface for each phase. One such solving strategies is the ghost fluid method (Fedkiw, Aslam, Merriman, & Osher, 1999), which uses the cells on the other side of the interface as ghost cells for the fluid on this side. These ghost cells are filled with the same fluid as the one under simulation, hence the name ghost fluid method. The variables in the ghost cells are determined by the jump conditions and based on the solution of the other phase. For example, if I am now calculating the liquid velocity, I will use the cells next to the interface but on the gas phase side as ghost cells and set the "liquid" velocity in such cells to the real gas velocity value at those places plus the velocity jump (see Figure 7.38).

Here I will use a similar approach to solve the temperature field, but I will still employ the one-fluid methodology for the velocity and pressure calculations.

When phase change like evaporation happens, the interface is not only a partition between phases, it is also the location where phase transition takes place. As we know from thermodynamics, under a given pressure the phase change of a pure substance occurs at a fixed saturation temperature. For this reason, we know that the temperature right at the interface must be the saturation temperature corresponding to the fluid pressure at the interface

$$T_I = T_{sat}(p_I) \tag{7.106}$$

If the fluid pressure does not vary too much along the interface, the interface temperature can be treated as a constant. And this is the boundary condition that has to be applied to the interface.

7.4.2 TEMPERATURE FIELD COMPUTATION

The temperature field in the ith phase is governed by the energy equation

$$\frac{\partial T_i}{\partial t} + \vec{u}_i \cdot \nabla T_i = \alpha_i \nabla^2 T_i \tag{7.107}$$

where α_i is the thermal diffusivity of the ith phase.

We have a few options to apply the boundary condition $T_I = T_{sat}$. Say we are solving the temperature field in the liquid phase cells which are denoted by hollow circles in Figure 7.39. The filled circles denote the gas phase cells. The hatched circles denote the interface-containing cells.

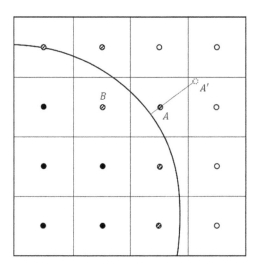

FIGURE 7.39 Phase interface illustrating temperature boundary condition enforcement.

The first option is regarding the interface-containing cells as boundary cells and set their temperature values with interpolation. For example, cell A in Figure 7.39 is such a cell. To determine T_A, we can draw a line passing A, normal to the interface and ends at a point, say point A' in the liquid phase. $T_{A'}$ may be interpolated from the temperature values of the four nearby points surrounding it based on the distance between point A' and these four points. Then assuming the temperature gradient is constant, we have

$$\frac{T_A - T_I}{T_{A'} - T_I} = \frac{T_A - T_{sat}}{T_{A'} - T_{sat}} = \frac{|\phi_A|}{|\phi_{A'}|} = \frac{|\phi_A|}{|\phi_A| + |AA'|} \qquad (7.108)$$

Temperatures at boundary points like T_A serve as boundary conditions for Equation (7.107) when we solve for the liquid phase temperatures.

The second option is taking the gas cells that contain interface as ghost cells of the liquid phase. The temperature in the ghost cell can be set with a similar interpolation procedure by drawing a line from the ghost point into the liquid phase. The line should be perpendicular to the interface. With this option all points in the liquid phase are now interior points and their temperature values are solved from Equation (7.107).

The third option is first populating the temperature gradient from the liquid phase into the gas region, then directly using such gradients to establish the ghost cell temperatures. For instance, the "liquid" temperature in ghost cell B in Figure 7.39 is

$$T_B = T_{sat} + \theta_B * \phi_B \qquad (7.109)$$

where θ_B is the temperature gradient populated from the liquid zone.

The question is, how should we outspread the temperature gradients from the liquid phase to the other side of the interface? You may go back to Section 7.3.3 for a hint.

Yes, we have done something similar before. When we reinitialize the level-set function, we advect ϕ contours away from the interface if they are too close to it. We can do the same thing here. We first calculate the temperature gradient in each cell in the liquid phase with e.g. central differences

$$\left.\frac{\partial T_l}{\partial x}\right|_{i,j} = \frac{T_{l,i+1,j} - T_{l,i-1,j}}{2\Delta x}, \quad \left.\frac{\partial T_l}{\partial y}\right|_{i,j} = \frac{T_{l,i,j+1} - T_{l,i,j-1}}{2\Delta y} \qquad (7.110)$$

and the temperature gradient is

$$\theta = \frac{\partial T_l}{\partial n} = \hat{n}\cdot\nabla T_l = \frac{\nabla\phi}{|\nabla\phi|}\cdot\nabla T_l = \frac{1}{|\nabla\phi|}\left(\frac{\partial\phi}{\partial x}\frac{\partial T_l}{\partial x} + \frac{\partial\phi}{\partial y}\frac{\partial T_l}{\partial y}\right) \qquad (7.111)$$

After we find $\theta_{i,j}$ for all liquid cells, we solve

$$\frac{\partial\theta}{\partial\tau} + S(\phi)\hat{n}\cdot\nabla\theta = 0 \tag{7.112}$$

within the gas cells. $S(\phi)$ is the sign of ϕ. Since this is an unsteady convection equation with "advection velocity" being unity in magnitude and pointing to the gas side, liquid θ values will be advected into the gas phase. You should have no problem with solving either Equation (7.112) or Equation (7.107).

We can calculate the temperature field in the gas phase with exactly the same procedure.

7.4.3 FLOW FIELD COMPUTATION

Once we have had the temperature values in both phases, we can compute the mass flow rate (per unit interface area) $\dot{m}$ using Equation (7.103). Obviously such values only make sense in those interface-containing cells. However, when we simulate the interface motion by advecting the level-set function ϕ, we need the interface velocity $\vec{u_I}$:

$$\frac{\partial\phi}{\partial t} + \vec{u_I}\cdot\nabla\phi = 0 \tag{7.113}$$

which demands the $\dot{m}$ information (see Equation (7.105)) everywhere. Fortunately we have populated temperature gradients of each phase into the whole flow domain which allows us to calculate a fictitious $\dot{m}$ in each phase.

Once the interface is displaced, we can update the physical properties and then solve the Navier–Stokes equations. Now due to the mass transfer between two phases, the continuity equation has to change a little.

Fluid velocity experiences a step change at the interface (see Figure 7.38), so that

$$\vec{u} = \vec{u_l}C + \vec{u_g}(1 - C) \tag{7.114}$$

where C is the step function that assumes constant value 0 in the gas phase while 1 in the liquid. Since $\nabla\cdot\vec{u_l} = \nabla\cdot\vec{u_g} = 0$, we have

$$\begin{aligned}
\nabla\cdot\vec{u} &= (\vec{u_l} - \vec{u_g})\cdot\nabla C \approx (\vec{u_l} - \vec{u_g})\cdot\nabla H = \dot{m}(\frac{1}{\rho_g} - \frac{1}{\rho_l})\hat{n}\cdot\nabla H \\
&= \dot{m}(\frac{1}{\rho_g} - \frac{1}{\rho_l})\frac{\partial H}{\partial n} = \dot{m}(\frac{1}{\rho_g} - \frac{1}{\rho_l})\delta
\end{aligned} \tag{7.115}$$

where the approximate step function H and Dirac delta function δ are given in Equations (7.88) and (7.94).

Notice that Equation (7.115) is valid everywhere since $\nabla C = 0$ and in turn the familiar $\nabla \cdot \vec{u} = 0$ is recovered in each individual phase.

The pressure Poisson equation, Equation (7.59), hence becomes

$$
\frac{\partial}{\partial x}\left(\frac{1}{\rho^m}\frac{\partial p^{m+1}}{\partial x}\right) + \frac{\partial}{\partial y}\left(\frac{1}{\rho^m}\frac{\partial p^{m+1}}{\partial y}\right)
$$
$$
= \frac{1}{\Delta t}\left[\left(\frac{\partial u^*}{\partial x} + \frac{\partial v^*}{\partial y}\right) - \dot{m}\left(\frac{1}{\rho_g} - \frac{1}{\rho_l}\right)\delta\right]
\tag{7.116}
$$

The flowchart of using level-set method to solve a multiphase flow with phase change is shown in Figure 7.40. If you decide to use the VOF instead of the level-set method, you only need to advect the C field rather than the ϕ field.

7.4.4   EXAMPLE: 1-D STEFAN PROBLEM

Consider the 1-D Stefan problem illustrated in Figure 7.41.

A certain fluid is in contact with a wall whose temperature T_{wall} is greater than the fluid saturation temperature T_{sat}. After a certain amount of time the whole fluid reaches the saturation temperature and then some liquid close to the wall begins to evaporate. The gas phase keeps growing and pushes the liquid phase to the right. We want to find out the gas–liquid interface location as well as the liquid speed versus time.

In this simulation we use the properties of saturated water under room pressure. The density of the gas (vapor) phase of water is then $\rho_g = 0.5974$ kg/m^3; the liquid water density is $\rho_l = 958.4$ kg/m^3. Viscosities are $\mu_g = 1.26 \times 10^{-5}$ $Pa{\cdot}s$ and $\mu_l = 2.8 \times 10^{-4}$ $Pa{\cdot}s$. Thermal conductivities are $k_g = 0.0246$ $W/m{\cdot}K$ and $k_l = 0.6828$ $W/m{\cdot}K$. Thermal diffusivities are $\alpha_g = 1.98 \times 10^{-5}$ m^2/s and $\alpha_l = 1.69 \times 10^{-7}$ m^2/s. The latent heat of vaporization is $h_{lg} = 2.257 \times 10^6$ J/kg. Saturation temperature is $T_{sat} = 373$ K and we set $T_{wall} = 398$ K.

The computational domain is 1mm long and we use 250 interior control volumes in this 1-D domain. The initial gas–liquid interface is put at the east face of the very first interior control volume next to the wall. We assume that the initial temperature at the center of the first interior control volume is the average of T_{wall} and T_{sat}. We will continue our calculation until the final time moment $t = 1$ s.

You may find that for this simple 1-D case many steps we discussed in previous sections can be simplified. For example, the reinitialization of the level-set function now can be really easily done by shifting all ϕ values for the same value, namely the displacement of the interface.

The key to the success of this simulation is the accurate estimate of temperature and velocity distributions in both phases. In this example in fact the temperature of the liquid phase is always a constant (T_{sat}) and the velocity of the gas phase is always zero. We therefore only need to worry about the temperature in the gas phase and velocity in the liquid. With using the ghost fluid method we can have a sharp temperature distribution (i.e. it is not diffused close to the interface), as shown

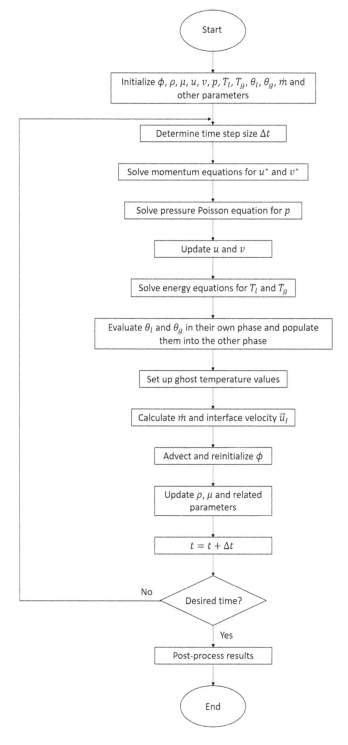

FIGURE 7.40 Flowchart of the level-set method solving problems with phase change.

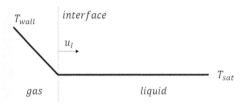

FIGURE 7.41 1-D Stefan problem.

in the left panel of Figure 7.42. This assures us a precise temperature gradient at the interface and in turn an accurate mass flow rate $\dot{m}$. However, the velocity is smeared in proximity to the interface due to our one-fluid methodology when solving the Navier–Stokes equations, as you can observe from the right panel of Figure 7.42. This gives us a little trouble when we evaluate the interface speed $u_I = u_{g,I} + \dot{m}/\rho_g = u_{l,I} + \dot{m}/\rho_l$ as the fluid velocity values at the interface, $u_{g,I}$ and $u_{l,I}$, are not exact. For example, the point that is closest to the interface yet still in the gas phase is point A marked in Figure 7.42, whose velocity is most likely used as $u_{g,I}$. This velocity is nonetheless not very good since the gas phase velocity should be zero.

For this specific example, this problem can be easily fixed by using the exact gas phase velocity, zero, in calculating the interface speed. In more complicated situations, one may estimate $u_{g,I}$ or $u_{l,I}$ by extrapolating the velocity of the corresponding phase from $1.5\Delta x$ distance away onto the interface. Or just use ghost fluid method to solve the flow field.

The results of liquid velocity and interface position versus time in one second period are shown in Figure 7.43.

The exact solution to this problem is given by the following equations (Alexiades & Solomon, 1993)

$$\lambda e^{\lambda^2} \mathrm{erf}(\lambda) = C_{p,g} \frac{T_{wall} - T_{sat}}{h_{lg}\sqrt{\pi}} \tag{7.117}$$

where *erf* is the error function and $C_{p,g} = 2079.7\ J/kg{\cdot}K$ is the specific heat of the water vapor; the temperature distribution in the gas phase is

$$T_g = T_{wall} - \frac{T_{wall} - T_{sat}}{\mathrm{erf}(\lambda)} \mathrm{erf}\!\left(\frac{x}{2\sqrt{\alpha t}} \right) \tag{7.118}$$

And the position of the interface is

$$x_I = 2\lambda\sqrt{\alpha t} \tag{7.119}$$

The liquid phase speed can be derived from the velocity jump condition

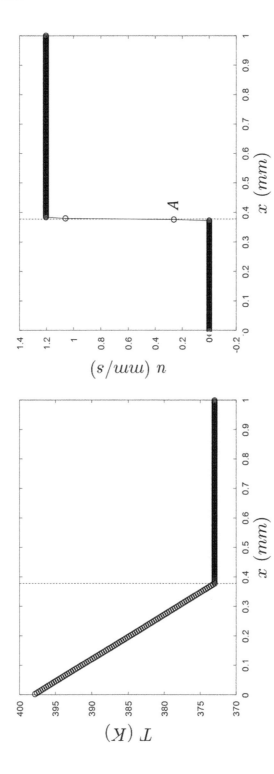

FIGURE 7.42 Temperature and velocity profiles.

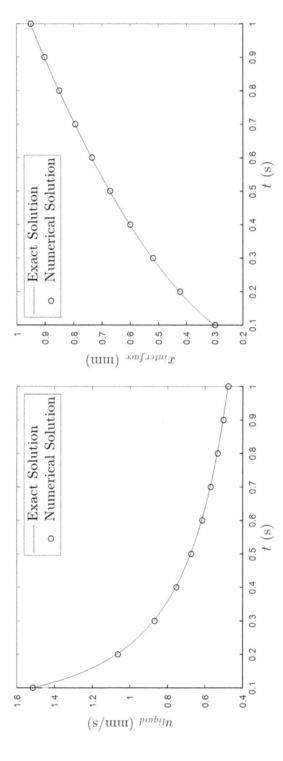

FIGURE 7.43 Liquid velocity and interface position versus time.

$$u_l = u_g + \dot{m}\left(\frac{1}{\rho_g} - \frac{1}{\rho_l}\right) = \dot{m}\left(\frac{1}{\rho_g} - \frac{1}{\rho_l}\right) \qquad (7.120)$$

And the exact value of $\dot{m}$ may be estimated with

$$\dot{m} = -\frac{k_g \frac{dT_g}{dx}\Big|_I}{h_{lg}} \approx \frac{k_g(T_{wall} - T_{sat})}{x_l h_{lg}} \qquad (7.121)$$

since the temperature distribution in the gas phase is very close to a linear one.

Exercises

1. Write a program to calculate the C values for a circular air bubble in liquid water on a 7×7 mesh. Radius of the bubble is 0.3 and the bubble is at the center of the domain.
2. Once you obtained the C values for problem 1, calculate the unit normal of the interface in each cell in Figure 7.44 that contains a piece of interface. Use the MATLAB built-in function "quiver" to display these unit normal vectors.
3. Most likely you used Equations (7.9) and (7.10) when you solved problem 2. It is stated in the text that these two equations are nothing but the simple average of central differences at four corners of the cells. Justify this statement.
4. Once you finished problems 1 and 2, you can try to write a program to reconstruct the bubble based on the C values you obtained for problem 1.

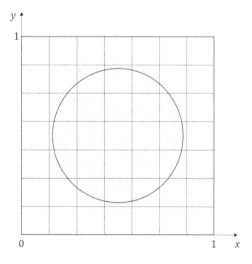

FIGURE 7.44 Air bubble in liquid water on a 7×7 mesh.

5. Write a program to advect the bubble. Suppose the flow field is $u = v = 1$ and $\Delta t = 0.125$. Calculate the bubble position after one time step.

6. Write a program to simulate the motion of a bubble. The bubble is a circle of radius 0.2π initially and is centered at $(0.5\pi, 0.75)$ in a box which spans from 0 to π in both x and y directions. The velocity field in the box is $u = -\sin x \cos y$, $v = \cos x \sin y$. Use a 35 × 35 mesh, find how the bubble looks like at $t = 3$.

7. After you finish problem 6, reverse the flow, resume your calculation until $t = 6$. Does the bubble return to its initial state?

8. Write a program to simulate the Rayleigh–Taylor instability. As Figure 7.45 shows, a heavy fluid ($\rho = 1.255$ kg/m³, $\mu = 1.78 \times 10^{-5}$ Pa·s) overlays a lighter one ($\rho = 0.1694$ kg/m³, $\mu = 1.94 \times 10^{-5}$ Pa·s). Each fluid occupies half of a 0.01 m × 0.04 m tank. The interface is perturbed by a cosine wave $\delta y = 0.0005 \cos(200\pi x)$ to initialize the process. Continue your calculation to the moment $t = 0.118$s. Show the interface.

9. Write a function to realize the smoothing of C field with solving Equation (7.75) and (7.76).

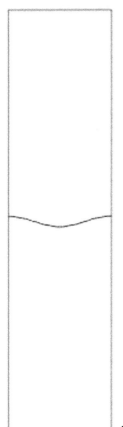

FIGURE 7.45 Rayleigh–Taylor instability problem setup.

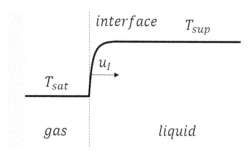

FIGURE 7.46   1-D sucking interface problem setup.

10. Write a VOF program to simulate the evolution of a square water drop in a box filled with air in outer space. The side length of the drop is 0.0375m. Use physical properties of water and air at 25°C. Show results at t = 0.2s and t = 0.4s.

11. Redo problem 6 with the level-set method.

12. Simulate a gas bubble rising in a quiescent viscous liquid. Using the same parameters adopted in Section 7.3.5 except that Eo = 17.7, and Mo = 711 now. Calculate until $\tilde{t}$ = 3. What is the shape of the bubble? What is the Reynolds number based on the bubble terminal speed?

13. Let us use something like Equation (7.112) to extrapolate values defined in a region, say within a circle $(x - 1)^2 + y^2 = 1$ onto a broader zone, say the rectangle whose both sides span from $-\pi$ to π. The values defined in the circle are given by $f(x, y) = \cos(x)\sin(y)$.

14. Solve the 1-D sucking interface problem. The setup is shown in Figure 7.46. A liquid is superheated to a temperature T_{sup} greater than the saturation temperature T_{sat}, which is not a very stable state. The existence of dusts in the liquid, or a solid surface in contact with the liquid can cause the liquid to vaporize.

We will simulate such a case when the vapor (gas) phase forms on the left wall and keeps growing, thereby pushing the liquid away from the wall. Using the same physical properties employed in Section 7.4.4, except that now the wall temperature is T_{sat} and $T_{sup} = T_{sat} + 5$ K. The flow domain is between $x = 0$ and $x = 6$mm. The interface is initially Δx away from the wall. You need to show the liquid speed and interface location versus time. Continue calculations till $t = 1$s. You probably have to use relatively fine mesh (e.g. 600 interior cells) for this problem. And to obtain good results, we may need to "cheat" a little by using the exact solution at a certain time moment e.g. at $t = 0.1$s to initialize the temperature distribution in the liquid phase. The exact solution can be found in Welch & Wilson (2000).

8 Turbulent Flow

8.1 INTRODUCTION

From what we learned in Section 2.5 in Chapter 2, we understand that a turbulent flow is always unsteady and the flow field variables including velocity and pressure fluctuate with time constantly and randomly. This is due to a great many eddies of different sizes, life spans, orientations, and energy levels in the flow. These eddies are the outcome of amplification of inevitable disturbances (environmental noise, lab setup vibrations, etc.) introduced into the flow. Since such disturbances are random in nature and their growth is a complex nonlinear process, the resultant eddies also have a random aspect: it is impossible to predict their exact characteristics like size, life span, orientation, and energy level, etc. Based on this brief description, you can see that the instantaneous flow field of a turbulent flow is always 3-D and unsteady, as we cannot force the disturbances or eddies to stay on a specific plane, although the time-averaged flow field can be 2-D, and statistically steady.

These eddies interact with the mean flow field as well as with each other. As we know, such eddy motions have a diffusive effect to the mean flow field in the sense that they can smooth out differences in the mean flow field; from another perspective, the differences within the mean flow field ("mean shear" as it is called) in fact can stretch the eddies and make them spin faster. That is, kinetic energy can transfer from the mean flow to the eddies. This transfer is most effective for the largest eddies. The same energy transfer also occurs between eddies. Since larger eddies usually are also more energetic, energy typically flows from the large eddies to the smaller ones, and then from the smaller ones to the even smaller ones. This is the so-called energy cascade in turbulent flows. Does this cascade process continue without limit? Or is there a certain minimum eddy scale at the end of this chain of energy transfer? Well, in fact there exists such a smallest eddy scale. The key to understand this fact is the momentum equation, say Equation (1.17). The diffusion term in this equation becomes more and more significant as the Reynolds number becomes less and less. At the smallest scale the eddy velocity is also the lowest, therefore the Reynolds number, being proportional to length scale and velocity scale, is very small (of order 1) at the smallest eddy scale. And the viscous diffusion can dissipate the kinetic energy of the smallest eddies instantly (into heat) and no smaller eddies can survive. This energy cascade process was expressed in a famous poem by Lewis F. Richardson (1922):

> Big whirls have little whirls
> that feed on their velocity,
> And little whirls have lesser whirls
> and so on to viscosity

DOI: 10.1201/9781003138822-8

One should understand that although energy cascade is a net kinetic energy transfer from large to small scales, it does not mean there is no opposite transfer from small to large scales (which is called "backscatter"). In fact, energy transfer in both directions is present in real turbulent flows but the forward-scatter typically dominates.

There are some significant differences between the large and the small eddies. The size of the largest eddies in a turbulent flow usually is comparable to the characteristic length scale of the flow field; while the size of the smallest eddies is independent of such scales. The large eddies usually have some preferred directions, which are the directions along which they can be most effectively stretched by the mean flow. The smaller eddies gradually lose such direction preferences and become more and more isotropic as the eddy size decreases. The largest eddies obviously have the most significant contribution to the flow field fluctuations i.e. the turbulent kinetic energy. The smallest eddies on the other hand contribute the most to the viscous dissipation of such fluctuations.

Based on these considerations, we might estimate the length, velocity, and time scales of the eddies at the two ends of the eddy size spectrum. Define the (specific) turbulent kinetic energy as

$$k = \frac{1}{2}(\overline{u'^2} + \overline{v'^2} + \overline{w'^2}) \tag{8.1}$$

in which the overbar denotes time-average or ensemble-average (refer to Section 2.5 for definition of these terms), and the prime stands for fluctuations:

$$u_i' = u_i - \bar{u}_i \tag{8.2}$$

Obviously the SI unit of k is m^2/s^2.

Suppose the time rate of the turbulent kinetic energy dissipation per unit mass of fluid is ϵ, whose SI unit is W/kg = m^2/s^3.

Since the largest eddies contribute the most to k, their velocity scale u_L should be of the order of $k^{0.5}$, and their time scale τ_L should be of the order of k/ϵ and their length scale l_L should be of the order of $u_L \tau_L \sim k^{1.5}/\epsilon$.

The smallest eddies dissipate energy at the rate of ϵ by means of viscosity. It is reasonable to relate scales of the smallest eddies with ϵ and the kinematic viscosity ν, whose SI unit is m^2/s. By requiring the unit of $\nu^a \epsilon^b$ to be the same as the unit of the desired scales, we can find the length scale of the smallest eddies, l_η, which is of the order of $\nu^{0.75}/\epsilon^{0.25}$; their velocity scale u_η is of the order of $\nu^{0.25}\epsilon^{0.25}$ and their time scale τ_η is of the order of $\nu^{0.5}/\epsilon^{0.5}$. Such scales are called Kolmogorov scales.

If we define a Reynolds number based on the largest eddy scales

$$Re_L = \frac{u_L l_L}{\nu} \tag{8.3}$$

you may find

$$\frac{l_L}{l_\eta} \sim Re_L^{0.75}, \quad \frac{u_L}{u_\eta} \sim Re_L^{0.25}, \quad \frac{\tau_L}{\tau_\eta} \sim Re_L^{0.5} \tag{8.4}$$

Since Reynolds number is typically very high in a turbulent flow, the scales of the smallest eddies are usually way smaller than the largest scales. For example, if $Re_L = 10^5$, and the largest eddy size is $l_L \sim 1$m, the size of the smallest eddies is only $l_\eta \sim 10^{-4}$m, which is roughly the same as the thickness of a piece of paper. Also you can see if the time scale of the largest eddy is $\tau_L \sim 1$s, the time scale of the smallest eddies is only $\tau_\eta \sim 10^{-3}$s at the same Reynolds number.

One way to simulate turbulent flows is simply solving the Navier–Stokes equations with very fine mesh and very small time step sizes so that even the eddy motion at the Kolmogorov scale is calculated. In this way no turbulent models are needed since every flow detail is resolved and nothing is left for modeling. This method is called direct numerical simulation (DNS). DNS is very accurate. In fact DNS data are usually used to calibrate turbulent flow measuring techniques. DNS however is very expensive. The reason is because DNS must use a computational domain big enough to contain the largest eddies, yet with a grid spacing small enough to capture the smallest eddies. The number of mesh cells in each spatial direction should be of the order of $l_L/l_\eta \sim Re_L^{0.75}$ and the total number of mesh cells should be $N \sim (Re_L^{0.75})^3 = Re_L^{2.25}$ since turbulence are intrinsically 3-D. Moreover, the total flow time a DNS simulates should be longer than the time scale of the largest eddies, but the time step size Δt should be less than the time scale of the smallest eddies so that behavior of all eddies can be discerned. In fact usually an even more stringent criterion is applied to Δt in DNS i.e. Δt has to be proportional to the time required for a small-scale eddy to pass a fixed point when advected by the large-scale eddy i.e. $\Delta t \sim l_\eta/u_L$. The total number of time steps needed for such a simulation is therefore of the order of $t_L/\Delta t \sim l_L/l_\eta \sim Re_L^{0.75}$. Each time step involves solving equations whose number is of the order of $Re_L^{2.25}$. The total number of equations is thus of the order of Re_L^3. For example, if we use DNS to simulate a flow with a Reynolds number $Re_L = 10^5$, and if 100 floating point operations have to be executed to solve one equation, the total work load is of the order of 10^{17} flops. It takes at least 290 days to do such a DNS on a personal computer with a 4 GHz CPU.

For this reason DNS is almost exclusively done on super computers and is restricted to relatively simple flows e.g. the fully developed turbulent channel flow and duct flow at relatively low Reynolds numbers. For most turbulent flows of practical engineering interest, which are usually much more complicated than such simple flows and with much higher Reynolds numbers, DNS is still not a realistic option at least for now. Instead we will have to ignore certain, even most of the eddy scales and only calculate the flow field of the rest scales so that the simulation can be accomplished in reasonable time frame with affordable resources. The effects of those neglected scales on the resolved scales then have to be modeled.

We have discussed one of the oldest Reynolds-averaged Navier–Stokes (RANS) models, namely the Prandtl's mixing length model in Section 2.5.2 in Chapter 2. Within the RANS framework, only the flow field at the longest time scale i.e. the mean flow is solved directly from its governing equations. The effects of all turbulent eddies are modeled in a RANS simulation.

The time-average of a flow variable, say u_i is defined as

$$\bar{u}_i = \frac{\int_0^T u_i dt}{T} \tag{8.5}$$

in which T is a time period much longer than the time scale of the largest eddies: $T \gg t_L$. For a statistically steady flow, $\bar{u}_i$ is time-independent. For a statistically unsteady flow, it is better to interpret $\bar{u}_i$ as the ensemble average of the velocity. That is, if we repeat the same flow experiment many times, then the average of the flow fields of all these repetitions is the ensemble-averaged flow field. The ensemble-averaged flow field can vary with time. We may give time and ensemble averages one single terminology called Reynolds average, named after Osborne Reynolds who proposed these averaging techniques.

One may then decompose the instantaneous flow field into two parts, namely the Reynolds-averaged flow field and the flow fluctuations:

$$u_i = \bar{u}_i + u_i' \tag{8.6}$$

which has been introduced already in Equation (8.2). The Reynolds average of the fluctuation vanishes i.e. $\overline{u_i'} = 0$ because $\bar{u}_i = \overline{\bar{u}_i + u_i'} = \bar{\bar{u}}_i + \overline{u_i'} = \bar{u}_i + \overline{u_i'}$.

If we carry out Reynolds averaging to the continuity equation, we have

$$\overline{\frac{\partial u}{\partial x}} + \overline{\frac{\partial v}{\partial y}} + \overline{\frac{\partial w}{\partial z}} = \frac{\partial \bar{u}}{\partial x} + \frac{\partial \bar{v}}{\partial y} + \frac{\partial \bar{w}}{\partial z} = 0 \tag{8.7}$$

The Reynolds-averaged velocity field, therefore, still satisfies the continuity equation.

Similarly the Reynolds-averaged momentum equation is

$$\frac{\partial \bar{u}_i}{\partial t} + \frac{\partial(\bar{u}\bar{u}_i)}{\partial x} + \frac{\partial(\bar{v}\bar{u}_i)}{\partial y} + \frac{\partial(\bar{w}\bar{u}_i)}{\partial z}$$

$$= -\frac{1}{\rho}\frac{\partial \bar{p}}{\partial x_i} + \nu\left(\frac{\partial^2 \bar{u}_i}{\partial x^2} + \frac{\partial^2 \bar{u}_i}{\partial y^2} + \frac{\partial^2 \bar{u}_i}{\partial z^2}\right) + \frac{\partial(-\overline{u'u_i'})}{\partial x} \tag{8.8}$$

$$+ \frac{\partial(-\overline{v'u_i'})}{\partial y} + \frac{\partial(-\overline{w'u_i'})}{\partial z}$$

These equations are called RANS equations.

Notice that although the average of fluctuations is zero, the average of products of fluctuations are typically not zero. For example, suppose u' takes the following five values: $- 0.25, 0.125, 0.125, 0.2$, and $- 0.2$. The average of these values is zero but the average of $u'u' = u'^2$ is not zero but 0.03475.

Such averages of velocity fluctuation products are the so-called Reynolds stresses[1]:

$$\tau_{ij} = -\overline{u'_i u'_j} \tag{8.9}$$

They present in the momentum equations representing the effects of the turbulent eddies on the mean flow. Reynolds stresses have to be modeled. One of the most commonly used modeling methodology is the Boussinesq approximation:

$$\tau_{ij} - \frac{1}{3}\tau_{kk}\delta_{ij} = 2\nu_T\left(\bar{S}_{ij} - \frac{1}{3}\bar{S}_{kk}\delta_{ij}\right) \tag{8.10}$$

in which

$$\bar{S}_{ij} = \frac{1}{2}\left(\frac{\partial \bar{u}_i}{\partial x_j} + \frac{\partial \bar{u}_j}{\partial x_i}\right) \tag{8.11}$$

is the Reynolds-averaged strain rate tensor. δ_{ij} is the Kronecker delta which is zero except when $i = j$ it is one (see Equation (5.198)). ν_T is the so-called turbulent or eddy viscosity, a parameter at the very heart of any RANS models. Notice that in Equation (8.10) the Einstein convention is used.

Therefore, $\tau_{kk} = -(\overline{u'^2} + \overline{v'^2} + \overline{w'^2}) = -2k$ and $\bar{S}_{kk} = \partial\bar{u}/\partial x + \partial\bar{v}/\partial y + \partial\bar{w}/\partial z = 0$. With Boussinesq approximation the Reynolds-averaged momentum equation becomes

$$\frac{\partial \bar{u}_i}{\partial t} + \frac{\partial(\bar{u}_j\bar{u}_i)}{\partial x_j} = -\frac{1}{\rho}\frac{\partial \bar{p}}{\partial x_i} + \nu\frac{\partial}{\partial x_j}\left(\frac{\partial \bar{u}_i}{\partial x_j}\right) + \frac{\partial}{\partial x_j}\left[\nu_T\left(\frac{\partial \bar{u}_i}{\partial x_j} + \frac{\partial \bar{u}_j}{\partial x_i}\right)\right] - \frac{2}{3}\frac{\partial k}{\partial x_i} \tag{8.12}$$

Again Einstein convention applies. This equation can be further simplified by absorbing the last term on the right side into the pressure gradient and combining the two diffusion terms

$$\frac{\partial \bar{u}_i}{\partial t} + \frac{\partial(\bar{u}_j\bar{u}_i)}{\partial x_j} = -\frac{1}{\rho}\frac{\partial \bar{\Psi}}{\partial x_i} + \frac{\partial}{\partial x_j}\left[(\nu + \nu_T)\left(\frac{\partial \bar{u}_i}{\partial x_j} + \frac{\partial \bar{u}_j}{\partial x_i}\right)\right] \tag{8.13}$$

in which $\bar{\Psi} = \bar{p} + (2/3)\rho k$ behaves just like a "pressure." From now on we will still use $\bar{p}$ in place of $\bar{\Psi}$.

The Reynolds-averaged or mean velocities $\bar{u}$, $\bar{v}$, and $\bar{w}$ can be solved from the Reynolds-averaged momentum equations if we know how to evaluate the turbulent or eddy viscosity ν_T. As Section 2.5.2 reasons, turbulent viscosity ν_T should be of the order of $u_{eddy}l_{eddy}$. And obviously the largest eddies contribute the most to ν_T from the perspective of the mean flow field, so $u_{eddy} \sim u_L$ and $l_{eddy} \sim l_L$.

In the mixing length model, l_L and u_L are directly related to the mean flow characteristic scales like the distance from the nearest solid wall: $l_L = \kappa y_{wall}$ (van Driest damping function has to be added approaching the wall, see Section 2.5) and the mean shear: $u_L \sim l_L |d\bar{u}/dy|$. If a flow is not bounded by solid walls e.g. a mixing layer due to a fluid flowing away from a solid surface into an open space as depicted in Figure 8.1, the mixing length model assumes a different eddy length scale: $l_L = \alpha\delta$ in which δ is the mixing layer thickness and α an empirical constant.

One might wonder what turbulent length scale should be used at a point that is somewhat in between a solid wall and a mixing layer, say point A in Figure 8.1. Similar problems always arise if one calculates complex turbulent flows e.g. the turbulent flow over a backward-facing step with the mixing length model.

A more satisfactory way to assess u_L and l_L is solving two differential equations (since we have two unknowns) which includes more physics. Such two-equation models are probably the most commonly used turbulent models for turbulent flows of engineering and industrial importance in the past few decades. We will introduce some widely used two- and four-equation RANS models in the next several sections.

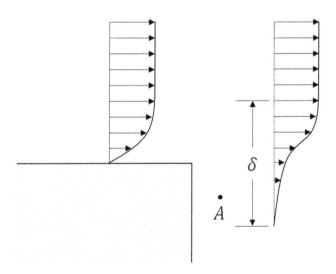

FIGURE 8.1   Sketch of a mixing layer.

8.2   TWO- AND FOUR-EQUATION RANS MODELS

8.2.1 $k - \epsilon$ MODEL

According to the order of magnitude estimates of u_L and l_L, we may conclude that $\nu_T \sim u_L l_L \sim k^2/\epsilon$. In the $k - \epsilon$ model (Jones & Launder, 1972), two differential equations are solved for the turbulent kinetic energy k and turbulent energy dissipation rate ϵ, which are then used to evaluate the eddy viscosity. The k equation can be derived from the instantaneous and RANS equations. The deduction is (I am afraid) not very interesting, so I will simply put the final equation here (Einstein convention applies):

$$
\frac{\partial k}{\partial t} + \frac{\partial \left(\bar{u}_j k \right)}{\partial x_j} = -\overline{u_i' u_j'} \frac{\partial \bar{u}_i}{\partial x_j} + \frac{\partial}{\partial x_j}\left(-\frac{1}{2}\overline{u_j' u_i' u_i'} - \frac{1}{\rho}\overline{p' u_j'} + \nu \frac{\partial k}{\partial x_j} \right)
$$
$$
- \nu \overline{\frac{\partial u_i'}{\partial x_j} \frac{\partial u_i'}{\partial x_j}}
$$
(8.14)

Although looks terrible, this equation indeed is still a convection–diffusion equation. The left side has, of course, the transient and convection terms; the second term on the right side is the diffusion term and the first and third terms are the the turbulent kinetic energy production and dissipation terms, respectively. The turbulent kinetic energy production term is almost always positive and represents the time rate at which turbulent kinetic energy is produced due to the stretching of large eddies by the mean shear. The turbulent kinetic energy dissipation term is always negative and performs as a drain on the turbulent kinetic energy. This term in fact is the definition of ϵ. Something unexpected is that in the k equation we have many terms more frightening than the Reynolds stress $-\overline{u_i' u_j'}$ we try to model with the help of this equation.

The solution to this so-called closure problem is more modeling. For example, the triple correlation term is modeled as

$$
-\frac{1}{2}\overline{u_j' u_i' u_i'} = \sigma_k \nu_T \frac{\partial k}{\partial x_j}
$$
(8.15)

in which σ_k is a modeling constant. The rationale behind this practice is regarding this triple correlation term as the diffusion flux of $k = \overline{u_i' u_i'}/2$ due to large eddies and a formula similar to the Fourier law for heat flux thus is assumed to hold. The pressure diffusion term $-\overline{p' u_j'}$ is simply neglected as it is usually pretty small.

The modeled k equation then reads

$$
\frac{\partial k}{\partial t} + \frac{\partial \left(\bar{u}_j k \right)}{\partial x_j} = \nu_T \left(\frac{\partial \bar{u}_i}{\partial x_j} + \frac{\partial \bar{u}_j}{\partial x_i} \right)\frac{\partial \bar{u}_i}{\partial x_j} + \frac{\partial}{\partial x_j}\left[(\nu + \sigma_k \nu_T)\frac{\partial k}{\partial x_j} \right] - \epsilon
$$
(8.16)

Notice that according to the Boussinesq approximation, the production term is equal to

$$P = -\overline{u_i' u_j'} \frac{\partial \bar{u}_i}{\partial x_j} = \nu_T \left(\frac{\partial \bar{u}_i}{\partial x_j} + \frac{\partial \bar{u}_j}{\partial x_i} \right) \frac{\partial \bar{u}_i}{\partial x_j} = \nu_T \frac{1}{2} \left(\frac{\partial \bar{u}_i}{\partial x_j} + \frac{\partial \bar{u}_j}{\partial x_i} \right) \left(\frac{\partial \bar{u}_i}{\partial x_j} + \frac{\partial \bar{u}_j}{\partial x_i} \right) = \nu_T S^2$$

$$(8.17)$$

where the magnitude of the strain rate tensor is

$$S = \sqrt{2 \left[\frac{1}{2} \left(\frac{\partial \bar{u}_i}{\partial x_j} + \frac{\partial \bar{u}_j}{\partial x_i} \right) \frac{1}{2} \left(\frac{\partial \bar{u}_i}{\partial x_j} + \frac{\partial \bar{u}_j}{\partial x_i} \right) \right]} = \sqrt{2 \bar{S}_{ij} \bar{S}_{ij}} \qquad (8.18)$$

You may want to take a look at Equation (8.11) for the definition of the strain rate tensor. So with the Boussinesq approximation the turbulent kinetic energy production term is always positive.

The ϵ equation again can be derived from the instantaneous and RANS equations. And the derived ϵ equation again contains many correlation terms that have to be modeled. The modeled ϵ equation is

$$\frac{\partial \epsilon}{\partial t} + \frac{\partial \left(\bar{u}_j \epsilon \right)}{\partial x_j} = \frac{\epsilon}{k} \left[C_{\epsilon 1} \nu_T \left(\frac{\partial \bar{u}_i}{\partial x_j} + \frac{\partial \bar{u}_j}{\partial x_i} \right) \frac{\partial \bar{u}_i}{\partial x_j} - C_{\epsilon 2} \epsilon \right] + \frac{\partial}{\partial x_j} \left[(\nu + \sigma_\epsilon \nu_T) \frac{\partial \epsilon}{\partial x_j} \right]$$

$$(8.19)$$

in which those Cs and σ_ϵ are modeling constants.

With k and ϵ solved, the eddy viscosity ν_T is assumed to be

$$\nu_T = C_\mu \frac{k^2}{\epsilon} \qquad (8.20)$$

according to the order of magnitude estimate of ν_T. Again C_μ is another model constant.

The model constants are obtained by matching the model predictions with experimental data of some well-studied simple turbulent flows and the values are

$$C_{\epsilon 1} = 1.44; \quad C_{\epsilon 2} = 1.92; \quad \sigma_k = 1; \quad \sigma_\epsilon = 0.77; \quad C_\mu = 0.09 \qquad (8.21)$$

In summary, when we use $k - \epsilon$ model to solve turbulent flow problems, we have to solve the RANS equations, Equations (8.7) and (8.13), together with the k Equation (8.16) and ϵ Equation (8.19), which then determine the eddy viscosity according to Equation (8.20).

Let us solve the fully developed turbulent channel flow (cf. Section 2.5) using the k-ϵ model.

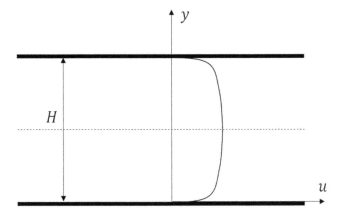

FIGURE 8.2 Turbulent channel flow example setup.

The problem setup is repeated here:

The fluid is water, density is $\rho = 1000$ kg/m³, viscosity $\mu = 0.00112$ Pa·s. The height of channel is $H = 0.2$ m. The pressure gradient is $d\bar{p}/dx = -0.2$ Pa/m. See Figure 8.2. Due to flow symmetry we will only simulate the flow in the lower half of the channel.

We want to determine

1. The distribution of mean velocity $\bar{u}$, including u^+ versus y and u^+ versus y^+ distributions. The relevant definitions are

$$u^+ = \frac{\bar{u}}{u_\tau}, \quad y^+ = \frac{y_w u_\tau}{\nu}, \quad u_\tau = \sqrt{\frac{\tau_w}{\rho}} = \sqrt{\nu \frac{d\bar{u}}{dy}\bigg|_{y=0}} \qquad (8.22)$$

y_w is the distance from the nearest channel wall. τ_w is the wall shear stress and u_τ is the so-called friction velocity, which is frequently used as reference velocity in turbulent flow simulations due to its close relationship with many flow parameters.

2. The distribution of Reynolds stress τ_{xy}, which is

$$\tau_{xy} = -\overline{u'v'} = \nu_T \left(\frac{\partial \bar{u}}{\partial y} + \frac{\partial \bar{v}}{\partial x} \right) = \nu_T \frac{d\bar{u}}{dy} \qquad (8.23)$$

Notice $\bar{v} = 0$.

3. The distribution of the turbulent kinetic energy k.

4. The distribution of the production term and dissipation term in the k equation.

The governing equations can be greatly simplified for this example. The continuity equation is

$$\frac{\partial \bar{u}}{\partial x} + \frac{\partial \bar{v}}{\partial y} = 0 \qquad (8.24)$$

which infers $\bar{v} = 0$ since $\partial \bar{u}/\partial x = 0$ (fully developed flow), so $\partial \bar{v}/\partial y = 0$ i.e. $\bar{v}$ does not change in the y-direction; now that $\bar{v} = 0$ at $y = 0$, it must be so at the other y values. Obviously $\bar{u}$ is only a function of y: $\bar{u} = \bar{u}(y)$.

The $\bar{u}$ momentum equation is reduced to

$$0 = -\frac{1}{\rho}\frac{\partial \bar{p}}{\partial x} + \frac{d}{dy}\left[(\nu + \nu_T)\frac{d\bar{u}}{dy}\right] \qquad (8.25)$$

The k and ϵ equations become

$$0 = \nu_T \frac{d\bar{u}}{dy}\frac{d\bar{u}}{dy} + \frac{d}{dy}\left[(\nu + \sigma_k \nu_T)\frac{dk}{dy}\right] - \epsilon \qquad (8.26)$$

and

$$0 = \frac{\epsilon}{k}\left(C_{\epsilon 1}\nu_T \frac{d\bar{u}}{dy}\frac{d\bar{u}}{dy} - C_{\epsilon 2}\epsilon\right) + \frac{d}{dy}\left[(\nu + \sigma_\epsilon \nu_T)\frac{d\epsilon}{dy}\right] \qquad (8.27)$$

You may find these three equations are ordinary differential equations and can be written in the same form

$$0 = \frac{d}{dy}\left(\Gamma_\phi \frac{d\phi}{dy}\right) + S_c - S_p \phi \qquad (8.28)$$

Here we want to make sure S_p is positive so that the final finite difference formulae are stable. The interpretation of terms of this general form equation are listed in Table 8.1.

If we use a uniform mesh as shown in Figure 8.3 , Equation (8.28) can be discretized by integration over each control volume, say the jth control volume:

$$0 = \left[(\nu + \sigma_\phi \nu_T)\frac{d\phi}{dy}\right]_{J-1}^{J} + S_c \Delta y - S_p \phi_j \Delta y \qquad (8.29)$$

where Δy is the grid spacing. Then using central differences for the derivatives at the cell surfaces, we have

TABLE 8.1

Terms in the General Form Equation for the k-ε Model

ϕ		S_c	S_p
$\bar{u}$	$\nu + \nu_T$	$-\frac{1}{\rho}\frac{\partial p}{\partial x}$	0
k	$\nu + \sigma_k \nu_T$	$\nu_T\left(\frac{d\bar{u}}{dy}\right)^2$	$\frac{\varepsilon}{k}$
ε	$\nu + \sigma_\varepsilon \nu_T$	$C_{\varepsilon 1}\nu_T\left(\frac{d\bar{u}}{dy}\right)^2\frac{\varepsilon}{k}$	$C_{\varepsilon 2}\frac{\varepsilon}{k}$

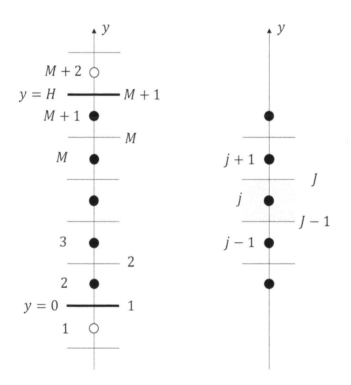

FIGURE 8.3 Finite volume mesh for the turbulent channel flow example.

$$0 = (\nu + \sigma_\phi \nu_T)_J\frac{\phi_{j+1}-\phi_j}{\Delta y} - (\nu + \sigma_\phi \nu_T)_{J-1}\frac{\phi_j-\phi_{j-1}}{\Delta y} + S_c\Delta y$$
$$- S_p\phi_j\Delta y \tag{8.30}$$

These formulas form a tridiagonal system of equations

$$AS_j\phi_{j-1} + AP_j\phi_j + AN_j\phi_j = B_j \tag{8.31}$$

in which

$$AS_j = -\frac{(\nu + \sigma_\phi \nu_T)_{J-1}}{\Delta y^2}, \quad AN_j = -\frac{(\nu + \sigma_\phi \nu_T)_J}{\Delta y^2}, \quad AP_j$$

$$= -\left(AS_j + AN_j\right) + S_p, \quad B_j = S_c \tag{8.32}$$

The viscosity at the surfaces can be evaluated with harmonic averages

$$\nu_J^{tot} = \frac{2}{\frac{1}{\nu_j^{tot}} + \frac{1}{\nu_{j+1}^{tot}}}, \quad \nu_{J-1}^{tot} = \frac{2}{\frac{1}{\nu_j^{tot}} + \frac{1}{\nu_{j-1}^{tot}}} \tag{8.33}$$

where $\nu^{tot} = \nu + \sigma_\phi \nu_T$.

The procedure of solving this problem is

1. Give an initial guess of mean velocity $\bar{u}$, turbulent kinetic energy k, turbulent kinetic energy dissipation rate ϵ and turbulent eddy viscosity ν_T distributions. We may use the log-law, Equation (2.143) to derive such initial guesses.
2. Using TDMA to solve the $\bar{u}$, k and ϵ equations, then update ν_T according to the new k and ϵ values.
3. Update the coefficients of the $\bar{u}$, k and ϵ equations.
4. Repeat steps (2) and (3) until convergence.

Let us see what the log-law is, why we have the log-law, and what we may infer from the log-law.

The log-law typically refers to the following mean velocity distribution

$$u^+ = \frac{1}{\kappa} \ln(y^+) + C \tag{8.34}$$

in which κ (Greek letter kappa, not k) and C are the so-called von Kármán constants. From experiments it is found $\kappa \approx 0.41$ and $C \approx 5$. This law is found to be a good approximation to the mean velocity in a certain region of wall-bounded turbulent flows with a negative pressure gradient (which is also called favorable pressure gradient). The region that log-law holds is

$$30 \lesssim y^+ \lesssim 0.2Re_\tau \tag{8.35}$$

where Re_τ is the flow Reynolds number based on the friction velocity u_τ and the half height of the channel h. The upper limit of the log-law layer can also be written as $y \lesssim 0.2h$ which means the log-law region in fact is pretty close to the wall. This can be observed from Figure 8.4, which shows the mean velocity distribution in half of a fully developed channel flow with $Re_\tau = 2000$.

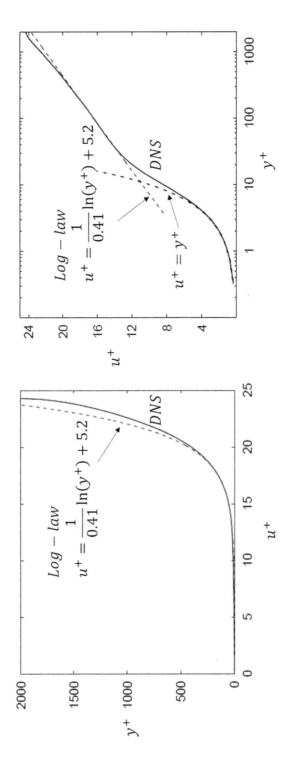

FIGURE 8.4 Log-law.

The log-law can be derived from the Reynolds-averaged momentum equation with the help of the mixing length model. Let us use the statistically steady, fully developed turbulent channel flow as an example. The momentum equation is

$$0 = -\frac{1}{\rho}\frac{\partial \bar{p}}{\partial x} + \frac{d}{dy}\left[(\nu + \nu_T)\frac{d\bar{u}}{dy}\right] \tag{8.36}$$

Notice the mean pressure gradient is a constant and can be related to the (mean) wall shear stress with simple force balance consideration: the pressure force acting on the fluid in the channel is balanced by the shear force at the wall

$$(-\Delta\bar{p})H\,(\Delta z) = 2\tau_w\,(\Delta x)(\Delta z) \tag{8.37}$$

Here Δx and Δz are the dimensions of the fluid segment we consider, τ_w is the wall shear stress. Refer to Figure 8.5 for a better idea of the force balance.

From the force balance we find

$$-\frac{1}{\rho}\frac{\partial \bar{p}}{\partial x} = \frac{1}{h}u_\tau^2 \tag{8.38}$$

where $h = H/2$ is the half height of the channel and $u_\tau = \sqrt{\tau_w/\rho}$ is the friction velocity. Since the pressure gradient is a constant, u_τ is also a constant. Using the given values in the problem statement, we can calculate $u_\tau = 0.00448$ m/s. The Reynolds number based on u_τ and h is $Re_\tau = \rho u_\tau h/\mu \approx 400$.

Substituting Equation (8.38) into Equation (8.36) we have

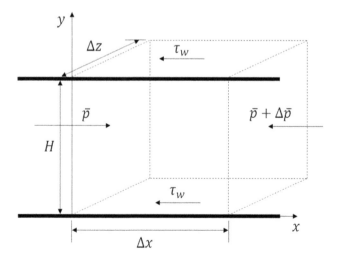

FIGURE 8.5 Force balance in the turbulent channel flow.

$$0 = \frac{1}{h}u_\tau^2 + \frac{d}{dy}\left[(\nu + \nu_T)\frac{d\bar{u}}{dy}\right] \tag{8.39}$$

Now use the mixing length model to evaluate ν_T:

$$\nu_T = u_{eddy}l_{eddy} = \left(l_{eddy}\left|\frac{d\bar{u}}{dy}\right|\right)l_{eddy} = l_{eddy}^2\left|\frac{d\bar{u}}{dy}\right| = \kappa^2 y^2\left|\frac{d\bar{u}}{dy}\right| \tag{8.40}$$

Then we drop the diffusion term due to the molecular motion, then the momentum equation becomes

$$0 = \frac{1}{h}u_\tau^2 + \frac{d}{dy}\left(\nu_T\frac{d\bar{u}}{dy}\right) = \frac{1}{h}u_\tau^2 + \frac{d}{dy}\left(\kappa^2 y^2\left|\frac{d\bar{u}}{dy}\right|\frac{d\bar{u}}{dy}\right) \tag{8.41}$$

If you are still following this deduction, you might wonder why I can drop a diffusion term. This is because the diffusion caused by the turbulent eddy motion is usually much stronger than that caused by the molecular motion, as long as it is not super close to the wall. Since Equation (8.41) is not valid as it is super close to the wall, let alone at the wall itself, we cannot integrate it from the wall up to a certain y location. Let us integrate it from y up to the channel half-height h:

$$\left(\kappa y\frac{d\bar{u}}{dy}\right)^2\Bigg|_y^h + \frac{1}{h}u_\tau^2(h - y) = 0 \tag{8.42}$$

Since at $y = h$, $d\bar{u}/dy = 0$ due to symmetry, we have

$$\kappa y\frac{d\bar{u}}{dy} = \sqrt{u_\tau^2\left(1 - \frac{y}{h}\right)} \tag{8.43}$$

Now let me drop y/h inside the square root, then we have

$$\kappa y\frac{d\bar{u}}{dy} = u_\tau \rightarrow du^+ = \frac{1}{\kappa}\frac{1}{y}dy \rightarrow du^+ = \frac{1}{\kappa}\frac{1}{y^+}dy^+ \rightarrow u^+ = \frac{1}{\kappa}\ln(y^+) + C \tag{8.44}$$

Notice that $u^+ = \bar{u}/u_\tau$. The dimensionless distance $y^+ = yu_\tau/\nu$ only differs from y by a constant factor. I dropped the y/h in Equation (8.43) because, as you have already observed from Figure 8.4, the log-law is only valid in a region pretty close (although not super close) to the wall i.e. y/h is small compared with unity.

In this log-law region,

1. $\bar{u} = u_\tau [\ln(y u_\tau / \nu)/\kappa + C]$ and $u_\tau = \sqrt{-(h/\rho)\partial \bar{p}/\partial x}$. See Equation (8.38).
2. $\nu_T = \kappa y u_\tau$. See Equations (8.40) and (8.44). $\nu_T = \kappa y^+ \nu$ is indeed much greater than ν since $y^+ \gtrsim 30$ in this region. This further justifies our practice of dropping the ν term in the above deduction.
3. Experiments (Townsend, 1976) show that the Bradshaw assumption (Bradshaw, Ferriss, & Atwell, 1967) is valid in many thin turbulent shear flows. This assumption states that the principal turbulent shear stress is proportional to the turbulent kinetic energy. This assumption can be derived from the mixing length model as well. For our channel flow, according to the mixing length model the large eddy velocity scale is

$$u_{eddy} \sim l_{eddy} \left| \frac{d\bar{u}}{dy} \right| \tag{8.45}$$

Notice that

$$k = \frac{\overline{u_i' u_i'}}{2} \sim u_{eddy}^2 \tag{8.46}$$

since large eddies contribute the most to the velocity fluctuations. The Reynolds stress

$$\tau_{xy} \sim \nu_T \left(\frac{d\bar{u}}{dy} \right) \sim \left(u_{eddy} l_{eddy} \right) \left(\frac{d\bar{u}}{dy} \right) \tag{8.47}$$

Therefore

$$\left| \tau_{xy} \right| \sim u_{eddy} \left(l_{eddy} \left| \frac{d\bar{u}}{dy} \right| \right) \sim u_{eddy} u_{eddy} \sim k \tag{8.48}$$

That is

$$\left| \tau_{xy} \right| = Ck \tag{8.49}$$

It is found from experiments that the constant $C \approx \sqrt{C_\mu} = 0.3$. Hence in the lower half of channel we have

$$- \overline{u'v'} \approx \sqrt{C_\mu} k \tag{8.50}$$

i.e. $\nu_T (d\bar{u}/dy) \approx \sqrt{C_\mu} k$. Since $\nu_T = \kappa y u_\tau$ and $d\bar{u}/dy = u_\tau/(\kappa y)$ in the log-law region, $k \approx u_\tau^2/\sqrt{C_\mu}$ in this region, which is roughly a constant;

4. Since k is a constant in this region, we must have $\epsilon = \nu_T (d\bar{u}/dy)^2 = u_\tau^3/(\kappa y)$ due to Equation (8.26). That is, the production and dissipation of turbulent kinetic energy reach an equilibrium in the log-law region.

Now we have the initial guesses of $\bar{u}$, ν_T, κ, and ϵ.

Before leaving this topic (log-law), let us see what happens in the region $y^+ < 30$. In the layer super close to the wall where $y^+ \lesssim 5$, turbulence is so suppressed due to the walls blocking eddies and strong viscous dissipation, $\nu_T \ll \nu$, therefore the wall shear stress is solely due to the familiar laminar diffusion term:

$$\frac{\tau_w}{\rho} = \nu \frac{d\bar{u}}{dy} \tag{8.51}$$

from which we find

$$u^+ = y^+ \tag{8.52}$$

in this so-called viscous sublayer. The mean velocity therefore first changes linearly with distance from the wall in the very thin viscous sublayer then transitions to the log-law via a buffer layer ($5 \lesssim y^+ \lesssim 30$).

Another issue that has to be addressed before we can really work on our CFD code is the boundary conditions of $\bar{u}$, k, and ϵ. At the channel centerline i.e. $y = h$, the y derivatives of these variables vanish because of symmetry; at the channel wall ($y = 0$), $\bar{u} = 0$ due to no-slip condition. Since fluid velocity is zero at the wall all the time, there cannot be any velocity fluctuations either. So at the wall the turbulent kinetic energy $k = (\overline{u'^2} + \overline{v'^2} + \overline{w'^2})/2$ is also zero. Moreover, velocity fluctuations very close to the wall may be expressed as Taylor's series

$$u' = a_1 y + a_2 y^2 + \ldots; \quad v' = b_1 y + b_2 y^2 + \ldots; \quad w' = c_1 y + c_2 y^2 + \ldots \tag{8.53}$$

whence $k \sim y^2$ as $y \sim 0$. For this reason not only is $k = 0$ at the wall, dk/dy is also zero.

Although both $\bar{u}$ and k vanish at the channel wall, ϵ does not. In fact ϵ reaches a local maximum at the wall. Notice that (refer to Equation (8.14))

$$\epsilon = \nu \overline{\frac{\partial u_i'}{\partial x_j} \frac{\partial u_i'}{\partial x_j}} \tag{8.54}$$

While $u_i' = 0$ at the wall, the derivative of u_i' (except v', which will be clear later) along the wall-normal (y) direction is not. Therefore $\epsilon \neq 0$ at the wall. To derive the boundary condition of ϵ at the wall, we can use the k equation, Equation (8.26), which reduces to

$$0 = \nu \frac{d^2 k}{dy^2} - \epsilon \tag{8.55}$$

at the wall. This may serve as the wall boundary condition of ϵ. However, numerical tests show it is better to go one step further. Since $k \sim y^2$ as $y \sim 0$, $d^2 k/dy^2 \sim 2k/y^2$, we have

$$\epsilon = 2\nu \lim_{y \to 0} \left(\frac{k}{y^2} \right) \tag{8.56}$$

which is not infinity and can be easily approximated by using the k and y values at the first mesh point off the wall.

The turbulent channel flow is solved by using the $k - \epsilon$ model with 2000 grid points. The results are compared with DNS data of Mansour, Kim, and Moin (Mansour, Kim, & Moin, 1988). The mean velocity distribution is shown in Figure 8.6.

Notice that $y^+ = yu_\tau/\nu$, so that at the channel center $y = h$, $y^+ = hu_\tau/\nu = Re_\tau$.

The distributions of dimensionless turbulent shear stress $\tau_{xy}^+ = \tau_{xy}/u_\tau^2$ and turbulent kinetic energy $k^+ = k/u_\tau^2$ are shown in Figure 8.7 .

The distribution of the normalized turbulent kinetic energy production term $P^+ = P\nu/u_\tau^4$ and dissipation term $\epsilon^+ = \epsilon\nu/u_\tau^4$ are shown in Figure 8.8 . You can see the production is indeed in balance with dissipation in the log-law region $(30 \lesssim y^+ \lesssim 0.2Re_\tau)$. In the immediate proximity to the wall the dissipation, however, prevails over production.

Using the $k - \epsilon$ model, we find that under the given pressure drop the Reynolds number based on the bulk velocity U_b and channel height H is about 1.12×10^4 and the exact value is 1.398×10^4 (Abe, Kawamura, & Matsuo, 2001). The bulk velocity is

$$U_b = \frac{1}{H} \int_0^H \bar{u}\, dy \tag{8.57}$$

The calculated friction coefficient $c_f = 2\tau_w/(\rho U_b^2) \approx 0.01$ which is almost 1.5 times the exact value 0.0065.

Obviously the agreement between the $k - \epsilon$ model and DNS is pretty bad close to the wall but becomes better away from it. It is interesting to note the opposite trend showed by the mixing length model, which performs well close to the wall but getting worse away from it (see Figure 2.15).

Let us see why $k - \epsilon$ does a poor job close to the wall. As mentioned, we may expand the velocity fluctuations very close to a solid wall into Taylor's series

$$u' = a_1 y + a_2 y^2 + \ldots; \quad v' = b_1 y + b_2 y^2 + \ldots; \quad w' = c_1 y + c_2 y^2 + \ldots \tag{8.58}$$

The continuity equation is always valid:

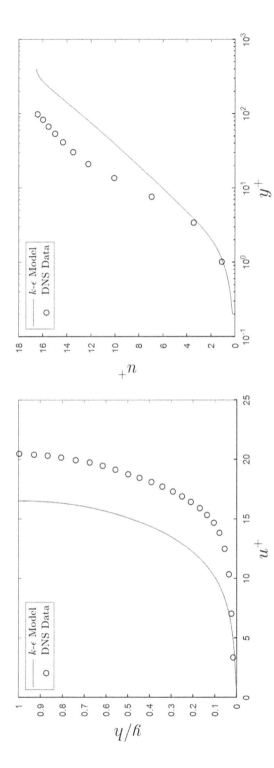

FIGURE 8.6 Mean velocity distribution with the k-ϵ model.

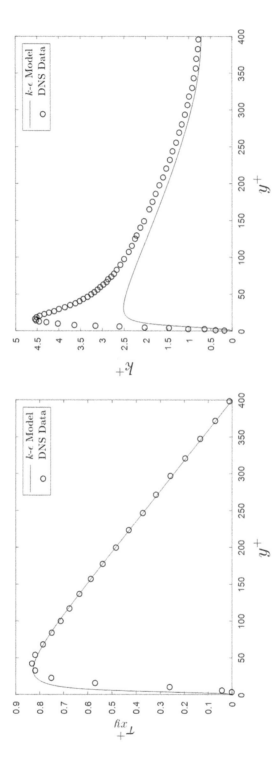

FIGURE 8.7   Turbulent shear stress and turbulent kinetic energy distributions with the k-ϵ model.

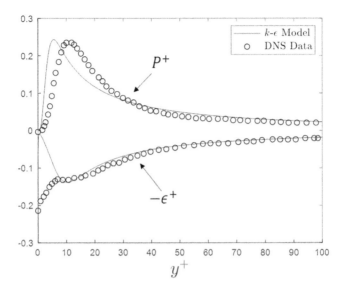

FIGURE 8.8   Distributions of the production term and dissipation term with the k-ϵ model.

$$\frac{\partial u}{\partial x} + \frac{\partial v}{\partial y} + \frac{\partial w}{\partial z} = \frac{\partial \bar{u}}{\partial x} + \frac{\partial \bar{v}}{\partial y} + \frac{\partial \bar{w}}{\partial z} + \frac{\partial u'}{\partial x} + \frac{\partial v'}{\partial y} + \frac{\partial w'}{\partial z} = 0 \qquad (8.59)$$

Since we know that the mean velocity alone satisfies the continuity equation as well (see Equation (8.7)), we have to conclude that the velocity fluctuations also satisfy the continuity equation:

$$\frac{\partial u'}{\partial x} + \frac{\partial v'}{\partial y} + \frac{\partial w'}{\partial z} = 0 \qquad (8.60)$$

Substituting Equation (8.58) into Equation (8.60), we find $\partial v'/\partial y = 0$ very close to a wall ($y{\sim}0$). That is $b_1 = 0$. So very close to a solid wall

$$u'{\sim}y; \quad v'{\sim}y^2; \quad w'{\sim}y \qquad (8.61)$$

Since u' and w' are tangent to the wall, they should not contribute to the turbulent mixing between fluid layers very close to the wall; therefore, the turbulent mixing should be most influenced by v'. As a result the appropriate velocity scale that should be used to evaluate eddy viscosity close to a wall is $v'{\sim}y^2$ and in turn the eddy viscosity $\nu_T {\sim} y^3$ approaching a wall. On the other hand, the turbulent kinetic energy $k{\sim}y^2$ close to a wall so that $\sqrt{k}{\sim}y \gg y^2{\sim}v'$ as $y{\sim}0$. In the $k - \epsilon$ model the velocity scale for eddy viscosity is always $\sqrt{k}$, which causes over-estimation of eddy viscosity close to solid walls as $\nu_T {\sim} y^2$ according to this scaling.

8.2.2 $k - \epsilon$ Model with Wall Models

There are a few ways to fix the scaling problem close to wall in the $k - \epsilon$ model. For example, one may modify the eddy viscosity equation of the $k - \epsilon$ model by adding a damping function to take into account the wall effects

$$\nu_T = C_\mu \frac{k^2}{\epsilon} f\,(y^+) \tag{8.62}$$

Such modifications give the so-called low-Reynolds-number $k - \epsilon$ models. This terminology is a little misleading as such models, like other turbulent models, are used to solve flows at high Reynolds numbers. The term "low-Reynolds-number" refers to the Reynolds number based on local turbulent eddy velocity and length scales, like the one defined in Equation (8.3), which is low approaching the wall.

Another idea is using a different turbulent model in a zone close to the wall and only apply the $k - \epsilon$ model outside this zone, as shown in the left panel of Figure 8.9.

For example, since the mixing length model works well close to walls at least for the channel flow example, it may be used as the wall model. Other popular wall models include the $k - l$ model and the $k - \omega$ model. An even simpler means of wall treatment is directly prescribing the correct turbulence quantities in the wall layer. For example, because the log-law is found to be valid in many wall-bounded turbulent flows, one may specify the $\bar{u}$, k, and ϵ values provided by the log-law at the first node of the $k - \epsilon$ model region, as shown in the right panel of Figure 8.9. One should make sure this first node is located in the log-law region i.e. $30 \lesssim y_1^+ \lesssim 0.2Re_\tau$.

When we use log-law as the wall function, we have

$$\bar{u}_1 = u_\tau \left[\frac{1}{\kappa} \ln\left(y_1^+\right) + C \right]; \quad k_1 = \frac{u_\tau^2}{\sqrt{C_\mu}}; \quad \epsilon_1 = \frac{u_\tau^3}{\kappa y_1} \tag{8.63}$$

which serve as the "wall" boundary conditions for the $k - \epsilon$ model calculation.

Now let us redo the turbulent channel flow calculation with using the mixing length model in the region $0 \le y^+ \le 40$ and $k - \epsilon$ model in $40 \le y^+ \le Re_\tau = 400$. Basically, we will use these two models to evaluate eddy viscosity in their own

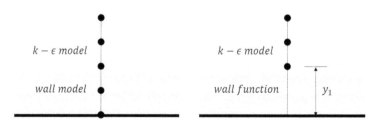

FIGURE 8.9 k-ϵ model with wall model or wall function.

zones. We may use k_1 and ϵ_1 in Equation (8.63) as boundary conditions at $y^+ = 40$ for $k - \epsilon$ model since this grid point is located in the log-law region.

The mean velocity distribution is shown in Figure 8.10. Although still not perfect, the current result is much better than that predicted by $k - \epsilon$ model alone.

The Reynolds stress and turbulent kinetic energy distributions are shown in Figure 8.11 . The turbulent kinetic energy production and dissipation terms are shown in Figure 8.12. Since the mixing length model does not assess k or ϵ, only the k^+ and ϵ^+ values in the $k - \epsilon$ model region are displayed. Again the results improve compared with those obtained from the $k - \epsilon$ model alone.

If we want to find k and ϵ in the wall-adjacent layer, we can use a wall model that calculates these two parameters e.g. the $k - l$ model (Wolfshtein, 1969; Chen & Patel, 1988). This model uses the same k equation as $k - \epsilon$ model but replaces the ϵ equation with an algebraic formula (for this reason $k - l$ model is a one-equation model as it only solves one differential equation)

$$\epsilon = \frac{k^{1.5}}{l_\epsilon} \tag{8.64}$$

And the length scale l_ϵ is evaluated as

$$l_\epsilon = C_l y \left(1 - e^{-\frac{y\sqrt{k}}{\nu A_\epsilon}} \right) \tag{8.65}$$

in which y is the distance from the wall; C_l and A_ϵ are two model constants. The eddy viscosity is calculated by

$$\nu_T = C_\mu \sqrt{k}\, l_\nu \tag{8.66}$$

where the length scale l_ν is found by

$$l_\nu = C_l y \left(1 - e^{-\frac{y\sqrt{k}}{\nu A_\nu}} \right) \tag{8.67}$$

Again A_ν is a model constant. $C_\mu = 0.09$ is the same constant used in the $k - \epsilon$ model.

The similarity between the formulas of length scales l_ϵ and l_ν and the mixing length in the mixing length model is obvious. One may, therefore, conjecture that $C_l \approx \kappa = 0.41$. Unfortunately, this is not the case. Let us see why. In the log-law layer, $\nu_T = \kappa y u_\tau$ and $k \approx u_\tau^2/\sqrt{C_\mu}$. To give the same eddy viscosity with $k - l$ model, we should have

$$\nu_T = C_\mu \sqrt{k}\, l_\nu \approx C_\mu \frac{u_\tau}{\sqrt[4]{C_\mu}} C_l y \approx \kappa y u_\tau \tag{8.68}$$

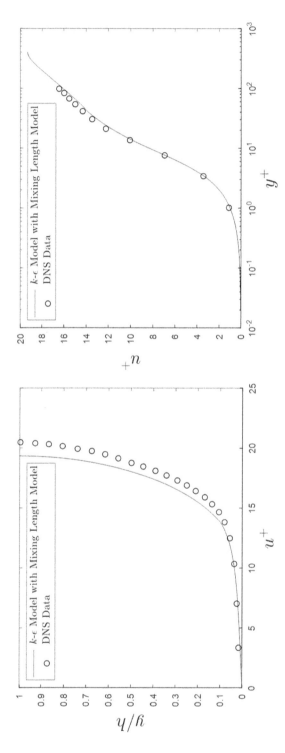

FIGURE 8.10 Mean velocity distribution with the k-ϵ model and mixing length wall model.

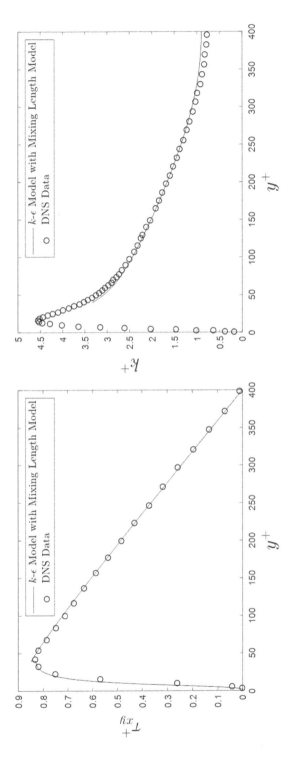

FIGURE 8.11 Turbulent shear stress and turbulent kinetic energy distributions with the k-ϵ model and mixing length wall model.

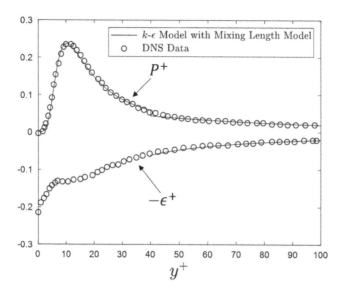

FIGURE 8.12 Distributions of the production term and dissipation term with the k-ϵ model and mixing length wall model.

which gives $C_l \approx \kappa/C_\mu^{0.75} = 2.5$, not 0.41. Notice that in the log-law layer $y\sqrt{k}/(\nu A_\nu)$ is large and the exponential term in l_ν is negligible. The constant A_ν is estimated by fitting the model prediction with experimental data and $A_\nu \approx 50.5$.

The model constant A_ϵ has to be twice C_l i.e. $A_\epsilon = 5$. The reason is as follows. Very close to wall, $k \approx Cy^2$. So based on the k equation, we have

$$\epsilon = \nu \frac{d^2 k}{dy^2} = 2\nu \frac{k}{y^2} \tag{8.69}$$

as $y \sim 0$. To reproduce this relationship, using the ϵ formula of $k - l$ model, we have

$$\epsilon = \frac{k^{1.5}}{C_l y \left(1 - e^{-\frac{y\sqrt{k}}{\nu A_\epsilon}} \right)} = 2\nu \frac{k}{y^2} \tag{8.70}$$

Since $y \sim 0$,

$$e^{-\frac{y\sqrt{k}}{\nu A_\epsilon}} \approx 1 - \frac{y\sqrt{k}}{\nu A_\epsilon} \tag{8.71}$$

Then you may find A_ϵ indeed is equal to $2C_l$ by substituting Equation (8.71) into Equation (8.70).

One problem of the hybrid of $k - l$ and $k - \epsilon$ model is its sensitivity to the location where the two models meet. A good choice is patching two models at $y\sqrt{k}/(\nu A_\nu) = 3$. That means you will have to adjust the two model regions on the fly as k is being updated.

The results of the turbulent channel flow with using the hybrid of $k - l$ and $k - \epsilon$ model are shown in Figures 8.13, 8.14, and 8.15.

Most results agree with DNS data very well, especially the mean velocity and Reynolds stress distributions, which are usually the most interested parameters for engineers. Since Reynolds stress is equal to eddy viscosity times mean velocity gradient, good Reynolds stress and mean velocity results imply good eddy viscosity estimation. The predicted Reynolds number based on the bulk velocity and channel height is 1.42×10^4 which is close to the exact value 1.398×10^4 and so is the friction coefficient which is 0.0064 versus the exact value 0.0065. The prediction of k and ϵ is, however, not satisfactory. In fact the calculated ϵ behavior shows a local minimum instead of maximum at the wall. The predicted ϵ maximum is closer to the real value than the normal $k - \epsilon$ model though.

Another popular option of wall model is the $k - \omega$ model (Wilcox, 1988), which will be discussed next.

8.2.3 $k - \omega$ Model

$k - \omega$ model is a two-equation model used to determine eddy viscosity. It solves two differential equations, one for turbulent kinetic energy k and one for vorticity frequency $\omega = \epsilon/(C_\mu k)$. As a frequency its unit is s^{-1}.

In the standard $k - \omega$ model (Wilcox, 1988), the k equation is of the same form as that used in the $k - \epsilon$ model except the dissipation term ϵ is replaced by $C_\mu k\omega$

$$\frac{\partial k}{\partial t} + \bar{u}_j\frac{\partial k}{\partial x_j} = \nu_T\left(\frac{\partial \bar{u}_i}{\partial x_j} + \frac{\partial \bar{u}_j}{\partial x_i}\right)\frac{\partial \bar{u}_i}{\partial x_j} + \frac{\partial}{\partial x_j}\left[(\nu + \sigma_k\nu_T)\frac{\partial k}{\partial x_j}\right] - C_\mu k\omega \quad (8.72)$$

The ω equation is

$$\frac{\partial \omega}{\partial t} + \bar{u}_j\frac{\partial \omega}{\partial x_j} = \alpha\frac{\omega}{k}\nu_T\left(\frac{\partial \bar{u}_i}{\partial x_j} + \frac{\partial \bar{u}_j}{\partial x_i}\right)\frac{\partial \bar{u}_i}{\partial x_j} + \frac{\partial}{\partial x_j}\left[(\nu + \sigma_\omega\nu_T)\frac{\partial \omega}{\partial x_j}\right] - \beta\omega^2 \quad (8.73)$$

The eddy viscosity is evaluated by

$$\nu_T = C_\mu\frac{k^2}{\epsilon} = \frac{k}{\omega} \quad (8.74)$$

By virtue of Equation (8.74), the ω equation becomes

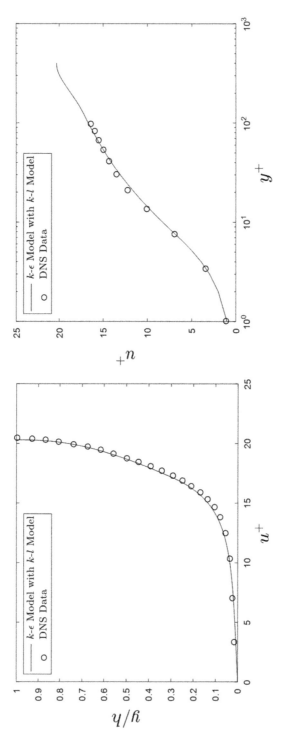

FIGURE 8.13 Mean velocity distribution with the k-ϵ model and k-l wall model.

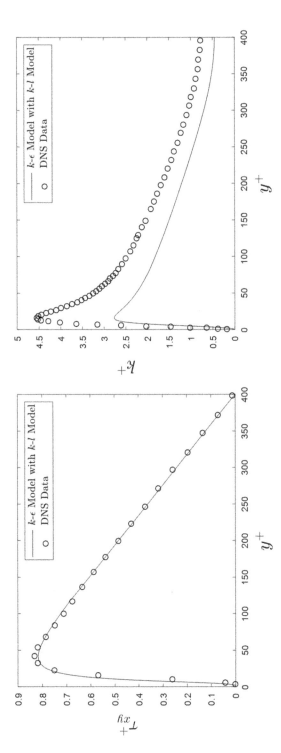

FIGURE 8.14 Turbulent shear stress and turbulent kinetic energy distributions with the k-ϵ model and k-l wall model.

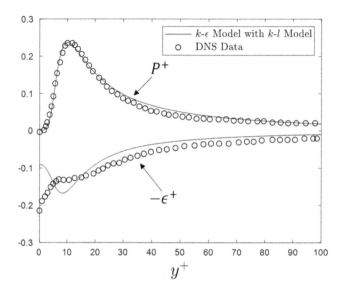

FIGURE 8.15 Distributions of the production term and dissipation term with the k-ϵ model and k-l wall model.

$$\frac{\partial \omega}{\partial t} + \bar{u}_j \frac{\partial \omega}{\partial x_j} = \alpha \left(\frac{\partial \bar{u}_i}{\partial x_j} + \frac{\partial \bar{u}_j}{\partial x_i} \right) \frac{\partial \bar{u}_i}{\partial x_j} + \frac{\partial}{\partial x_j} \left[(\nu + \sigma_\omega \nu_T) \frac{\partial \omega}{\partial x_j} \right] - \beta \omega^2 \quad (8.75)$$

The model constants are

$$\alpha = 0.56, \quad \beta = 0.075, \quad \sigma_k = \sigma_\omega = 0.5, \quad C_\mu = 0.09 \quad (8.76)$$

Since wall gives us most trouble in our simulations of the turbulent channel flow, it is desirable to explore the behavior of the $k - \omega$ model near the wall. Let us first apply the $k - \omega$ model alone to the turbulent channel flow example.

For a statistically steady, fully developed turbulent channel flow, the k equation can be simplified to

$$0 = \nu_T \left(\frac{d\bar{u}}{dy} \right)^2 + \frac{d}{dy} \left[(\nu + \sigma_k \nu_T) \frac{dk}{dy} \right] - C_\mu k\omega \quad (8.77)$$

the ω equation reduces to

$$0 = \alpha \left(\frac{d\bar{u}}{dy} \right)^2 + \frac{d}{dy} \left[(\nu + \sigma_\omega \nu_T) \frac{d\omega}{dy} \right] - \beta \omega^2 \quad (8.78)$$

These two equation are of the same general form

TABLE 8.2

Terms in the General Form Equation for the k-ω Model

ϕ	Γ_ϕ	S_c	S_p
$\bar{u}$	$\nu + \nu_T$	$-\dfrac{1}{\rho}\dfrac{\partial \bar{p}}{\partial x}$	0
k	$\nu + \sigma_k \nu_T$	$\nu_T \left(\dfrac{d\bar{u}}{dy}\right)^2$	$C_\mu \omega$
ω	$\nu + \sigma_\omega \nu_T$	$\alpha \left(\dfrac{d\bar{u}}{dy}\right)^2$	$\beta \omega$

$$0 = \frac{d}{dy}\left(\Gamma_\phi \frac{d\phi}{dy}\right) + S_c - S_p \phi \tag{8.79}$$

The parameters in three governing equations for $\bar{u}$, k, and ω are listed in Table 8.2.

The boundary condition of ω at the channel centerline is $d\omega/dy = 0$. Very close to the channel wall, say in the viscous sublayer, the ω equation can be further simplified to

$$0 = \nu \frac{d^2\omega}{dy^2} - \beta \omega^2 \tag{8.80}$$

We may find that in the viscous sublayer

$$\omega = \omega_{vis} = \frac{6\nu}{\beta y^2} \tag{8.81}$$

which becomes singular at the wall. This is not surprising because $\omega \sim \epsilon/k$ but $k = 0$ at the wall. Similarly one may find that in the log-law region

$$\omega = \omega_{log} = \frac{\sqrt{k}}{\sqrt[4]{C_\mu}\,\kappa y} = \frac{u_\tau}{\sqrt{C_\mu}\,\kappa y} \tag{8.82}$$

ω value can be automatically determined as long as y is inside the law of the wall region (from $y = 0$ to the upper limit of the log-law region) by blending the previous two formulae

$$\omega = \sqrt{\omega_{vis}^2 + \omega_{log}^2} \tag{8.83}$$

To avoid singularity at the wall, we may assume that the ω value at the wall is equal to the ω value at the first point above the wall, which is given by Equation (8.83). The results are shown in Figures 8.16, 8.17, and 8.18.

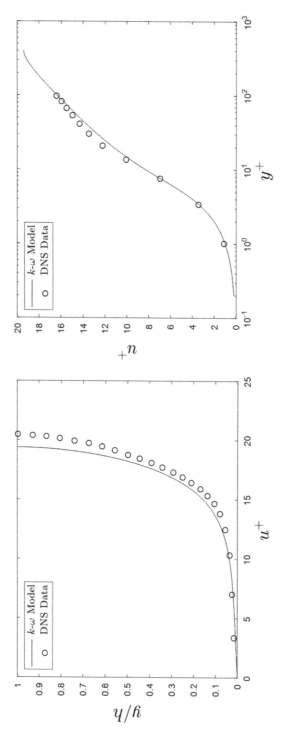

FIGURE 8.16 Mean velocity distribution with the k-ω model.

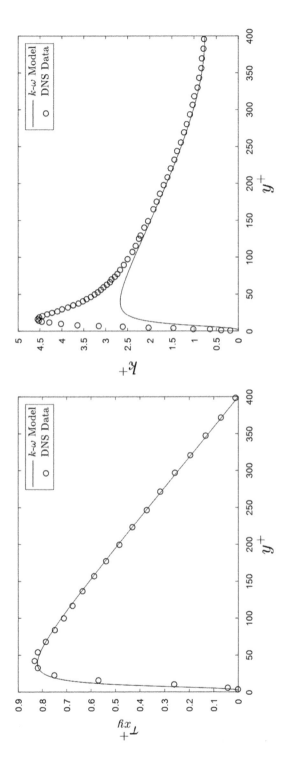

FIGURE 8.17 Turbulent shear stress and turbulent kinetic energy distributions with the k-ω model.

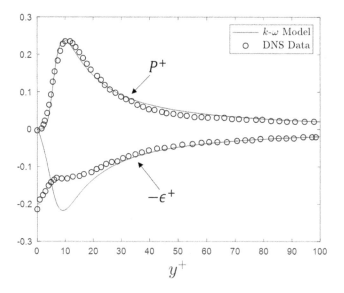

FIGURE 8.18 Distributions of the production term and dissipation term with the k-ω model.

The mean velocity and Reynolds stress profiles are well reproduced by the $k - \omega$ model, especially close to the wall. As a result, the Reynolds number based on bulk velocity and friction coefficient values are also very accurate, which are 1.37×10^4 and 0.0068, respectively. This success is quite impressive since the model itself does not have any damping functions in it.

The computed k and ϵ from $k - \omega$ model still show quite significant discrepancy from the DNS close to the wall, but the predicted ϵ maximum matches the exact value very well.

A hybrid model which transitions from the $k - \omega$ model close to the wall to the $k - \epsilon$ model far away from walls is the shear stress transport (SST) model (Menter, 1994).

8.2.4 SST MODEL

The SST model solves two differential equations. The k equation is

$$\frac{\partial k}{\partial t} + \bar{u}_j \frac{\partial k}{\partial x_j} = P + \frac{\partial}{\partial x_j}\left[(\nu + \sigma_k \nu_T)\frac{\partial k}{\partial x_j}\right] - C_\mu k\omega \qquad (8.84)$$

where the production term

$$P = \min\left[\nu_T\left(\frac{\partial \bar{u}_i}{\partial x_j} + \frac{\partial \bar{u}_j}{\partial x_i}\right)\frac{\partial \bar{u}_i}{\partial x_j},\ 10C_\mu k\omega\right] \qquad (8.85)$$

The ω equation is

$$\frac{\partial \omega}{\partial t} + \bar{u}_j \frac{\partial \omega}{\partial x_j} = \alpha \left(\frac{\partial \bar{u}_i}{\partial x_j} + \frac{\partial \bar{u}_j}{\partial x_i} \right) \frac{\partial \bar{u}_i}{\partial x_j} + 2(1 - F_1) \frac{\sigma_{\omega 2}}{\omega} \frac{\partial k}{\partial x_j} \frac{\partial \omega}{\partial x_j}$$
$$+ \frac{\partial}{\partial x_j} \left[(\nu + \sigma_\omega \nu_T) \frac{\partial \omega}{\partial x_j} \right] - \beta \omega^2$$

(8.86)

in which F_1 is a function that equals one at the wall and becomes zero far from walls. Notice that $\epsilon \sim k\omega$, it is therefore possible to convert the ϵ equation of the $k - \epsilon$ model into a form of an ω equation. This practice results in the $(\partial k/\partial x_j)(\partial \omega/\partial x_j)$ term. Such a term is absent in the standard $k - \omega$ model. So F_1 switches the behavior of the SST model from being like a $k - \omega$ model near the wall to a $k - \epsilon$ model far from the wall. The F_1 function is

$$F_1 = \tanh(\xi^4), \quad \xi = \min \left[\max \left(\frac{\sqrt{k}}{C_\mu \omega y}, \frac{500\nu}{\omega y^2} \right), \frac{4k\sigma_{\omega 2}}{CD_\omega y^2} \right],$$
$$CD_\omega = \max \left(\frac{2\sigma_{\omega 2}}{\omega} \frac{\partial k}{\partial x_j} \frac{\partial \omega}{\partial x_j}, 10^{-10} \right)$$

(8.87)

The eddy viscosity is

$$\nu_T = \frac{\sqrt{C_\mu} k}{\max \left(\sqrt{C_\mu} \omega, SF_2 \right)}$$

(8.88)

in which S is the magnitude of strain rate tensor (see Equation (8.18)) and F_2 is another function that transitions from one at the wall to zero far from walls

$$F_2 = \tanh(\eta^2), \quad \eta = \max \left(\frac{2\sqrt{k}}{C_\mu \omega y}, \frac{500\nu}{\omega y^2} \right)$$

(8.89)

Equation (8.88) is the regular eddy viscosity formula $\nu_T = k/\omega$ when S is small; but when S is large, so that the turbulent kinetic energy production $P = \tau_{ij} \bar{S}_{ij} = \nu_T S^2$ (see Equation (8.17)) is large e.g. when the pressure increases along the flow direction (the so-called adverse pressure gradient), experiments show it is more accurate to relate τ_{ij} with k by Bradshaw assumption $\left| \tau_{ij} \right| \approx \sqrt{C_\mu} k$, so that $\nu_T \approx \sqrt{C_\mu} k/S$.

The model constants are

$$\alpha_1 = 0.56; \quad \alpha_2 = 0.44; \quad \beta_1 = 0.075; \quad \beta_2 = 0.0828; \quad C_\mu = 0.09;$$
$$\sigma_{k1} = 0.85; \quad \sigma_{k2} = 1; \quad \sigma_{\omega 1} = 0.5; \quad \sigma_{\omega 2} = 0.856$$

(8.90)

Those constants with subscript "1" come from the $k - \omega$ model (σ_{k1} is tuned) and constants with subscript "2" essentially from the $k - \epsilon$ model. The model constants α, β, σ_k, and σ_ω in the k and ω equations should be evaluated by formulas like

$$\alpha = \alpha_1 F_1 + \alpha_2 (1 - F_1) \tag{8.91}$$

so that they are also blended between the $k - \omega$ and $k - \epsilon$ models.

Although this model seems incredibly complex, it is still of the same general form of equation

$$0 = \frac{d}{dy}\left(\Gamma_\phi \frac{d\phi}{dy}\right) + S_c - S_p \phi \tag{8.92}$$

The parameters in three governing equations for $\bar{u}$, k, and ω are listed in Table 8.3.

Under-relaxation is needed when you solve the k and ω equations to avoid possible solution instability due to the nonlinear terms. The same boundary conditions used for the $k - \omega$ model still applies. The numerical results of applying the SST model to the fully developed turbulent channel flow at $Re_\tau = 400$ are shown in Figures 8.19, 8.20, and 8.21.

As expected, the SST model improves the mean velocity prediction away from the wall compared with $k - \omega$ model (cf. Figure 8.16). The same k and ϵ profiles close to the wall show up as those given by the $k - \omega$ model, which is not surprising since SST reduces to $k - \omega$ close to the wall.

As we have discussed earlier, $\sqrt{k}$ is not the correct velocity scale for eddy viscosity close to walls. Instead v' contributes more to the mixing process near the wall than the other two velocity fluctuation components (u' and w'), which qualifies v' or more precisely its root-mean-square $\sqrt{\overline{v'^2}}$ as the proper velocity scale. This consideration with other thoughts like how the pressure brings $\overline{v'^2}$ information from the wall to the rest of the flow filed motivated the $\overline{v'^2} - f$ model (Durbin, 1993).

TABLE 8.3

Terms in the General Form Equation for the SST Model

ϕ	Γ_ϕ	S_c	S_p
$\bar{u}$	$\nu + \nu_T$	$-\dfrac{1}{\rho}\dfrac{\partial p}{\partial x}$	0
k	$\nu + \sigma_k \nu_T$	P	$C_\mu \omega$
ω	$\nu + \sigma_\omega \nu_T$	$\alpha\left(\dfrac{d\bar{u}}{dy}\right)^2 + \dfrac{2(1 - F_1)\sigma_{\omega 2}}{\omega}\dfrac{dk}{dy}\dfrac{d\omega}{dy}$	$\beta \omega$

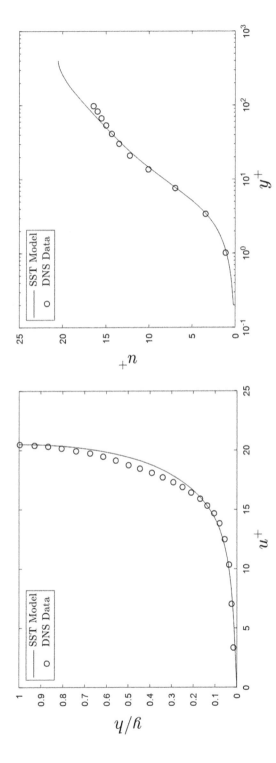

FIGURE 8.19   Mean velocity distribution with the SST model.

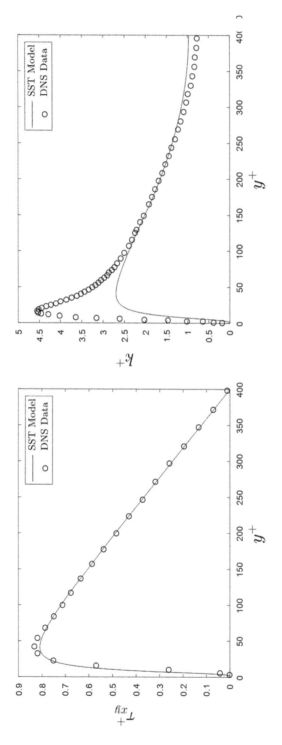

FIGURE 8.20 Turbulent shear stress and turbulent kinetic energy distributions with the SST model.

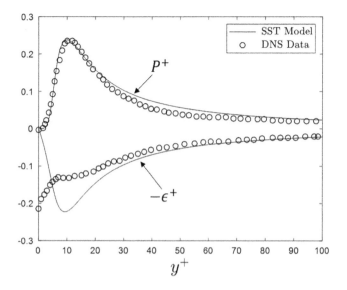

FIGURE 8.21 Distributions of the production term and dissipation term with the SST model.

8.2.5 V2F MODEL

The $\overline{v'^2} - f$ or V2F model is indeed a four-equation model as it has to solve four differential equations: the k equation, ϵ equation, and the $\overline{v'^2}$ as well as f equations. The k equation is the same as the one used in the standard $k - \epsilon$ model

$$\frac{\partial \epsilon}{\partial t} + \frac{\partial \left(\bar{u}_j k \right)}{\partial x_j} = \nu_T \left(\frac{\partial \bar{u}_i}{\partial x_j} + \frac{\partial \bar{u}_j}{\partial x_i} \right) \frac{\partial \bar{u}_i}{\partial x_j} + \frac{\partial}{\partial x_j} \left[(\nu + \sigma_k \nu_T) \frac{\partial k}{\partial x_j} \right] - \epsilon \quad (8.93)$$

And the ϵ equation now reads

$$\frac{\partial \epsilon}{\partial t} + \frac{\partial \left(\bar{u}_j \epsilon \right)}{\partial x_j} = \frac{1}{\tau} \left[C_{\epsilon 1} \nu_T \left(\frac{\partial \bar{u}_i}{\partial x_j} + \frac{\partial \bar{u}_j}{\partial x_i} \right) \frac{\partial \bar{u}_i}{\partial x_j} - C_{\epsilon 2} \epsilon \right] + \frac{\partial}{\partial x_j} \left[(\nu + \sigma_\epsilon \nu_T) \frac{\partial \epsilon}{\partial x_j} \right]$$

$$(8.94)$$

Compared with the ϵ equation (Equation (8.19)) used in the $k - \epsilon$ model, the time scale before the production and dissipation terms of the ϵ equation in the V2F model is not the large eddy time scale $\tau_L = k/\epsilon$ but

$$\tau = \max \left(\tau_L, 6\tau_\eta \right) = \max \left(\frac{k}{\epsilon}, 6\sqrt{\frac{\nu}{\epsilon}} \right) \quad (8.95)$$

which is bounded by the small eddy time scale τ_η (Kolmogorov scale). The idea is that very close to the wall, the "large" eddies are no longer large, and their behavior become more and more like small eddies approaching the wall. Hence the Kolmogorov scale is used as the lower limit in Equation (8.95).

The $\overline{v'^2}$ equation is

$$\frac{\partial \overline{v'^2}}{\partial t} + \frac{\partial \left(\bar{u}_j \, \overline{v'^2} \right)}{\partial x_j} = kf - 6\epsilon \frac{\overline{v'^2}}{k} + \frac{\partial}{\partial x_j}\left[(\nu + \nu_T)\frac{\partial \overline{v'^2}}{\partial x_j} \right] \tag{8.96}$$

where kf is the production term for the Reynolds stress $\overline{v'^2} = \overline{v'v'}$ and the $6\epsilon\overline{v'^2}/k$ is its dissipation term. f is solved from

$$f - l^2 \frac{\partial^2 f}{\partial x_k \partial x_k} = \frac{1}{\tau}\left[(5 - C_1)\frac{\overline{v'^2}}{k} + \frac{2}{3}C_1 \right] + C_2\frac{P}{k} \tag{8.97}$$

where P is the production term of the k equation

$$P = \nu_T\left(\frac{\partial \bar{u}_i}{\partial x_j} + \frac{\partial \bar{u}_j}{\partial x_i} \right)\frac{\partial \bar{u}_i}{\partial x_j} \tag{8.98}$$

and l is the length scale

$$l = 0.2 \max\left(l_L, 70 l_\eta \right) = 0.2 \max\left[\frac{k^{1.5}}{\epsilon}, 70\left(\frac{\nu^3}{\epsilon} \right)^{0.25} \right] \tag{8.99}$$

which again is limited by the Kolmogorov length scale l_η.

The eddy viscosity is evaluated by

$$\nu_T = C_\mu \overline{v'^2}\tau \tag{8.100}$$

The model constants are listed below

$$C_\mu = 0.22; \ C_{\epsilon 1} = 1.4\left(1 + 0.045\sqrt{\frac{k}{\overline{v'^2}}} \right); \ C_{\epsilon 2} = 1.9; \ \sigma_\epsilon = 0.77; \tag{8.101}$$
$$C_1 = 0.4; \ C_2 = 0.3$$

The boundary conditions at the wall are

$$k = \overline{v'^2} = f = 0; \; \epsilon = 2v\frac{k}{y^2} \tag{8.102}$$

At the channel center the y-derivatives of all these variables vanish. Notice that the model constant $C_{\epsilon 1}$ depends on k and $\overline{v'^2}$ values. We may simply use $\overline{v'^2} = 2k/3$ as an initial guess to start the calculation. The reason will be clear later. The order of variables we solve for is first k, then ϵ, then f and finally $\overline{v'^2}$ in every iteration.

The numerical results of the fully developed turbulent channel flow at $Re_\tau = 400$ with using the V2F model are shown in Figures 8.22, 8.23, and 8.24.

As we can see the predictions of V2F model are very good, especially the k and ϵ distributions which agree with DNS data very well. ϵ profile given by the V2F model now has a maximum at the wall as it should. The computed Reynolds number based on bulk velocity and channel height is 1.4×10^4 and the friction coefficient is 0.0065. These great improvements in results confirm the merit of rationale behind the V2F model e.g. using $\sqrt{\overline{v'^2}}$ as velocity scale and limiting the length scale and time scale by Kolmogorov scales.

All the turbulence models discussed hitherto make use of Boussinesq approximation which presumes that Reynolds stress is proportional to the corresponding component of strain rate tensor and the coefficient of proportionality v_T is the same for all Reynolds stress components at a specific location and time moment. None of these postulations are universal truth. In fact one may easily find their flaws based on physical reasoning. For example, since v_T is mainly contributed by the large eddies, which are usually affected by the mean flow field and hence are not isotropic, obviously we should expect v_T should also show anisotropy instead of being the same for all Reynolds stress components. Experimental observation also shows discrepancies from this assumption. For example, in the fully developed turbulent channel flow, since $\partial \bar{u}/\partial x = 0$, we have $\tau_{xx} = -\overline{u'u'} = -\overline{u'^2} = 2v_T (\partial \bar{u}/\partial x) - 2k/3 = -2k/3$; similarly we may find $-\overline{v'^2} = -\overline{w'^2} = -2k/3$ since $\bar{v} = \bar{w} = 0$ i.e. the mean-squares of u', v', and w' should the same at any given point in the channel according to Boussinesq approximation (this is why we use $2k/3$ as initial guess for $\overline{v'^2}$ to start the V2F model calculation). However this is not true. DNS (Mansour, Kim, & Moin, 1988) shows that $\overline{u'^2}$ is usually much higher than $\overline{v'^2}$ and $\overline{w'^2}$ at the same point in a turbulent channel flow. Another example is the effect of streamline curvature on turbulence. Since a very small streamline curvature only causes a little variation of mean velocity along the normal-to-streamline direction, it should not produce any appreciable change in Reynolds stress according to Boussinesq approximation. Yet it has been observed (Bradshaw, 1973; Thompson & Whitelaw, 1985) that even very small degrees of streamline curvature can significantly change the Reynolds stress level.

It is clear that in order to rectify such problems, one might have to make changes to or even give up the Boussinesq approximation. As a matter of fact, we have already done this a few times when we used the Bradshaw assumption to relate Reynolds stress with turbulent kinetic energy instead of mean strain rate (see e.g. Equation (8.88)). Following this path one may largely abandon the Boussinesq approximation and directly solve differential equations for the Reynolds stresses,

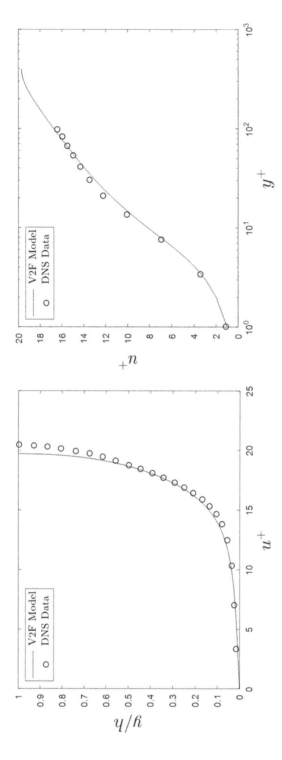

FIGURE 8.22 Mean velocity distribution with the V2F model.

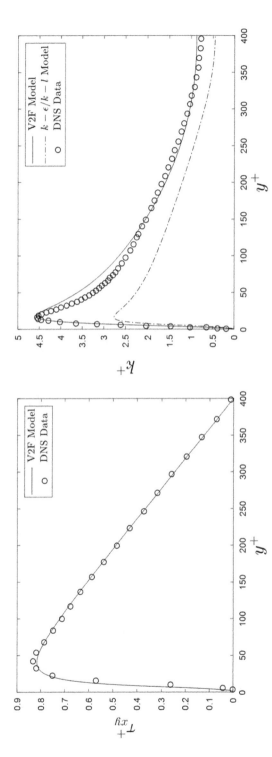

FIGURE 8.23 Turbulent shear stress and turbulent kinetic energy distributions with the V2F model.

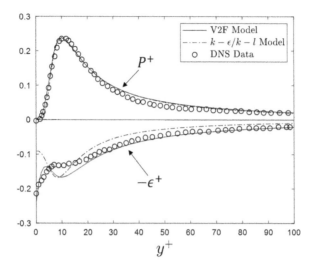

FIGURE 8.24 Distributions of the production term and dissipation term with the V2F model.

which can be derived from the instantaneous and RANS equations. In such differential equations we have terms more complicated than Reynolds stresses that have to be modeled (the recurrent closure problem). These models are called Reynolds stress models. For more details of such models, consult Pope (2000).

Before we close this section, let us study how turbulence affects the other transport phenomena e.g. heat transfer.

8.2.6 EXAMPLE: TURBULENT PIPE FLOW WITH HEAT TRANSFER

As we know one of the most outstanding features of turbulent flows is that the turbulent eddies enhance mixing processes in the flow. This applies to heat transfer as well. In a turbulent flow, the turbulent eddies transport fluid particles from one place to another continuously. If there exists a temperature difference between these two places, this mixing process will help annul the difference just like the normal heat diffusion does. This effect therefore can be taken into account by adding an extra diffusion term in the thermal energy equation

$$\frac{\partial \bar{T}}{\partial t} + \frac{\partial (u_j \bar{T})}{\partial x_j} = \frac{\partial}{\partial x_j}\left[(\alpha + \alpha_T)\frac{\partial \bar{T}}{\partial x_j}\right] \tag{8.103}$$

in which α is the thermal diffusivity of the fluid and α_T is the so-called turbulent thermal diffusivity. The situation is pretty alike the normal viscosity versus turbulent eddy viscosity in momentum equations. And just like turbulent eddy viscosity, α_T is also not a physical property of the fluid but being determined by the flow field.

Since the underlying mechanisms for turbulent eddy viscosity and turbulent thermal diffusivity is the same mixing caused by the eddy motion, we expect that these two parameters are proportional to each other

$$Pr_T = \frac{\nu_T}{\alpha_T} \approx constant \tag{8.104}$$

in which Pr_T is the turbulent Prandtl number and its value is about 0.9~1.0. Notice that there is a normal Prandtl number $Pr = \nu/\alpha$ which is a physical property of the fluid. For air, $Pr \approx 0.7$ at room temperature; for water $Pr \approx 6$ at room temperature.

In summary, when we solve a turbulent flow with heat transfer problem, we will have to solve the continuity equation, the momentum equations and the energy equation, in which additional diffusion terms due to turbulence should be included. These equations might be mutually dependent as temperature changes may result in physical property (ρ, μ, Pr, etc.) changes, which may prompt velocity changes through the momentum equations and velocity changes cause ν_T and Pr_T to change, which in turn induces temperature changes via energy equation. However, if the temperature change is not significant, we may treat all physical properties as constants then the velocity field is no longer affected by the temperature distribution. In such situations we can solve the continuity and momentum equations first for the velocity field. After we have a good velocity field, we solve the energy equation for temperature. Let us see one example.

Consider air at temperature T_{in} entering a long tube which is heated up by electrical heating units uniformly distributed over the pipe surface. Suppose the heat flux is a constant. The air mass flow rate $\dot{m}$ is also constant and high enough so that the flow is turbulent. The set up is shown in Figure 8.25.

Due to the constant heat flux at the surface, thermal energy of the fluid in the pipe increases linearly with x

$$q_w'' 2\pi R \Delta x = \dot{m} c_p \Delta \bar{T}_b \rightarrow \frac{d\bar{T}_b}{dx} = \frac{2\pi R q_w''}{\dot{m} c_p} \tag{8.105}$$

in which R is the pipe radius and $\bar{T}_b(x)$ is the bulk fluid temperature at location x.

After a certain distance, the air flow reaches a hydrodynamically and thermally fully developed state, in which the fluid mean velocity distribution as well as the shape of fluid mean temperature distribution stops changing along the flow direction, as shown in Figure 8.25.

In the fully developed region, the mean velocity $\bar{u}$ is only a function of radial distance r: $\bar{u} = \bar{u}(r)$; The pipe wall temperature T_w changes with x, and so is the fluid bulk temperature $\bar{T}_b$, but they change at the same rate as the mean fluid temperature in the fully developed region i.e.

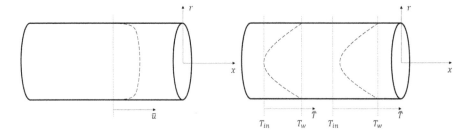

FIGURE 8.25 Sketches of the mean velocity and temperature distributions in the turbulent pipe flow.

$$\frac{dT_w}{dx} = \frac{d\bar{T}_b}{dx} = \frac{\partial\bar{T}}{\partial x} \tag{8.106}$$

Therefore the nondimensional temperature difference

$$\tilde{T} = \frac{T_w(x) - \bar{T}(r, x)}{T_w(x) - \bar{T}_b(x)} \tag{8.107}$$

does not change with x and is only a function of r.

The heat transfer characteristics of the flow may be condensed to the Nusselt number

$$Nu_D = \frac{D\left|\frac{d\tilde{T}}{dr}\right|_{wall}}{T_w - \bar{T}_b} = D\left|\frac{d\tilde{T}}{dr}\right|_{wall} \tag{8.108}$$

in which $D = 2R$ is the pipe diameter. Nu_D is a constant in the fully developed region. This constant is a function of the flow Reynolds number and the fluid Prandtl number. We will want to find out the Nusselt number in the turbulent pipe flow when Reynolds number $Re_D = U_b D/\nu = 10^5$ and Prandtl number $Pr = 0.7$.

Cylindrical coordinate system is used for this problem. In this system the x-momentum equation is

$$0 = -\frac{1}{\rho}\frac{d\bar{p}}{dx} + \frac{1}{r}\frac{d}{dr}\left[r(\nu + \nu_T)\frac{d\bar{u}}{dr}\right] \tag{8.109}$$

for this statistically steady fully developed turbulent flow. The pressure gradient is a constant. The energy equation reduces to

$$\frac{\partial(\bar{u}\bar{T})}{\partial x} = \bar{u}\frac{\partial \bar{T}}{\partial x} = \frac{1}{r}\frac{\partial}{\partial r}\left[r\left(\frac{\nu}{Pr} + \frac{\nu_T}{Pr_T}\right)\frac{\partial \bar{T}}{\partial r}\right] \tag{8.110}$$

The turbulent Prandtl number Pr_T is set to 0.97.

It will be beneficial to render these equations in the nondimensional form by defining

$$\tilde{u} = \frac{\bar{u}}{U_b}, \ \tilde{T} = \frac{T_w - \bar{T}}{T_w - \bar{T}_b}, \ \tilde{r} = \frac{r}{R}, \ \tilde{x} = \frac{x}{R}, \ \tilde{\nu}_T = \frac{\nu_T}{\nu}, \ \tilde{p} = \frac{\bar{p}}{\rho U_b^2} \tag{8.111}$$

The two governing equations then become

$$0 = -\frac{d\tilde{p}}{d\tilde{x}} + \frac{2}{Re_D}\frac{1}{\tilde{r}}\frac{d}{d\tilde{r}}\left[\tilde{r}(1 + \tilde{\nu}_T)\frac{d\tilde{u}}{d\tilde{r}}\right] \tag{8.112}$$

and

$$\tilde{u}\frac{dT_w}{d\tilde{x}} = -\frac{2}{Re_D}\frac{1}{\tilde{r}}\frac{d}{d\tilde{r}}\left[\tilde{r}\left(\frac{1}{Pr} + \frac{\tilde{\nu}_T}{Pr_T}\right)\frac{d\tilde{T}}{dr}\right](T_w - \bar{T}_b) \tag{8.113}$$

which is equivalent to

$$\frac{1}{T_w - \bar{T}_b}\frac{dT_w}{d\tilde{x}} = -\frac{2}{\tilde{u}Re_D}\frac{1}{\tilde{r}}\frac{d}{d\tilde{r}}\left[\tilde{r}\left(\frac{1}{Pr} + \frac{\tilde{\nu}_T}{Pr_T}\right)\frac{d\tilde{T}}{dr}\right] \tag{8.114}$$

Since the left side of Equation (8.114) is only dependent on x and the right side depends only on r, both sides must equal to a constant, say λ. We then find that

$$\frac{1}{\tilde{r}}\frac{d}{d\tilde{r}}\left[\tilde{r}\left(\frac{1}{Pr} + \frac{\tilde{\nu}_T}{Pr_T}\right)\frac{d\tilde{T}}{dr}\right] = -\frac{Re_D\lambda}{2}\tilde{u} \tag{8.115}$$

Similarly, Equation (8.112) can be written as

$$\frac{1}{\tilde{r}}\frac{d}{d\tilde{r}}\left[\tilde{r}(1 + \tilde{\nu}_T)\frac{d\tilde{u}}{d\tilde{r}}\right] = -\frac{Re_D}{2}\left(-\frac{d\tilde{p}}{d\tilde{x}}\right) \tag{8.116}$$

Now let us focus on Equation (8.116). To solve this equation for $\tilde{u}$, we will need to know $\tilde{\nu}_T$, pressure gradient $d\tilde{p}/d\tilde{x}$ and two boundary conditions. ν_T can be determined by a turbulent model; the pressure gradient can be related to $\tilde{u}$ by the overall force balance consideration as what we did in Section 8.2.1 (see Equations (8.37) and the paragraphs therein). You may find that

$$-\frac{d\tilde{p}}{d\tilde{x}} = \frac{4}{Re_D}\left.\frac{d\tilde{u}}{d\tilde{r}}\right|_{wall} \tag{8.117}$$

Therefore Equation (8.116) is

$$\frac{1}{\tilde{r}}\frac{d}{d\tilde{r}}\left[\tilde{r}(1+\tilde{v}_T)\frac{d\tilde{u}}{d\tilde{r}}\right] = -2\left.\frac{d\tilde{u}}{d\tilde{r}}\right|_{wall} \tag{8.118}$$

And the boundary conditions are

$$\frac{d\tilde{u}}{d\tilde{r}} = 0 \ \ at \ \ \tilde{r} = 0; \ \tilde{u} = 0 \ \ at \ \ \tilde{r} = 1 \tag{8.119}$$

The no-slip condition at the pipe wall may need modification per wall treatment adopted by turbulent models, as indicated in Section 8.2.2.

You might be a little concerned because Reynolds number drops out of Equation (8.118) so it seems that we will have the same velocity distribution no matter what Reynolds numbers may be. This is of course not true. In fact $\tilde{v}_T$ is a function of Reynolds number. For example if we use the mixing length model

$$v_T = l_{eddy}^2\left|\frac{du}{dr}\right| \rightarrow \tilde{v}_T = \frac{v_T}{v} = \frac{1}{v}\left(R^2\tilde{l}_{eddy}^2\right)\left|\frac{U_b d\tilde{u}}{R d\tilde{r}}\right| = \frac{Re_D}{2}\tilde{l}_{eddy}^2\left|\frac{d\tilde{u}}{d\tilde{r}}\right| \tag{8.120}$$

Nikuradse (Nikuradse, 1950) suggested the following expression for the mixing length

$$\tilde{l}_{eddy} = \frac{l_{eddy}}{R} = 0.14 - 0.08\left(1 - \frac{y}{R}\right)^2 - 0.06\left(1 - \frac{y}{R}\right)^4 \tag{8.121}$$

where y is the distance from the nearest pipe surface. The van Driest damping function has to be added to this formula near the pipe wall. The procedure to calculate the mean velocity distribution in the pipe is

1. Guess an initial $\tilde{u}$ distribution.
2. Use a turbulent model to evaluate the eddy viscosity v_T.
3. Solve Equation (8.118) and its boundary conditions for $\tilde{u}$.
4. Repeat steps (2) and (3) until convergence is reached.

Once the velocity is found, we may go ahead to solve the temperature distribution.

To solve Equation (8.115), we have to know λ. To this end let us define a new variable θ

$$\tilde{T} = \lambda\theta \tag{8.122}$$

Equation (8.115) then becomes

$$\frac{1}{\tilde{r}}\frac{d}{d\tilde{r}}\left[\tilde{r}\left(\frac{1}{Pr}+\frac{\tilde{\nu}_T}{Pr_T}\right)\frac{d\theta}{dr}\right] = -\frac{Re_D}{2}\tilde{u} \tag{8.123}$$

which is subject to boundary conditions

$$\frac{d\theta}{d\tilde{r}} = 0 \ at \ \tilde{r} = 0; \ \theta = 0 \ at \ \tilde{r} = 1 \tag{8.124}$$

Now we do not have λ in both governing equation and its boundary conditions. So we should have no difficulty to solve for θ. Yet we still need λ to really calculate the temperature distribution $\tilde{T}$ after we obtain θ. This is done with the aid of the definition of bulk temperature $\bar{T}_b$.

For an incompressible flow with constant specific heat, the bulk temperature is

$$\bar{T}_b = \frac{1}{U_b \pi R^2} \int_0^R \bar{u}\bar{T}2\pi r dr \tag{8.125}$$

By using the definitions of $\tilde{T}$ and θ we can derive the following expression for λ from Equation (8.125)

$$\lambda = \frac{1}{2\int_0^1 \tilde{u}\theta\tilde{r}d\tilde{r}} \tag{8.126}$$

Therefore, we can directly compute the λ value once we have the $\tilde{u}$ and θ distributions. After we have λ and in turn $\tilde{T}$, we can easily evaluate the Nusselt number using Equation (8.108), which is equivalent to

$$Nu_D = 2\left|\frac{d\tilde{T}}{d\tilde{r}}\right|_{wall} \tag{8.127}$$

The Nusselt numbers calculated by using the mixing length model is shown in Figure 8.26. The numerical result is compared with an empirical formula, namely the Petukhov–Gnielinski equation (Incropera & DeWitt, 2007)

$$Nu_D = \frac{\left(\frac{f}{8}\right)(Re_D - 1000)Pr}{1 + 12.7\left(\frac{f}{8}\right)^{\frac{1}{2}}\left(Pr^{\frac{2}{3}} - 1\right)}; f = \frac{1}{[1.82\log_{10}(Re_D) - 1.64]^2} \tag{8.128}$$

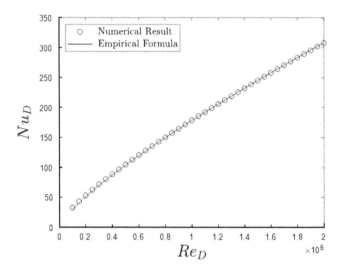

FIGURE 8.26 Nusselt numbers at different Reynolds numbers.

8.3 LARGE EDDY SIMULATION

RANS methods are popular choices for solving complex industrial turbulent flows due to their very competitive performance to cost ratio. However, the theoretical foundation of these methods is not without defects. With using Reynolds averaging, all turbulent fluctuations due to turbulent eddies of all scales are smoothed out from the RANS equations and their effects are modeled. The efficiency and economy of RANS methods rely on their effort to simulate the effects of all turbulent eddies with one universal model. One may, nonetheless, argue that such a universal model does not exist because the characteristics of large eddies are significantly influenced by physical boundaries of the flow field. It is hard to believe one model can cover the effects of myriads of different physical boundary conditions in the real world on the large eddies.

On the other hand, the small eddies in a turbulent flow usually do not interact with the fluid motion at the largest scale directly, therefore they are quite insensitive to different physical boundary conditions, which may justify a universal model for them.

Based on these considerations, large eddy simulation (LES) was proposed (Smagorinsky, 1963; Deardorff, 1970) to simulate turbulent flows. In LES, the large-scale motions are computed directly by solving filtered Navier–Stokes equations in a time-accurate manner while the small-scale motions are modeled. Since the large scale motion is intrinsically 3-D and unsteady, LES is also 3-D, unsteady simulations.

8.3.1 FILTERING

To separate the large eddies and small eddies, a filtering operation is applied to the flow field, so that the flow motion on scales smaller than the filter width Δ is smoothed out. The filtering operation is formally defined as (Leonard A., 1974)

$$\bar{\phi}(\vec{x}, t) = \int G(\vec{r}; \Delta)\phi(\vec{x} - \vec{r}, t)d\vec{r} \qquad (8.129)$$

where ϕ is the original flow field variable, $\bar{\phi}$ the filtered variable, and G is the filter function. The integration is carried out over the entire flow domain. The G function has to satisfy the normalization condition

$$\int G(\vec{r}; \Delta)d\vec{r} = 1 \qquad (8.130)$$

Let us see a specific filter function, namely the top-hat filter, which is

$$G(\vec{r}; \Delta) = \begin{cases} \frac{1}{\Delta^3} & if \ |\vec{r}| \le \frac{\Delta}{2} \\ 0 & if \ |\vec{r}| > \frac{\Delta}{2} \end{cases} \qquad (8.131)$$

The filtered variable $\bar{\phi}$ at a certain point in the flow field with using the top-hat filter is simply the average of ϕ over a volume Δ^3 around this point. You can see the filtering operation is indeed a local spatial averaging with a weighting function G. Other frequently adopted filters in LES studies include the sharp spectral filter and Gaussian filter (Pope, 2000).

As a result of the filtering operation, any flow field can be decomposed as

$$\phi = \bar{\phi} + \phi_{sgs} \qquad (8.132)$$

in which $\bar{\phi}$ is the filtered or locally averaged field and ϕ_{sgs} is the residual, or the so-called subgrid-scale (SGS) component. Although this decomposition looks analogous to the Reynolds decomposition, cf. Equation (8.6), they do have important differences: LES filtering is a spatial averaging rather than a temporal or ensemble averaging. For a statistically stable turbulent flow, the Reynolds average of any flow variable is time independent; on the contrary the filtered flow variable $\bar{\phi}$ fluctuates continuously. Moreover, the Reynolds average of fluctuations at any given point is zero, but the filtered subgrid-scale component is typically not zero: $\bar{\phi}_{sgs} \ne 0$. Unlike the Reynolds average, the twice filtered variable is also in general not equal to the one-time filtered variable: $\bar{\bar{\phi}} \ne \bar{\phi}$. This is because filtering is a local spatial averaging procedure, so it does not iron out all fluctuations at once. If you filter a field twice, it becomes smoother than the field filtered for the first time. Let me use a metaphor to explain this point. Suppose we ask all residents of a community to stand side by side in a row and have each of them calculate the average age of the two persons immediately next to them and her/himself. This average age is assigned as the filtered age of that person. Obviously, we can ask them to do the same calculation again, but this time averaging their filtered ages. The new average each one finds, which is the twice filtered age, should be different from the previous one. With the same analogy you may verify that $\bar{\phi}_{sgs} \ne 0$.

The top-hat filter is most commonly used in conjunction with finite volume method because if we choose the cubic root of the volume of the computational cell as the filter width Δ

$$\Delta^3 = \Delta x \Delta y \Delta z \tag{8.133}$$

the top-hat filtered variable $\bar{\phi}$ is nothing but the volume average of ϕ in the control volume, which is exactly our tacit understanding of the ϕ value at a control volume center when we use finite volume method. Therefore, whenever we use the finite volume method we have already applied an implicit top-hat filter to the flow field without probably noticing it. No more explicit filtering is needed.

If we apply the filtering operation to Navier–Stokes equations, we obtain the filtered governing equations

$$\frac{\partial \bar{u}_i}{\partial x_i} = 0$$

$$\frac{\partial \bar{u}_i}{\partial t} + \frac{\partial \overline{u_j} \bar{u}_i}{\partial x_j} = -\frac{\partial \bar{p}}{\partial x_i} + \nu \frac{\partial^2 \bar{u}_i}{\partial x_j \partial x_j} + \frac{\partial (u_j \bar{u}_i - \overline{u_j u_i})}{\partial x_j} + \bar{f}_i \tag{8.134}$$

in which f_i is the component of force (per unit fluid volume) acting along the ith direction. Also note that $\bar{p}$ in fact is pressure divided by fluid density. The extra term $\overline{u_j} \bar{u}_i - \overline{u_j u_i}$ on the right side is the subgrid-scale stress

$$\tau_{ij}^{sgs} = \overline{u_j} \bar{u}_i - \overline{u_j u_i} = \overline{u_j} \bar{u}_i - \overline{(\bar{u}_j + u_j^{sgs})(\bar{u}_i + u_i^{sgs})} \tag{8.135}$$

which obviously requires modeling because we do not directly solve the subgrid-scale variables like u_i^{sgs}. Notice that the filtering operation is a spatial averaging, so it does not affect a time derivative. That is, the filtered time derivative of a variable ϕ is equal to the time derivative of the filtered ϕ. This is also true for spatial derivatives as long as the filter width is constant.

As stated at the beginning of this section, if the grid we use in LES is small enough, there may exist a universal model to represent the effects of the subgrid scales i.e. the subgrid-scale stress.

8.3.2 Subgrid-Scale Stress Models

Since the pioneering work of Smagorinsky (1963), Deardorff (1970), and Schumann (1975), numerous subgrid-scale or SGS stress models have been proposed. A review can be found in Sagaut (2006). The most widely used SGS modeling strategy is still the eddy viscosity concept we have been familiar with in RANS models (cf. Equation (8.10))

$$\tau_{ij}^{sgs} - \frac{1}{3}\tau_{kk}^{sgs}\delta_{ij} = 2\nu_{sgs}\left(\bar{S}_{ij} - \frac{1}{3}\bar{S}_{kk}\delta_{ij}\right) \tag{8.136}$$

in which

$$\bar{S}_{ij} = \frac{1}{2}\left(\frac{\partial \bar{u}_i}{\partial x_j} + \frac{\partial \bar{u}_j}{\partial x_i}\right) \tag{8.137}$$

is the filtered (grid-scale) strain rate tensor. Again we may absorb the τ_{kk}^{sgs} term into the pressure gradient. $\bar{S}_{kk}$ vanishes for an incompressible flow. Then the only question left is how to determine the SGS viscosity ν_{sgs}. The simplest model is that of Smagorinsky (1963), which is an analogy for the mixing length model

$$\nu_{sgs} = u_{sgs}l_{sgs} = l_{sgs}^2 S = (C_s\Delta)^2 S \tag{8.138}$$

where C_s is the Smagorinsky coefficient and Δ is the filter width, which is usually simply set to the grid size. In the standard Smagorinsky model, $C_s = 0.18$. However, it has been found that this value is too large based on numerical tests. For example, Moin and Kim (1982) found that a lower value $C_s = 0.065$ in simulation of fully developed turbulent channel flows gave better agreement with experimental data.

Although looks just like the mixing length model, the Smagorinsky model uses grid size instead of distance from the wall as the length scale. A problem of the Smagorinsky model is that, like the mixing length model, it cannot reproduce the correct behavior of ν_{sgs} near the wall. Very close to the wall, we should have $\nu_{sgs} \sim y^3$ (see the last paragraph of Section 8.2.1). The Smagorinsky model, on the other hand leads, incorrectly, to a ν_{sgs} being independent of y near the wall. One way to alleviate this problem is adding a damping function e.g. the van Driest damping function to the length scale

$$\nu_{sgs} = u_{sgs}l_{sgs} = l_{sgs}^2 S = \left[C_s\Delta\left(1 - e^{-\frac{y^+}{A^+}}\right)\right]^2 S \tag{8.139}$$

where $A^+ \approx 26$.

Let us simulate the fully developed turbulent channel flow with using LES and Smagorinsky model.

An incompressible fluid flows through a channel formed by two parallel flat plates. The flow domain is shown in Figure 8.27. The mean flow is along the x direction. Notice that the flow domain size is not arbitrary. In fact we have to make sure the domain is larger than the size of the biggest eddies. One way to tell whether this is the case is to check if the flow becomes statistically uncorrelated at a separation of one half-size of the channel in the two homogeneous directions (the x and z directions). That is the two-point correlation of velocity fluctuations between

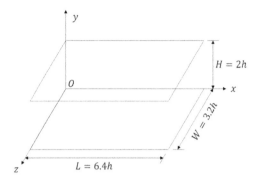

FIGURE 8.27 Turbulent channel flow setup for the LES.

such two points should be zero. The two-point correlation is defined as the mean (i.e. Reynolds-averaged, denoted by $<<\cdot>>$) product of velocity fluctuations at two points:

$$R_{ij}(\vec{x},\,\vec{r},\,t) = <\bar{u}_i'(\vec{x},\,t)\bar{u}_j'(\vec{x}+\vec{r},\,t)> \tag{8.140}$$

The velocity fluctuations at two points, no matter how far apart, along the same eddy are usually correlated i.e. $R_{ij} \neq 0$. A fluctuation is simply the deviation from the mean value:

$$\bar{u}_i'(\vec{x},\,t) = \bar{u}_i(\vec{x},\,t) - <\bar{u}_i(\vec{x},\,t)> \tag{8.141}$$

The flow domain size in fact is determined based on the DNS two-point correlation data of the same flow (Kim, Moin, & Moser, 1987).

We will simulation the case with the Reynolds number $Re_\tau = 395$ based on the friction velocity and channel half-height h, which is almost the same as the the the one we have calculated with RANS models.

The governing equations are the filtered Navier–Stokes equations, Equation (8.134) .

Since the flow is statistically invariant in the x and z directions, we may apply the periodic boundary condition along these directions. For example, if there are N interior control volumes along the x direction, we will set the flow variables of the virtual volume at the inlet to the corresponding variable values of the Nth volume; similarly, we use variable values of the first interior volume as the virtual volume variable values at the outlet. There is a small problem with pressure though, because pressure drops along the flow direction. This problem is resolved by separating the mean pressure gradient from the pressure term

$$\frac{\partial\bar{p}}{\partial x} = \left\langle\frac{\partial\bar{p}}{\partial x}\right\rangle + \frac{\partial\bar{\psi}}{\partial x} \tag{8.142}$$

This mean (i.e. Reynolds-averaged) pressure gradient $\partial \bar{p}/\partial x$ is a constant, see Equation (8.38) . $\bar{\psi}$ is then periodic at the inlet and outlet of the flow domain. The momentum equation then reads

$$\frac{\partial \bar{u}_i}{\partial t} + \frac{\partial \overline{u_j} \bar{u}_i}{\partial x_j} = -\frac{\partial \bar{\psi}}{\partial x_i} + \nu \frac{\partial^2 \bar{u}_i}{\partial x_j \partial x_j} + \frac{\partial \tau_{ij}^{sgs}}{\partial x_j} + f\delta_{i1} \tag{8.143}$$

in which $f = \langle \partial \bar{p}/\partial x \rangle$ and δ_{ij} is the Kronecker delta (see Equation (5.198)).

No-slip condition is implemented at the channel surfaces ($y = 0$ and $2h$) for velocities. Neumann condition is therefore enforced for pressure

$$\bar{u} = \bar{v} = \bar{w} = 0; \frac{\partial \bar{\psi}}{\partial y} = 0 \tag{8.144}$$

It is desired to recast the equations in dimensionless form. This is done by defining

$$\tilde{u}_i = \frac{\bar{u}_i}{u_\tau}; \tilde{x}_i = \frac{x_i}{h}; \tilde{t} = \frac{tu_\tau}{h}; \tilde{p} = \frac{\bar{\psi}}{u_\tau^2}; \tilde{\nu}_{sgs} = \frac{\nu_{sgs}}{\nu}; \tilde{f} = \frac{fh}{u_\tau^2} \tag{8.145}$$

The dimensionless governing equations are

$$\frac{\partial \tilde{u}_i}{\partial \tilde{x}_i} = 0$$

$$\frac{\partial \tilde{u}_i}{\partial \tilde{t}} + \frac{\partial \tilde{u}_j \tilde{u}_i}{\partial \tilde{x}_j} = -\frac{\partial \tilde{p}}{\partial \tilde{x}_i} + \frac{1}{Re_\tau} \frac{\partial}{\partial \tilde{x}_j} \left[\left(1 + \tilde{\nu}_{sgs} \right) \left(\frac{\partial \tilde{u}_i}{\partial \tilde{x}_j} + \frac{\partial \tilde{u}_j}{\partial \tilde{x}_i} \right) \right] + \tilde{f} \, \delta_{i1} \tag{8.146}$$

where $Re_\tau = hu_\tau/\nu$ is the Reynolds number based on friction velocity and channel half-height. The SGS viscosity is

$$\tilde{\nu}_{sgs} = Re_\tau \left[C_s \tilde{\Delta} \left(1 - e^{-\frac{y^+}{A^+}} \right) \right]^2 \tilde{S} \tag{8.147}$$

in which $\tilde{\Delta} = \Delta/h$ and

$$\tilde{S} = \sqrt{2\tilde{S}_{ij}\tilde{S}_{ij}} ; \tilde{S}_{ij} = \frac{1}{2} \left(\frac{\partial \tilde{u}_i}{\partial \tilde{x}_j} + \frac{\partial \tilde{u}_j}{\partial \tilde{x}_i} \right) \tag{8.148}$$

The nondimensional mean pressure gradient

$$\tilde{f} = \left\langle \frac{\partial \bar{p}}{\partial x} \right\rangle \frac{h}{u_\tau^2} = 1 \tag{8.149}$$

by virtue of Equation (8.38). From now on we will drop the tilde from all variables.

We will use a co-located $64 \times 64 \times 64$ structured mesh for this simulation. Uniform mesh is used along the x and z-directions; the mesh along the y direction is refined toward the channel walls by using a hyperbolic function

$$yf_j = 1 - \frac{\tanh\left\{a\left[1 - \frac{2(j-1)}{(NJ-1)}\right]\right\}}{\tanh a} \tag{8.150}$$

in which yf_j is the y coordinate of the jth surface (north or south surfaces of control volumes), NJ is the total number of such surfaces and a is a constant that controls the degree of mesh stretching. You may find $yf_1 = 0$ and $yf_{NJ} = 2$. The mesh on the $x - y$ plane with using $a = 2$ is shown in Figure 8.28.

So how fine the mesh should be? Well, in LES we only model the relatively small eddies which are more or less isotropic, which means the grid we use should be down to the size of such eddies. As we know eddies close to walls can be very small and anisotropic due to wall effects, we hence need very fine mesh at the wall. Usually we require the first 3~4 grid points above the wall fall in the viscous sublayer i.e. $y^+ \lesssim 5$. This is a quite stringent requirement indeed. For example, we will need $N = 2R_\tau \approx 800$ uniform cells along the y direction to reach a mesh resolution $\Delta y^+ = 1$ in the viscous sublayer. The computational cost of using such a mesh is prohibitive. That is why we use a nonuniform mesh clustering toward walls. When Reynolds number becomes really high, one may have to use a near-wall model to avoid an overly dense mesh.

The cell dimensions along the streamwise (x) and spanwise (z) directions can be relatively large but still subject to some constraints. As a rule of thumb, LES mesh

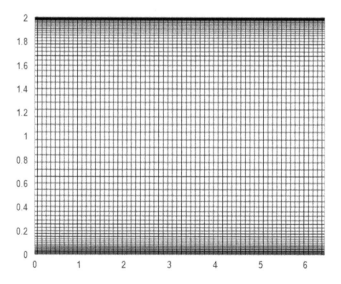

FIGURE 8.28   Nonuniform mesh used for the LES.

should be sufficiently fine to resolve about 80% of total turbulent kinetic energy (Pope, 2000). We can only check if this guideline is followed, however, *a posteriori* after we carry out an LES. With using the calculated SGS viscosity and length scale (see (8.139)), we can estimate the level of SGS turbulent kinetic energy at each control volume

$$k_{sgs} \sim u_{sgs}^2 = \left(\frac{u_{sgs} l_{sgs}}{l_{sgs}}\right)^2 = \left(\frac{\nu_{sgs}}{l_{sgs}}\right)^2 \tag{8.151}$$

By comparing with the resolved turbulent kinetic energy $k = 0.5 \langle \bar{u}'^2 + \bar{v}'^2 + \bar{w}'^2 \rangle$, we may find out if the mesh resolution is fine enough.

Both velocities $\tilde{u}_i$ and pressure $\tilde{p}$ are stored at the center of control volumes as Figure 8.29 shows.

A second-order backward difference scheme is used for the transient term in the momentum equations

$$\left(\frac{\partial \phi}{\partial t}\right)^{n+1} \approx \frac{3\phi^{n+1} - 4\phi^n + \phi^{n-1}}{\Delta t} \tag{8.152}$$

in which ϕ stands for any of the three velocity components. $n + 1$, n and $n - 1$ denote three consecutive time moments. The other terms are all treated implicitly. For this reason, iterations are needed in each time step since the convective terms are nonlinear. To start the calculation, we can use a simple implicit scheme for the first time step i.e.

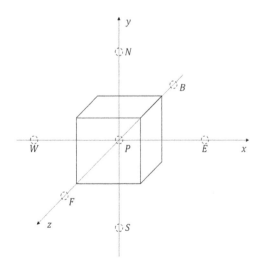

FIGURE 8.29   Control volume and node arrangement.

$$\left(\frac{\partial \phi}{\partial t}\right)^1 \approx \frac{\phi^1 - \phi^0}{\Delta t} \tag{8.153}$$

in which ϕ^0 is the initial guess of the flow field. Such a practice usually introduces an error of first order in time and this error stays for a long time before the second-order temporal accuracy is recovered. It is, however, not a very serious problem in LES since the initial guess of the flow field is often not accurate unless we use DNS or experimental data as initial field, so we have to run the code for many steps anyway before the simulated flow field reaches a statistically steady state.

Central difference is employed to discretize the spatial derivatives. For instance

$$\frac{\partial (u\phi)}{\partial x} \approx \frac{u_e\phi_e - u_w\phi_w}{\Delta x} \approx \frac{1}{\Delta x}\left[u_e\left(\frac{\phi_P + \phi_E}{2}\right) - u_w\left(\frac{\phi_P + \phi_W}{2}\right)\right] \tag{8.154}$$

in which u_e and u_w are obtained with linear interpolation.

One advantage of using a co-located mesh is all discretized momentum equations are of the same form

$$\begin{aligned}
\frac{3\phi_P^* - 4\phi_P^n + \phi_P^{n-1}}{\Delta t} &+ \frac{1}{\Delta x}\left[u_e^m\left(\frac{\phi_P^* + \phi_E^*}{2}\right) - u_w^m\left(\frac{\phi_P^* + \phi_W^*}{2}\right)\right] + \ldots \\
&= \frac{1}{Re_\tau}\frac{1}{\Delta x}\left[\left(1 + \nu_{sgs}\right)_e^m\left(\frac{\phi_E^* - \phi_P^*}{\Delta x}\right)\right. \\
&\left. - \left(1 + \nu_{sgs}\right)_w^m\left(\frac{\phi_P^* - \phi_W^*}{\Delta x}\right)\right] + \ldots + S_\phi^m
\end{aligned} \tag{8.155}$$

where m and $m + 1$ (temporarily denoted by $*$) are the two successive iteration steps in each time step from n to $n + 1$ time moment. S_ϕ^m is the source term which includes the pressure gradient and a part of the diffusion term. Iterations begin when we use values at the nth step as values at $m = 1$ iteration.

Such finite difference equations complemented by boundary conditions can be solved with SIP method for an intermediate velocity field $\vec{u}^*$, which has to be adjusted so that the continuity equation is satisfied. We will use a projection method to this end.

We first decouple the velocity field from the pressure field by subtracting the pressure term from the intermediate velocity field. For example

$$\hat{u}_P = u_P^* - \frac{\Delta t}{3}\left(-\frac{p_e^m - p_w^m}{\Delta x}\right) \tag{8.156}$$

Then we form the pseudo face velocities, e.g. $\hat{u}_e$, $\hat{u}_w$, $\hat{v}_n$, and $\hat{v}_s$ with linear interpolation. Afterwards we re-couple the velocity and pressure fields by adding the correct pressure gradient term back to the pseudo face velocity. As an example

$$u_e^* = \hat{u}_e + \frac{\Delta t}{3}\left(-\frac{p_E^m - p_P^m}{\Delta x}\right) \tag{8.157}$$

This step removes the possible odd–even decoupling between velocity and pressure. Now we are ready to adjust the flow field. Writing symbolically

$$u_{fi}^{m+1} - u_{fi}^* = \frac{\Delta t}{3}\left[-\frac{\partial p^{m+1}}{\partial x_i} - \left(-\frac{\partial p^m}{\partial x_i}\right)\right] = -\frac{\Delta t}{3}\frac{\partial(\delta p)}{\partial x_i} \tag{8.158}$$

where u_{fi} stands for the face velocity along the ith direction, we have

$$\frac{\partial u_{fi}^{m+1}}{\partial x_i} - \frac{\partial u_{fi}^*}{\partial x_i} = -\frac{\Delta t}{3}\frac{\partial^2(\delta p)}{\partial x_i \partial x_i} \tag{8.159}$$

By requiring that the flow field is divergence free at $m + 1$ iteration, we obtain the Poisson equation for pressure correction δp

$$\frac{\partial^2(\delta p)}{\partial x_i \partial x_i} = \frac{3}{\Delta t}\frac{\partial u_{fi}^*}{\partial x_i} \tag{8.160}$$

The boundary conditions are Neumann conditions at the channel walls and periodic conditions at the other boundaries. This equation can be very efficiently solved by a FFT method, see Section 4.4.

Once we have δp, we can update pressure and velocity

$$p^{m+1} = p^m + \delta p; \quad u_i^{m+1} = u_i^* - \frac{\Delta t}{3}\frac{\partial(\delta p)}{\partial x_i} \tag{8.161}$$

in which $\partial(\delta p)/\partial x_i$ has to be interpreted differently for cell face and cell center velocities.

The above procedure is summarized as follows:

1. Solve momentum Equation (8.155) based on solutions of the nth and $(n-1)^{th}$ time steps.
2. Form the pseudo cell center and intermediate face velocities.
3. Solve Poisson Equation (8.160) for pressure corrections.
4. Update velocity and pressure.
5. Repeat steps (2) through (4) until convergence, which is the flow field at $(n+1)^{th}$ time step.

6. Calculate the SGS viscosity.

7. Repeat steps (1) through (6) until desired flow time is reached.

The flow field becomes statistically steady after $\tilde{t} = tu_{tau}/h \approx 50$. So it takes about 10,000 steps before we can begin taking turbulent flow field statistics if we choose $\Delta t = 0.005$. These statistics include the mean velocity, root-mean-square (r.m.s) of velocity fluctuations, and Reynolds stress. The mean velocity u is obtained by adding up the velocity field every 10 time steps for about 10,000 steps then taking the average of them. The r.m.s of the streamwise velocity fluctuations is defined as

$$u'_{rms} = \sqrt{\langle u'^2 \rangle} = \sqrt{\langle ((u - \langle u \rangle)^2 \rangle} = \sqrt{\langle u^2 \rangle - \langle u^2 \rangle} \qquad (8.162)$$

v'_{rms} and w'_{rms} are defined in the same way. The Reynolds stress of interest is $-\langle u'v' \rangle$.

The mean velocity distribution $\langle \bar{u}^+ \rangle$ versus y^+ is shown in Figure 8.30. Pretty significant difference between LES and DNS (Moser, Kim, & Mansour, 1999) in the log-law region can be observed although their agreement in the viscous sublayer is very good.

The r.m.s values are shown in Figure 8.31. The u'_{rms} is grossly over-predicted by LES while v'_{rms} and w'_{rms} are underestimated.

The Reynolds stress distribution is shown in Figure 8.32 .

The relatively unsatisfactory results are mainly due to two reasons: the numerical scheme is of low-order accuracy and the SGS model is not very ideal. These two sources of numerical error are somehow entangled in LES. The central difference we use is of the second-order accuracy, which introduces an error of order $\mathcal{O}(\Delta^2)$. The Smagorinsky SGS viscosity is also proportional to Δ^2 (see (8.139)). Therefore

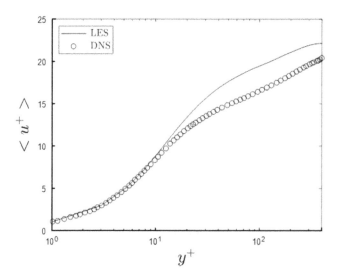

FIGURE 8.30 Mean velocity distribution with LES.

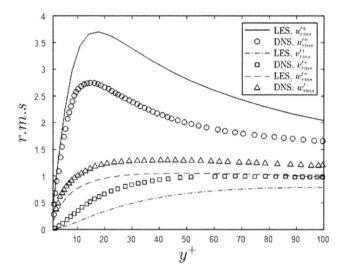

FIGURE 8.31 Root-mean-square values of velocity fluctuations with LES.

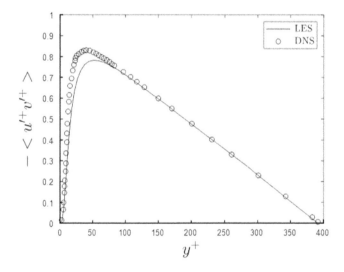

FIGURE 8.32 Reynolds stress distribution with LES.

the error resulted from our numerical scheme may perform just like an extra SGS stress, which might interfere with the Smagorinsky model in an unpredictable way. For this reason, it is highly desirable to use high-order schemes like compact schemes (Kobayashi, 1999) in LES.

The Smagorinsky model itself has some shortcomings. The Smagorinsky coefficient C_s should depend on factors like flow regime, distance to the wall, etc., so it should not be a constant. Another well-known problem of this model is the excess dissipation it puts into the flow (Pope, 2000), which leads to inaccurate flow prediction, even re-laminarization of relatively low Reynolds number turbulent flows. For a thorough discussion of SGS models, refer to Sagaut (2006).

So, this is the ultimate challenge for you: can you compose a LES code with proper high-order accurate, low-dissipation numerical schemes, and a well-chosen SGS model to simulate the turbulent channel flow?

Good luck.

Exercises

1. Show the velocity, length, and time scales of the smallest eddies are given by the Kolmogorov scales.
2. Show that $u^+ = y^+$ in the viscous sublayer.
3. Show that in the log-law region $\omega = u_\tau/(\sqrt{C_\mu}\kappa y)$. That is, Equation (8.82).
4. Reproduce the results shown in Figure 8.6 through Figure 8.8 using the $k - \epsilon$ model.
5. Reproduce the turbulent channel flow results shown in Figure 8.13 through Figure 8.15 using the $k - \epsilon$ model with $k - l$ wall model.
6. Reproduce the results shown in Figure 8.16 through Figure 8.18 using the $k - \omega$ model.
7. Reproduce the results shown in Figure 8.19 through Figure 8.21 using the SST model.
8. Reproduce the results shown in Figure 8.22 through Figure 8.24 using the V2F model.
9. Reproduce the heat transfer in a pipe flow example results shown in Figure 8.26.

NOTE

1 Strictly speaking, $\tau_{ij} = -\rho\overline{u_i' u_j'}$ is Reynolds stress which has the same unit as stress, but we will not struggle with this detail.

Appendix

A MATLAB® FUNCTIONS

A.1 Assemble Diagonal Vectors to Form Coefficient Matrix

```
function A=v2m(P,N,S)
% the inputs are all 1 x n row vectors which are coefficients
% ap, an and as for the n points
% the output is A which is an (n x n) matrix
n=length(P);
A = spdiags([N(:),P(:),S(:)],[-1,0,1],n,n)';
end
```

A.2 TDMA Algorithm (Thomas Algorithm)

```
function phi = TDMA(a,b,c,d)
n = length(d);
c(1) = c(1) / a(1);
d(1) = d(1) / a(1);
for i = 2:n-1
temp = a(i) - b(i) * c(i-1);
c(i) = c(i) / temp;
d(i) = (d(i) - b(i) * d(i-1))/temp;
end
d(n) = (d(n) - b(n) * d(n-1))/(a(n) - b(n) * c(n-1));
phi(n) = d(n);
for i = n-1:-1:1
phi(i) = d(i) - c(i) * phi(i + 1);
end
```

A.3 Stone's Strongly Implicit Procedure (SIP)

```
function [phi,res] = SIP(AP,AW,AS,AN,AE,b,phi,a,tol)
[iMax,jMax] = size(phi); iMaxM = iMax-1; jMaxM = jMax-1;
R = zeros(iMax,jMax);
% LU decomposition
LP = AP; LW = AW; LS = AS; UN = AN./LP; UE = AE./LP;
for j = 1: jMax
for i = 1: iMax
if i>1
LW(i,j) = AW(i,j)/(1+a*UN(i-1,j));
end
if j>1
```

```
LS(i,j) = AS(i,j)/(1+a*UE(i,j-1));
end
if i>1 && j>1
LP(i,j) = AP(i,j)+a*(LW(i,j)*UN(i-1,j)+LS(i,j)*UE(i,j-1))...
-LW(i,j)*UE(i-1,j)-LS(i,j)*UN(i,j-1);
elseif i>1 && j==1
LP(i,j)=AP(i,j)+a*(LW(i,j)*UN(i-1,j))-LW(i,j)*UE
(i-1,j);
elseif j>1 && i==1
LP(i,j)=AP(i,j)+a*(LS(i,j)*UE(i,j-1))-LS(i,j)*UN
(i,j-1);
end
if i>1
UN(i,j) = (AN(i,j)-a*LW(i,j)*UN(i-1,j))/LP(i,j);
else
UN(i,j)=AN(i,j)/LP(i,j);
end
if j>1
UE(i,j) = (AE(i,j)-a*LS(i,j)*UE(i,j-1))/LP(i,j);
else
UE(i,j)=AE(i,j)/LP(i,j);
end
end
end
% Iterations
res = tol + 1;
while (res > tol)
% Calculate Residual
ii = 2: iMaxM; jj = 2: jMaxM;
R(ii,jj) = b(ii,jj)-AE(ii,jj).*phi(ii+1,jj)-AW(ii,jj).*phi
(ii-1,jj)...
-AN(ii,jj).*phi(ii,jj+1)-AS(ii,jj).*phi(ii,jj-1)...
-AP(ii,jj).*phi(ii,jj);
ii = 1; jj = 2: jMaxM;
R(ii,jj) = b(ii,jj)-AE(ii,jj).*phi(ii+1,jj)...
-AN(ii,jj).*phi(ii,jj+1)-AS(ii,jj).*phi(ii,jj-1)...
-AP(ii,jj).*phi(ii,jj);
ii = iMax; jj = 2: jMaxM;
R(ii,jj) = b(ii,jj)-AW(ii,jj).*phi(ii-1,jj)...
-AN(ii,jj).*phi(ii,jj+1)-AS(ii,jj).*phi(ii,jj-1)...
-AP(ii,jj).*phi(ii,jj);
ii = 2: iMaxM; jj = 1;
R(ii,jj) = b(ii,jj)-AE(ii,jj).*phi(ii+1,jj)-AW(ii,jj).*phi
(ii-1,jj)...
-AN(ii,jj).*phi(ii,jj+1)...
-AP(ii,jj).*phi(ii,jj);
```

```
ii = 2: iMaxM; jj = jMax;
R(ii,jj) = b(ii,jj)-AE(ii,jj).*phi(ii+1,jj)-AW(ii,jj).*phi
(ii-1,jj)...
-AS(ii,jj).*phi(ii,jj-1)...
-AP(ii,jj).*phi(ii,jj);
res = max(max(abs(R)));
% Forward substitution
R(1,1) = R(1,1)/LP(1,1);
ii = 2: iMax;
R(ii,1) = (R(ii,1)-LW(ii,1).*R(ii-1,1))./LP(ii,1);
jj = 2: jMax;
R(1,jj) = (R(1,jj)-LS(1,jj).*R(1,jj-1))./LP(1,jj);
for j = 2: jMax
for i = 2: iMax
R(i,j)= (R(i,j)-LW(i,j)*R(i-1,j)-LS(i,j)*R(i,j-1))/LP(i,j);
end
end
% Backward substitution
for j = jMax: -1: 1
for i = iMax: -1: 1
if i<iMax && j<jMax
R(i,j) = R(i,j)-UN(i,j)*R(i,j+1)-UE(i,j)*R(i+1,j);
elseif i<iMax && j==jMax
R(i,j) = R(i,j)-UE(i,j)*R(i+1,j);
elseif i==iMax && j<jMax
R(i,j) = R(i,j)-UN(i,j)*R(i,j+1);
end
end
end
phi = phi + R;
end
```

A.4   ASSEMBLE DIAGONAL MATRICES TO FORM COEFFICIENT MATRIX

```
function A=m2m(P,N,E,W,S)
% the inputs are all n x m matrices which are coefficients at
(n x m)
% mesh nodes
% the outputs are A which is an (n x m) x (n x m) matrix
[n,m]=size(P);
len = n*m;
A = spdiags([N(:),E(:),P(:),W(:),S(:)],[-n,-1,0,1,n],len,
len)';
end
```

A.5 Incomplete Cholesky Conjugate Gradient (ICCG) Method

```
function phi = ICCG(A,b,n,phi,tol)
% A is a n^2 X n^2 matrix
% Step 1: find the preconditioning matrix M = L*U using
Incomplete Cholesky Method
% notice that I have "absorbed" the matrix D into L and U
d = zeros(n^2,1); d(1) = A(1,1);
for j = 2:n
d(j) = A(j,j) -A(j,j-1)*A(j-1,j)/d(j-1);
end
for j = n+1:n^2
d(j) = A(j,j) - A(j,j-1)*A(j-1,j)/d(j-1)-A(j-n,j)*A(j,j-
n)/d(j-n);
end
L = (tril(A,-1)+ spdiags(d,0,n^2,n^2))...
*spdiags(1./sqrt(d),0,n^2,n^2);
U = spdiags(1./sqrt(d),0,n^2,n^2)*...
(triu(A,1) + spdiags(d,0,n^2,n^2));
% Step 2: Using Conjugate Gradient method for the precondi-
tioned system
x = zeros(n^2,1); z = x; d = sqrt(d);
r=b-A*phi;
% solve L*U*z = r
z = U\(L\r);
p=z;
zrold=z'*r;
for i=1:length(b)
Ap=A*p;
alpha=zrold/(p'*Ap);
phi=phi+alpha*p;
r=r-alpha*Ap;
if max(abs(r))<tol
break;
end
z = U\(L\r);
zrnew = z'*r;
p=z+(zrnew/zrold)*p;
zrold=zrnew;
end
```

A.6 2-D Poisson Solver

```
clear; clc;
%============================================%
% Use FFT to solve
```

```
% u_xx + u_yy = 10
% with B.C.'s:
% u(0,y) = 3*y^2; u(1,y) = 2 + 3*y^2;
% u(x,0) = 2*x^2, u(x,1) = 3 + 2*x^2
% The exact solution is u = 2*x^2 + 3*y^2
%=========================================%
L = 1;
N = 128;
N_2times = 2*N; % number of elements in extended vectors
m = N-1; % number of interior nodes
h = L/N; % mesh size
x = (0:N)*h;
y = (0:N)*h;
% change f to take into account BCs
f = 10*ones(N+1);
uw = 3*y.^2;
ue = 2 + 3*y.^2;
us = 2*x.^2;
un = 3 + 2*x.^2;
f(2,:) = f(2,:)-uw/h^2;
f(N,:) = f(N,:)-ue/h^2;
f(:,2) = f(:,2)-us'/h^2;
f(:,N) = f(:,N)-un'/h^2;
f_interior = f(2:N,2:N);
% extend the matrix to form a sine wave pattern in both
directions
F = [zeros(1,m); f_interior; zeros(1,m); -f_interior(m:-1:
1,:)];
G = [zeros(N_2times,1),F,zeros(N_2times,1),-F(:,m:-1:1)];
% double DFT
ghat = fft2(G);
% double IDFT
[n,p] = meshgrid(1:N_2times,1:N_2times);
mu_plus_nu    =   -(4/h^2)*(sin((n-1)*pi/N_2times).^2+sin(
(p-1)*pi/N_2times).^2);
mu_plus_nu(1,1) = 1; % you have to set it to a nonzero value
U = real(ifft2(ghat./mu_plus_nu)); % double IDFT; you have
to take its real part
u = U(1:N+1,1:N+1); % extract u(i,j) from the U matrix
obtained from IDFT
% find the error and plot the solution
[Y,X] = meshgrid(y,x);
u_exact = 2*X.^2 + 3*Y.^2;
err = max(max(abs(u(2:N,2:N)-u_exact(2:N,2:N))))
surf(X(2:N,2:N),Y(2:N,2:N),u(2:N,2:N));
xlabel('x');
```

```
ylabel('y');
zlabel('u');
```

A.7 TRIANGULAR MESH GENERATION PROGRAM OF PERSSON AND STRANG[1]

```
function [p,t]=distmesh2-D(fd,fh,h0,bbox,pfix,varargin)
dptol=.001; ttol=.1; Fscale=1.2; deltat=.2; geps=.001*h0;
deps=sqrt(eps)*h0;
% 1. Create initial distribution in bounding box (equilateral
triangles)
[x,y]=meshgrid(bbox(1,1):h0:bbox(2,1),bbox(1,2):h0*sqrt
(3)/2:bbox(2,2));
x(2:2:end,:)=x(2:2:end,:)+h0/2; % Shift even rows
p=[x(:),y(:)]; % List of node coordinates
% 2. Remove points outside the region, apply the rejection
method
p=p(feval(fd,p,varargin{:})<geps,:); % Keep only d<0 points
r0=1./feval(fh,p,varargin{:}).^2; % Probability to keep
point
p=[pfix; p(rand(size(p,1),1)<r0./max(r0),:)]; % Rejection
method
N=size(p,1); % Number of points N
pold=inf; % For first iteration
while 1
% 3. Retriangulation by the Delaunay algorithm
if  max(sqrt(sum((p-pold).^2,2))/h0)>ttol  % Any  large
movement?
pold=p; % Save current positions
t=delaunayn(p); % List of triangles
pmid=(p(t(:,1),:)+p(t(:,2),:)+p(t(:,3),:))/3; % Compute
centroids
t=t(feval(fd,pmid,varargin{:})<-geps,:); % Keep interior
triangles
% 4. Describe each bar by a unique pair of nodes
bars=[t(:,[1,2]);t(:,[1,3]);t(:,[2,3])]; % Interior bars
duplicated
bars=unique(sort(bars,2),'rows'); % Bars as node pairs
% 5. Graphical output of the current mesh
trimesh(t,p(:,1),p(:,2),zeros(N,1))
view(2),axis equal,axis off; drawnow
end
% 6. Move mesh points based on bar lengths L and forces F
barvec=p(bars(:,1),:)-p(bars(:,2),:); % List of bar vectors
L=sqrt(sum(barvec.^2,2)); % L = Bar lengths
hbars=feval(fh,(p(bars(:,1),:)+p(bars(:,2),:))/2,var-
argin{:});
```

```
L0=hbars*Fscale*sqrt(sum(L.^2)/sum(hbars.^2)); % L0 = Desired
lengths
F=max(L0-L,0); % Bar forces (scalars)
Fvec=F./L*[1,1].*barvec; % Bar forces (x,y components)
Ftot=full(sparse(bars(:,[1,1,2,2]),ones(size
(F))*[1,2,1,2],[Fvec,-Fvec],N,2));
Ftot(1:size(pfix,1),:)=0; % Force = 0 at fixed points
p=p+deltat*Ftot; % Update node positions
% 7. Bring outside points back to the boundary
d=feval(fd,p,varargin{:}); ix=d>0; % Find points out-
side (d>0)
dgradx=(feval(fd,[p(ix,1)+deps,p(ix,2)],varargin{:})-d
(ix))/deps; % Numerical
dgrady=(feval(fd,[p(ix,1),p(ix,2)+deps],varargin{:})-d
(ix))/deps; % gradient
p(ix,:)=p(ix,:)-[d(ix).*dgradx,d(ix).*dgrady]; % Project
back to boundary
% 8. Termination criterion: All interior nodes move less than
dptol (scaled)
if max(sqrt(sum(deltat*Ftot(d<-geps,:).^2,2))/h0)<dptol,
break; end
end
```

A.8   USEFUL ACCESSORY FUNCTIONS TO TRIANGULAR MESH GENERATION PROGRAM OF PERSSON AND STRANG [2]

```
function d=dcircle(p,xc,yc,r) % Circle
d=sqrt((p(:,1)-xc).^2+(p(:,2)-yc).^2)-r;
end
function d=drectangle(p,x1,x2,y1,y2) % Rectangle
d=-min(min(min(-y1+p(:,2),y2-p(:,2)),...
-x1+p(:,1)),x2-p(:,1));
end
function d=dunion(d1,d2) % Union
d=min(d1,d2);
end
function d=ddiff(d1,d2) % Difference
d=max(d1,-d2);
end
function d=dintersect(d1,d2) % Intersection
d=max(d1,d2);
end
function p=pshift(p,x0,y0) % Shift points
p(:,1)=p(:,1)-x0;
p(:,2)=p(:,2)-y0;
end
```

```
function p=protate(p,phi) % Rotate points around origin
A=[cos(phi),-sin(phi);sin(phi),cos(phi)];
p=p*A;
end
function d=dmatrix(p,xx,yy,dd,varargin) % Interpolate d
(x,y) in meshgrid matrix
d=interp2(xx,yy,dd,p(:,1),p(:,2),'*linear');
end
function h=hmatrix(p,xx,yy,dd,hh,varargin) % Interpolate h
(x,y) in meshgrid matrix
h=interp2(xx,yy,hh,p(:,1),p(:,2),'*linear');
end
function h=huniform(p,varargin) % Uniform h(x,y) distribution
h=ones(size(p,1),1);
end
```

B   VON NEUMANN ANALYSIS OF FTCS SCHEME

To explore the stability of the FTCS scheme for the 1-D unsteady convection–diffusion equation, we can substitute the presumed solution

$$\phi = e^{i(\xi x - \omega t)} \tag{B.1}$$

into the finite difference equation:

$$\phi_j^{n+1} - \phi_j^n + \frac{c}{2}\left(\phi_{j+1}^n - \phi_{j-1}^n\right) - r\left(\phi_{j-1}^n - 2\phi_j^n + \phi_{j+1}^n\right) \tag{B.2}$$

We end up with

$$k_{1,2} = \frac{1 - 2r - z \pm \sqrt{(1 - 2r - z)^2 + (c^2 - 4r^2)}}{c - 2r} \tag{B.3}$$

where

$$k = e^{i\xi \Delta x}; \ z = e^{-i\omega \Delta t} \tag{B.4}$$

Equation (B.3) defines the relationship $\omega = f(\xi)$. Notice that the waves indeed travel as wave packets at their group velocity (Trefethen, 1983)

$$c_g = \frac{\partial \omega_r}{\partial \xi_r} \tag{B.5}$$

So if we can find a point on the f function that has two group velocities of opposite signs and a positive imaginary part of ω (so that it grows with time), the scheme is unstable.

For example, for $r = 0.3$, $c = 2\sqrt{r} - 10^{-2}$, and $\omega_i = 10^{-5}$, the $\xi_r \Delta x - \omega_r \Delta t$ and $\xi_i \Delta x - \omega_r \Delta t$ graphs are shown in Figure B.1. The subscripts "r" and "i" denote the real and imaginary parts, respectively. Notice that it is enough to only explore the range $-\pi \leq \omega_r \Delta t \leq \pi$.

You can see although there are points like point A on the $\xi_r \Delta x$-$\omega_r \Delta t$ curve at which we have two group velocities (the slope of the curve) of opposite signs (e.g. along the k_2 branch), but this point is a discontinuity on the $\xi_i \Delta x$-$\omega_r \Delta t$ graph, so it is not a supported solution and the scheme is stable;

On the other hand, if $c = 2\sqrt{r} + 10^{-2}$, the graphs change (see Figure B.2):

Now we have such two wave-packet solutions (e.g. at point A, along either the k_1 or k_2 branch), which have exactly the same wave number ξ, and exactly the same frequency ω, but propagating in opposite directions, and growing with time. The scheme is thus unstable.

Another case: $r = 0.6$, as $c = 2\sqrt{2r - 1} + 10^{-2}$, the graphs are shown in Figure B.3. It is stable.

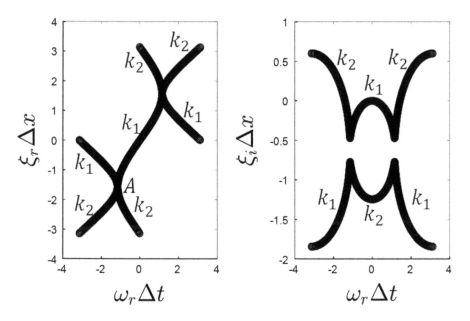

FIGURE B.1 ξ-ω graph as $c < 2\sqrt{r}$.

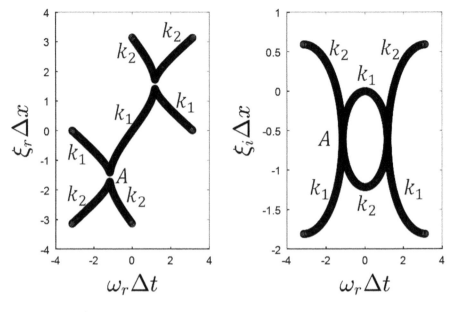

FIGURE B.2 ξ-ω graph as $c > 2\sqrt{r}$.

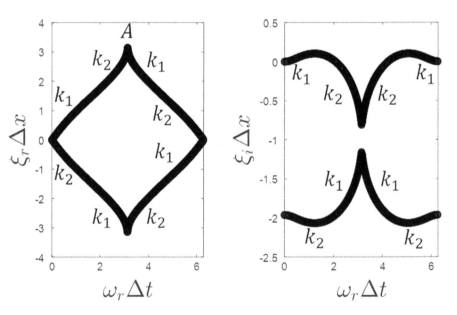

FIGURE B.3 ξ-ω graph as $c > 2\sqrt{2r-1}$.

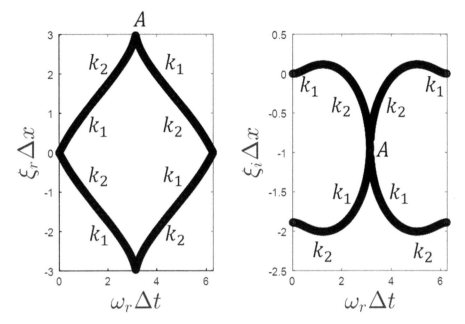

FIGURE B.4 ξ-ω graph as $c < 2\sqrt{2r-1}$.

On the other hand, if $c = 2\sqrt{2r-1} - 10^{-2}$, we have two solutions which share the same ω and ξ but with group velocity of opposite signs. See point A in Figure B.4. The scheme is thus unstable.

NOTES

1 This program is included here with kind permission from Dr. Per-Olof Persson
2 This program is included here with kind permission from Dr. Per-Olof Persson

References

Abe, H., Kawamura, H., & Matsuo, Y. (2001). Direct Numerical Simulation of a Fully Developed Turbulent Channel Flow with Respect to the Reynolds Number Dependence. *Journal of Fluids Engineering*, 123: 382–393.

Alexiades, V., & Solomon, A. D. (1993). *Mathematical Modeling of Melting and Freezing Processes*. Washington DC: CRC Press.

Armaly, B. F., Durst, F., Peireira, J., & Schonung, B. (1983). Experimental and Theoretical Investigation of Backward-Facing Step Flow. *Journal of Fluid Mechanics*, 127: 473–496.

Bhaga, D., & Weber, M. E. (1981). Bubbles in Viscous Liquids: Shapes, Wakes and Velocities. *Journal of Fluid Mechanics*, 105: 61–85.

Boltzmann, L. (2011). *Lectures on Gas Theory*. New York: Dover.

Brackbill, J. U., Kothe, D. B., & Zemach, C. (1992). A Continuum Method for Modeling Surface Tension. *Journal of Computational Physics*, 100: 335–354.

Bradshaw, P. (1973). *Effects of Streamline Curvature on Turbulent Flow*. Neuilly-sur-Seine, France: AGARD.

Bradshaw, P., Ferriss, D. H., & Atwell, N. P. (1967). Calculation of Boundary Layer Using the Turbulent Energy Equation. *Journal of Fluid Mechanics*, 28: 593–616.

Cabelli, A. (1977). Storage Tanks a Numerical Experiment. *Solar Energy*, 19: 45–54.

Chen, H., & Patel, V. (1988). Near-wall Turbulence Models for Complex Flows including Separation. *AIAA Journal*, 26: 641–648.

Chorin, A. J. (1968). Numerical Solution of the Navier-Stokes Equations. *Mathematics of Computation*, 22: 745–762.

De Vahl Davis, G. (1983). Natural Convection of Air in a Square Cavity: A Bench Mark Numerical Solution. *International Journal for Numerical Methods in Fluids*, 3: 249–264.

Deardorff, J. W. (1970). A Numerical Study of Three-dimensional Turbulent Channel Flow at Large Reynolds Numbers. *Journal of Fluid Mechanics*, 41: 453–480.

Deen, W. M. (1998). *Analysis of Transport Phenomena*. New York: Oxford University Press.

Dhiman, A. K., Chhabra, R. P., Sharma, A., & Eswaran, V. (2006). Effects of Reynolds and Prandtl Numbers on Heat Transfer across a Square Cylinder in the Steady Flow Regime. *Numerical Heat Transfer*, 49: 717–731.

Durbin, P. A. (1993). Application of a Near-wall Turbulence Model to Boundary Layers and Heat Transfer. *International Journal of Heat and Fluid Flow*, 14: 316–323.

Fedkiw, R., Aslam, T., Merriman, B., & Osher, S. (1999). A Non-oscillatory Eulerian Approach to Interfaces in Multimaterial Flows (the Ghost Fluid Method). *Journal of Computational Physics*, 152: 457–492.

Ferziger, J. H., & Peric, M. (2002). *Computational Methods for Fluid Dynamics*. New York: Springer.

Gaskell, P. H., & Lau, A. (1988). Curvature-compensated Convective Transport: SMART, a New Boundedness-preserving Transport Algorithm. *International Journal of Numerical Methods in Fluids*, 8: 617–641.

Ghia, U., Ghia, K. N., & Shin, C. T. (1982). High-Re Solutions for Incompressible Flow Using the Navier-Stokes Equations and a Multigrid Method. *Journal of Computational Physics*, 48: 387–411.

Gilat, A., & Subramaniam, V. (2008). *Numerical Methods for Engineers and Scientists*. Hoboken: John Wiley & Sons.

Gingold, R. A., & Monaghan, J. J. (1977). Smoothed Particle Hydrodynamics: Theory and Application to Non-Spherical Stars. *Monthly Notices of the Royal Astronomical Society*, 181: 375–389.

Goldstein, S. (1938). *Mordern Developments in Fluid Dynamics*. Oxford: Clarendon Press.

Golub, G. H., & Van Loan, C. F. (1989). *Matrix Computations*. Baltimore and London: The Johns Hopkins University Press.

Gottlieb, S., & Shu, C. (1998). Total Variation Diminishing Runge-Kutta Schemes. *Mathematics of Computation*, 67: 73–85.

Grace, J. R. (1973). Shapes and Velocities of Bubbles Rising in Infinite Liquids. *Transactions of the Institution of Chemical Engineers*, 51: 116–120.

Guillaument, R., Vincent, S., & Caltagirone, J.-P. (2015). An Original Algorithm for VOF Based Method to Handle Wetting Effect in Multiphase Flow Simulation. *Mechanics Research Communications*, 63: 26–32.

Harlow, F. H., & Welch, J. E. (1965). Numerical Calculation of Time-dependent Viscous Incompressible Flow of Fluid with a Free Surface. *Physics of Fluids*, 8: 2182–2189.

Harten, A. (1983). High Resolution Schemes for Hyperbolic Conservation Laws. *Journal of Computational Physics*, 49: 357–393.

Harten, A., Engquist, B., & Osher, S. (1987). Uniformly High Order Accurate Essentially Non-oscillatory Schemes. *Journal of Computational Physics*, 71, 231–303.

Hartmann, D., Meinke, M., & Schröder, W. (2008). Differential Equation Based Constrained Reinitialization for Level Set Methods. *Journal of Computational Physics*, 227: 6821–6845.

Hirt, C. W., & Nichols, B. D. (1981). Volume of Fluid (VOF) Method for the Dynamics of Free Boundaries. *Journal of Computational Physics*, 39: 201–225.

Hu, C., & Sueyoshi, M. (2010). Numerical Simulation and Experiment on Dam Break Problem. *Journal of Marine Science and Application*, 9: 109–114.

Incropera, F. P., & DeWitt, D. P. (2007). *Fundamentals of Heat and Mass Transfer*. Hoboken: Wiley.

Jacqmin, D. (1999). A Calculation of Two-phase Navier-Stokes Flows Using Phase-field Modeling. *Journal of Computational Physics*, 155: 96–127.

Jones, W. P., & Launder, B. E. (1972). The Prediction of Laminarization with a Two-equation Model. *International Journal of Heat and Mass Transfer*, 15: 301–314.

Khosla, P. K., & Rubin, S. G. (1974). A Diagonally Dominant Second Order Accurate Implicit Scheme. *Computers & Fluids*, 2: 207–209.

Kim, J., Moin, P., & Moser, R. (1987). Turbulence Statistics in Fully Developed Channel Flow at Low Reynolds Number. *Journal of Fluid Mechanics*, 117: 133–166.

Kobayashi, M. H. (1999). On a Class of Pade Finite Volume Methods. *Journal of Computational Physics*, 156: 137–180.

Kuehn, T. H., & Goldstein, R. J. (1976). An Experimental and Theoretical Study of Natural Convection in the Annulus between Horizontal Concentric Cylinders. *Journal of Fluid Mechanics*, 74: 695–719.

Launder, B. E., & Spalding, D. B. (1974). The Numerical Computation of Turbulent Flows. *Computer Methods in Applied Mechancs and Engineering*, 3: 269–289.

Leonard, A. (1974). Energy Cascade in Large-Eddy Simulations of Turbulent Fluid Flows. *Advances in Geophysics*, 18: 237–248.

Wark, K. (1983). *Thermodynamics*. New York: McGraw-Hill.

Wei, J. J., Yu, B., Tao, W. Q., Kawaguchi, Y., & Wang, H. S. (2003). A New High-order Accurate and Bounded Scheme for Incompressible Flow. *Numerical Heat Transfer*, 43: 19–41.

Welch, S. W., & Wilson, J. (2000). A Volume of Fluid Based Method for Fluid Flows with Phase Change. *Journal of Computational Physics*, 160: 662–682.

Wesseling, P. (1991). *An Introduction to Multigrid Methods*. New York: Wiley.

Wesseling, P. (2001). *Principles of Computational Fluid Dynamics*. New York: Springer.

Wilcox, D. C. (1988). Re-Assessment of the Scale-Determining Equation for Advanced Turbulence Models. *AIAA Journal*, 26: 1299–1310.

Wolfshtein, M. (1969). The Velocity and Temperature Distribution in One-Dimensional Flow with Turbulence Augmentation and Pressure Gradient. *International Journal of Mass and Heat Transfer*, 12: 301–318.

Yankovskii, A. P. (2017). Analysis of the Spectral Stability of the Generalized Runge-Kutta Methods Applied to Initial-Boundary-Value Problems for Equations of the Parabolic Type. I. Explicit Methods. *Journal of Mathematical Sciences*, 229: 227–240.

Youngs, D. L. (1982). Time-dependent Multimaterial Flow with Large Fluid Distortion. In K. Morton, & M. Baines, *Numerical Methods for Fluid Dynamics* (pp. 273–285). New York: Academic Press.

Yu, B., Ozoe, H., & Tao, W.-Q. (2005). A Collocated Finite Volume Method for Incompressible Flow on Unstructured Meshes. *Progress in Computational Fluid Dynamics*, 5: 181–189.

Index

Adams-Bashforth schemes, 74–75
ADI method, 121–22, 238
advection equation, 109, 112, 235–36;
     see also pure convection equation
all-time-wiggle-free condition, 72–74, 86–88
alternating direction implicit method *see* ADI
     method
amplification factor, 60, 65–66, 72, 77–78, 81–83

backward difference: first-order, 11–14, 18, 22, 24
    second-order, 74, 343
backward substitution: of LU decomposition
    method, 122
    of SIP method, 122–24
    of TDMA algorithm, 20
barrier function, 32
    and universal barrier function, 32–33
BiCGSTAB method, 131
bi-conjugate gradient stabilized method
    see BiCGSTAB method
bounded scheme, 26–29, 58, 66, 69, 93, 95, 109,
    236, 249, 264; *see also* positive
    scheme; TVD scheme
boundedness, 28; *see also* positiveness; TVD
boundedness condition, 28
    and of Euler explicit scheme, 63
    of FTCS scheme, 86
    of Runge-Kutta methods, 72–74
    of TVD schemes, 103, 106–09
Boussinesq approximation, 291, 294, 327
Boussinesq assumption, 173–74
Bradshaw assumption, 302, 321, 327
buffer layer, 303

CFL number, 77, 102, 106–9, 236–37, 241, 250
co-located mesh, 154
continuum surface force model *see* CSF model
convection-diffusion equation, 3, 199–206
    stability of FTCS scheme applied to unsteady
       1-D, 356
    steady 1-D, 89, 93–106, 109–111

and unsteady 1-D, 75, 86, 106;
    see also transport equation
convergence, 58–59
coupled method, 148
Courant number *see* CFL number
Crank-Nicolson scheme, 68–71
cross-diffusion term, 204
CSF model, 255, 267
cut-cell mesh, 220–21

dam-break problem, 250–252, 269
deferred-correction method, 99–101, 106,
    204, 210
Delaunay triangulation, 195–98
delta function: Dirac, 267, 277
    Kronecker, 187, 291, 341
DFT, 132–38, 353; *see also* FFT
diffusion, 2–6
diffusion coefficient, 3–6
    artificial, 258
    of heat transfer, 210
    of mass transfer, 75
    and of momentum, 6
    varying, 42
diffusion number, 59, 64, 108
diffusion term, 4, 57, 64–65, 78, 88, 172, 204–06,
    258, 293, 301, 303, 330, 343
    and artificial, 95
diffusive scheme, 95
direct numerical simulation *see* DNS
Dirichlet boundary condition, 16–17, 61, 109–12,
    133–38, 147, 151, 169, 182, 189, 207
Discrete Fourier transform *see* DFT
dispersive scheme, 95
dissipation: rate of, 293, 304, 313, 325–26
    of Smagorinsky model, 347
    of turbulent kinetic energy, 288, 303
dissipative scheme *see* diffusive scheme
divergence: of numerical methods, 58, 121,
    165, 219
    of a vector, 172–73, 200, 235, 345